Soil Physics

Agricultural and Environmental Applications

Soil Physics

Agricultural and Environmental Applications

H. Don Scott

Iowa State University Press / Ames

H. Don Scott received his doctoral degree in soil science from the University of Kentucky. Dr. Scott is university professor of soil physics in the Department of Crop, Soil, and Environmental Sciences at the University of Arkansas, Fayetteville. His research interests are primarily in soil/water/crop relationships, transport of agricultural chemicals in soil, and the applications of geographical information systems technologies to agriculture.

Iowa State University Press
2121 South State Avenue, Ames, Iowa 50014

Orders: 1-800-862-6657
Office: 1-515-292-0140
Fax: 1-515-292-3348
Web site (secure, interactive): www.isupress.edu

First edition, 2000

Library of Congress Cataloging-in-Publication Data

Scott, H. Don
Soil physics: agricultural and environmental applications / H. Don Scott—1st ed.
    p. cm.
Includes bibliographical references and index.
ISBN 0-8138-2087-1 (acid-free paper)
1. Soil physics. 2. Soil moisture. I. Title.
S592.3 .S36 2000
631.4'3—dc21   00-025759

The last digit is the print number: 9   8   7   6   5   4   3   2   1

To my parents, Hubert Cox and Rachel Mozelle Scott,
and to my wife, Janice Elizabeth Scott,
for their love, encouragement, and support.

To Dr. J. F. Lutz and Dr. R. E. Phillips,
who taught me more than soil physics.

Contents

Contents

Preface

This textbook is concerned with the physical properties of soil and how they affect other soil properties, the transport of water, heat, solutes, and oxygen in soil, and soil water and its impact on plant growth and development. This book will enable the student to understand how the soil, plant, and engineering sciences utilize knowledge of soil physical behavior and to develop the mathematical and quantitative skills needed to solve applied problems in soil science. The book is not intended to be an exhaustive treatise of soil physics but presents a subset of the current knowledge on the subject. The material may be covered in a single semester.

Emphasis is placed on understanding how soil physical properties impact agriculture, natural resources, and the environment. To achieve this goal, considerable use is made of elementary concepts of physics, mathematics, and statistics, which are needed to quantify amounts and rates of processes in soil systems. In most cases, conservation laws and rate equations are used to account for the spatial and temporal distributions of mass and energy in soil systems. These are the basic underlying threads throughout the book and should result in a greater appreciation and understanding of the physical processes that influence soil behavior.

Chapter 1 presents a perspective of soil physics in terms of soil science and of physics and provides a brief sketch of its history. Chapter 2 presents the parameters to be used in quantifying the physical properties of soil and illustrates their use. Chapters 3 and 4 provide the basics for characterization of the physical properties of the solid phase and emphasize the important role of the solid phase in the retention and transport of mass and energy. Chapter 5 serves as the key chapter for the remainder of the textbook. This chapter introduces the balance and steady-state rate equations of mass and energy transport. An understanding of these concepts and principles is necessary before proceeding to the remaining chapters, in which balances of mass and energy are emphasized. Chapters 6 and 7 present the principles associated with soil temperature and gas transport. Chapter 8 discusses the principles associated with the content and energy status of soil water. Chapters 9 and 10 review soil water flow processes. Chapter 11 presents the principles of solute transport in soil and emphasizes the importance of water flow and interactions between the solid, liquid, and gaseous phases and the solute. Chapter 12 examines soil-plant-water relationships and concludes that water and energy processes in soils, plants, and the atmosphere are analogous.

Each chapter includes definitions, essential terms and concepts, and an overview of the principles and practical importance of the topic. Each chapter also contains numerous examples and, at the end of the chapter, problems to be worked. The examples and problems present opportunities for students to review and apply their knowledge and mathematical skills such as algebra and calculus to real-world problems. Many of the problems are specifically designed to encourage applications of theory to field situations.

My students in soil physics have been using the text in various somewhat inconvenient approximations for several years, and most have been patient and invaluable in helping me shape it. Since the academic backgrounds of these students were diverse, I have included three fairly detailed appendixes to provide brief reviews of the concepts of statistics, mathematics, and physics. One measure of success will be achieved if students use the concepts and principles presented in this textbook in their work in the future. And one can hope that the work performed will be of benefit to all.

I am deeply indebted to my former students who labored through early drafts of the concepts and muddled through the theory and field applications. In particular, I am grateful to those students who offered helpful suggestions. Also, I appreciate those special colleagues who reviewed one or more chapters, especially Tina Udouj, Don Baker, Bernali Dixon, Dennis Brewer, Andy Mauromostakos, and Bob Sojka. Marty McKimmey and Evey Fisher were instrumental in developing the graphics. Michael Scott designed and developed the cover. All were extremely helpful in focusing the material and correcting errors in the original manuscript. I thank Dr. Jim Ice and Lynne Bishop of Iowa State University Press and Pam Bruton, the copyeditor, for their help in editing and publishing this text. As expected, I am solely responsible for the material and any errors in the final copy.

Soil Physics

Agricultural and Environmental Applications

Soil Physics in Perspective

Introduction

Humans are dependent on soils, and to a certain extent "good" soils are dependent upon the care given them and the uses made of them by humans. The physical properties of soil are important components in determining the quality, potential uses, sustainability, and productivity of soil. In this textbook, we will examine the physical properties of soil and show how these properties can be used to gain a greater understanding of soil behavior in agricultural and environmental applications. Emphases are given to the use of familiar concepts of physics, mathematics, and chemistry as applied to soils. We begin by presenting a perspective of soil physics as related to other disciplines in soil science and physics and a brief history of the discipline.

Definition of Soil Physics

Soil physics is the study of the physical properties of the soil and the relation of these soil physical properties to agricultural, environmental, and engineering uses. It has also been defined as the study of the state and transport of matter and energy in the soil. These definitions imply that soil physics is quantitative and mathematical in nature and is primarily concerned with the fundamental properties of soils. Historically, the emphasis has been on the application of soil physical principles to the solution of agricultural, engineering, and environmental problems.

Over the years, the primary motivation of soil physics has been to obtain a greater understanding of the physical processes of soil to ultimately aid in agriculture, particularly in the production of crops. Lately, this emphasis has been expanded and supplemented to develop a greater understanding of (1) transport of water, solutes, and heat in porous media, (2) hydrology and water resources, (3) various environmental and pollution concerns, and (4) spatial and temporal variability of soil properties in the landscape. The landscape is defined as the physical surroundings or landform within the viewing region.

Soil Physical Properties

Five soil physical properties have been identified as important to the maintenance of the physical fertility of the soil:

1. *Soil texture:* particle size distribution.
2. *Soil temperature and heat:* processes include heat transport and thermal regimes.
3. *Soil aeration:* exchanges of gases such as oxygen and carbon dioxide by plant roots and soil microorganisms with the atmosphere.
4. *Soil water:* processes include retention, infiltration, runoff, redistribution within the profile, evaporation, transpiration and plant response, groundwater, drainage, irrigation, and transport of contaminants.
5. *Soil structure:* particle and pore arrangement, compaction, consistency, and tilth.

Spatial and Temporal Influences

All of the soil physical properties vary with space and time. Mathematically, this relationship can be written as

$$SP = f(x,y,z,t) \qquad [1]$$

where SP can be any soil physical parameter such as soil water content, pressure, temperature, etc.; f is a symbol that means "is a function of" or "depends upon"; x, y, and z are the Cartesian coordinates; and t is the time.

Example

Suppose you were given the assignment to sample the soil water content of a soybean field. Since soil water content varies with time of sampling and with position in the field, you should be concerned with

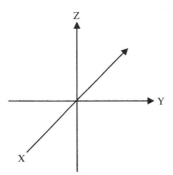

Figure 1.1 The Cartesian coordinate system.

- sampling time, t
- depth in the profile, z
- row in which sample is taken, x
- position along the row, y

Would time and position be as important for the measurement results if you were sampling for a fertilizer component or a pesticide instead of soil water? We will examine the answer to this question in greater detail in this textbook.

Amounts and rates of change are both important in soil physics. The amount of change of a soil physical property is expressed as the difference in its magnitude between two samplings. We will consider changes in the properties in both space and time.

The functional dependence of a soil physical property on x, y, and z is known as spatial variability. This indicates that where the sample of soil is taken is important. Graphically, the Cartesian coordinate system is shown in Figure 1.1.

Suppose that the soil water content in the Ap horizon was determined at two locations that were 10 meters apart. We designate the soil water content with the symbol θ_v along with a subscript designating the location. At location 1, θ_{v1} was 0.25 m^3/m^3, and at location 2, θ_{v2} was 0.30 m^3/m^3. The amount of change in soil water content between the two locations, designated as $\Delta\theta_v$, is $\theta_{v2} - \theta_{v1} = 0.30 - 0.25 = 0.05$ m^3/m^3. Since the distance between the two locations was 10 m, the average rate of change in soil water content, $\Delta\theta_v/\Delta x$, between the two locations is

$$\Delta\theta_v/\Delta x = 0.05 \text{ m}^3/\text{m}^3/10 \text{ m} = 0.005 \text{ m}^3/\text{m}^3/\text{m} \qquad [2]$$

Spatial variability of soil physical properties in many old, well-developed soils changes most rapidly in the vertical direction (the z direction) of the soil, which is known as the profile. Spatial variability of soil physical properties in the x and y directions is expressed as the landscape.

The functional dependence of soil physical properties on time indicates that we should also consider the time of sampling and the change in magnitude of the soil property during a designated time interval—that is, the time rate of change. This is known as temporal variability.

Suppose that at a given location the water content of 10 cm³ of soil was 0.30 m³/m³ at sampling time 1 and 0.25 m³/m³ at sampling time 2. Then the amount of change was –0.05 m³/m³ of soil water. Thus, water was lost from the soil by one or more consumption processes such as evaporation, plant uptake, or drainage. Mathematically, this change in soil water content is calculated from $\theta_{v2} - \theta_{v1}$, or $\Delta\theta_v$, where θ_{v2} and θ_{v1} are the soil water contents at sampling times 2 and 1, respectively. The change in soil water content can be positive or negative; no change is represented by a difference of zero. A brief review of statistics with an emphasis on spatial statistics is presented in Appendix A.

The time rate of change is the amount of change divided by the elapsed time. In the example above, if the sampling interval was 7 days, then the average time rate of change was –0.05 m³ of water per m³ of soil per 7 days, or –0.0071 m³/m³/d. Mathematically, the time rate of change is represented by the ratio

$$\Delta\theta_v/\Delta t = (\theta_{v2} - \theta_{v1})/(t_2 - t_1) \qquad\qquad [3]$$

where t_2 and t_1 are sampling times 2 and 1, respectively. For some applications, it may be of interest to know the rate of change for a shorter period of time. Therefore, it is often important to distinguish between the *average rate of change,* $\Delta\theta_v/\Delta t$, and the *instantaneous rate of change,* $d\theta_v/dt$. The differences between these two mathematical ways of expressing the rate of change of a soil physical quantity will be discussed in greater detail later.

Quantification of the rates of change of mass and energy in soil and the direction of these changes are extremely important concepts to the greater understanding of physical behavior of soil. A brief review of mathematics is given in Appendix B.

Scales of Observation

Investigations of soil physical behavior are conducted at the molecular, microscopic, local, and regional scales of observation. At the molecular scale, the molecules are the system, and the atomic particles (e.g., electrons and protons) are the system elements.

At the microscopic scale, the three phases—solid, liquid, and gas—form the soil system, and the atoms, molecules, and ions are the indivisible elements of the system. Due to the exceedingly large number of these elements, it is usually more convenient to choose a volume containing a sufficiently large number of atoms, molecules, or ions so that only their mean statistical behavior is relevant.

A volume enclosing such a continuum molecular mixture is called a *representative elementary volume* (REV). The REV must be large compared to the mean free path of molecules caused by Brownian motion. The concept of REV was developed because of the need to describe or lump the physical properties at a geometrical point. We say that we give to one point in space and time the value of the property of a certain volume surrounding this point. The REV is used to define and sometimes to measure the mean properties of the volume in question. Consequently, this concept involves an integration in space. According to de Marsily (1986) the size of the REV is determined by saying that it must be

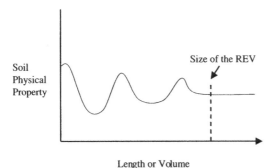

Figure 1.2. Definition of the REV.

1. sufficiently large to contain a soil volume large enough to allow the definition of a mean global property while ensuring that the effects of the fluctuations from one pore to another are negligible and

2. sufficiently small that the parameter variations from one domain to the next may be approximated by continuous functions, in order that we may use calculus, without in this way introducing any error that may be picked up by the measuring instruments at the macroscopic scale.

One example of the usefulness of the REV concept is to integrate from the Navier-Stokes equations of fluid flow in the microscopic realm to the less complicated Darcy's law in the macroscopic realm (Corey 1994).

Figure 1.2 illustrates how to choose the size of the REV. The size of the REV is generally linked to the existence of a flattening of the curve that connects the physical property with the dimension. It is an averaging of the soil physical properties within the volume. Obviously, the size of the REV varies widely with soil physical property, location, and time and is somewhat arbitrary.

At the local scale of investigation, it is necessary to define a REV. Representative in this sense refers to the necessity of finding an element that is uniform for all properties and variables. The two conditions given above must be met.

At the regional scale of observation, selection of the REV becomes even more important. Here the solid particles are replaced by zones of materials that have specified characteristics.

Areas of Important Soil Physical Research

Areas of current research interest in soil physics include the following:

1. Transport studies, including
   a. movement and retention of water, soluble salts, and pesticides in and through soil profiles,
   b. extraction and movement of plant nutrients and pesticides to seed and plant roots,
   c. movement of oxygen in soil,
   d. transfer of heat in soil.

2. Tillage of the soil, including minimum tillage and no-till.
3. Influence of soil compaction, including tillage pans, fragipans, and crusts.
4. Evapotranspiration, including drought effects on plants and soil-plant-water relations.
5. Soil and water conservation, including reductions in water and sediment transport off-site.
6. Irrigation and water quality studies.
7. Spatial and temporal variability of soil properties.
8. Development of new techniques to measure soil water such as remote sensing and time domain reflectance.
9. Use of geographical information systems of soil properties to summarize, model, and display soil properties.
10. Computer simulation modeling of dynamic processes in soil science.

Relation of Soil Physics to Physics

Soil physics concerns itself with the physical, physical-chemical, and physical-biological relationships among the solid, liquid, and gaseous phases of soil as they are affected by temperature, pressure, and energy.

Physics concerns itself mainly with the more fundamental aspects of energy and non-living matter. As an aid to study, physics has been divided into six areas: sound, light, atomics, mechanics, thermodynamics, and electricity. The memory aid for these areas is "slamte." Physics has been called the most basic of the natural sciences; it seeks to establish mathematical laws to explain and predict the behavior of mass and energy. In this course, we will emphasize the application of mechanics and thermodynamics to understanding and predicting soil physical behavior.

Physics often uses mathematics as its language. The results of many observations and experiments, called laws, can usually be written with algebraic equations. We will also use mathematics to represent fundamental concepts and will show similarities between these concepts in soil physics and those in other closely related scientific areas. A brief review of several elementary ideas in physics is given in Appendix C.

Relation of Soil Physics to Soil Science

Soil physics is considered one of the three fundamental areas of soil science. Currently, the Soil Science Society of America is divided into 11 areas of concentration:

S-1   Soil physics
S-2   Soil chemistry
S-3   Soil biology and biochemistry
S-4   Soil fertility and plant nutrition
S-5   Pedology
S-6   Soil and water management and conservation
S-7   Forest and range soils

S-8   Fertilizer technology
S-9   Soil mineralogy
S-10   Wetland soils
S-11   Soil and environmental quality

There are no definite boundaries between these areas of soil science; all areas are related. However, soil physics, soil chemistry, and soil biology and biochemistry are considered the most fundamental areas of Soil Science.

History of Soil Physics

Excellent reviews of the history of soil physics have been published by Philip (1974), Gardner (1977), and Swartzendruber (1977). They indicate that the physical aspects of the soil have been manipulated since the early use of irrigation and tillage. These basic activities are estimated to have been practiced for at least 6000 years. However, soil physics as a scientific area of emphasis had to wait until the emergence and development of physics itself, which occurred primarily during the latter half of the 17th century.

The systematic study of agriculture did not get under way until the first half of the 19th century, and soil physical studies were a large part of the early work. The intensity of this effort to study the physical properties of soil soon decreased after the "new" chemical theory of soil fertility emerged. Therefore, the development of other branches of soil science, namely, pedology and soil chemistry, preceded soil physics by a generation or two.

Early soil physicists were mainly concerned with the physical description of soils. They measured the particle size distribution, water content, water potential, pore space, and temperature of the soil. To aid in these descriptions, a number of terms were defined, particularly with regard to soil water. Such terms as hygroscopic water, capillary water, gravitational water, etc., were used to qualitatively describe the soil water. As the science progressed, however, the qualitative descriptions of the physical state of the soil system gave way to more quantitative descriptions, and soil system dynamics began to be considered. Over time, greater use was made of mathematics to describe the status and dynamics of soil physical properties.

Since the beginning of this century, and particularly since the 1930s, soil physics has become well established and increasingly recognized as a vital field of the earth sciences, as a separate discipline, and as a meeting place for the natural sciences. In his review of the status of soil physics, Philip (1974) estimated that (1) about 90% of all work ever accomplished in soil physics had been done in the preceding 50 years, and (2) the soil physicists living in 1974 constituted more than 80% of those who ever lived. Other textbooks that provide a brief history of soil physics include Baver et al. (1972) and Hillel (1982).

A discussion of the historical impact of mismanagement of soils has been given by Hillel (1991). He provides examples of empires that have vanished due to the impairment of soil properties, including the soil physical processes of erosion, runoff, sedimentation, salinity, drainage, and evapotranspiration. Several of these processes are discussed in detail in this book. The approach taken here will be to initially characterize the physical status of soils emphasizing quantification of the soil physical properties in the profile and the landscape and then to discuss the principles of dynamic processes such as water, heat, oxygen, and solute transport.

Cited References

Baver, L. D., W. H. Gardner, and W. R. Gardner. 1972. *Soil Physics*, pp. vii–xv. 4th ed. John Wiley and Sons. New York.

Corey, A. T. 1994. *Mechanics of Immiscible Fluids in Porous Media.* Water Resources Publications. Highlands Ranch, CO.

de Marsily, G. 1986. *Quantitative Hydrogeology: Groundwater Hydrology for Engineers.* Academic Press. New York.

Gardner, W. H. 1977. Historical highlights in American soil physics, 1176–1976. *Soil Science Society of America Journal* 41:221–229.

Hillel, D. 1982. *Introduction to Soil Physics,* chaps. 1–2. Academic Press. New York.

Hillel, D. J. 1991. *Out of the Earth: Civilization and the Life of the Soil.* Free Press, Macmillan. New York.

Philip, J. R. 1974. Fifty years of progress in soil physics. *Geoderma* 12:265–280.

Swartzendruber, D. 1977. *Soil Physics: Reflections and Perspectives.* American Society of Agronomy Special Publication 30-53-71. Madison.

Additional Reference

Medawar, P. B. 1979. *Advice to a Young Scientist.* Harper's. New York.

Problems

1. Define the following:
    a. soil physics
    b. physics
    c. mechanics
    d. thermodynamics

2. Read the article by W. H. Gardner published in the *Soil Science Society of America Journal* in 1977. Who was selected as the "father of soil physics" in the United States? Why?

3. Read Daniel Hillel's book published in 1990. Write a summary of his conclusions on the historical importance of soil and water to civilization. Choose a continent and describe the effects of civilization on the conservation of its soil and water resources.

4. Distinguish between physics and soil physics.

5. What are the differences between the rate of a solute transport process and the total amount of solute moved?

6. How can the position of a substance in a landscape be defined?

7. Thought-provoking ideas

   a. In a two-dimensional profile plot, how does one decide which variable is to be plotted on the x axis and which is to be plotted on the y axis?

   b. How does the texture of a soil influence soil temperature, soil aeration, infiltration, and runoff of soil water?

   c. How does the temperature of a soil affect the rates of microbial and chemical reactions?

   d. How does the water content of a soil affect the rates of microbial and chemical reactions?

8. What are the differences between average rate of change and instantaneous rate of change?

CHAPTER TWO

Calculations and Dimensions of Physical Quantities

Introduction

This chapter is a review of selected background information in soil physics. Many of the concepts encountered will have been discussed previously in introductory courses in soil science. Why then the need for a review? First, from experience, I have found it necessary to restate these familiar concepts in a condensed, clear fashion. Second, students generally need practice to develop their abilities to analyze and solve practical problems.

Many advanced undergraduates and graduate students can understand the fundamental principles fairly easily but often have difficulty in applying them to unfamiliar situations in the field. Scientists who become competent in their profession by mastering the techniques developed by their predecessors have the potential to become pioneers in developing new techniques or solving new, more complicated problems.

This chapter begins with a discussion of units, dimensions, and conversion factors and then reviews several soil physical quantities and their volume and mass relationships. Examples are presented in which the soil physical parameters are calculated from known dimensions.

Physical Dimensions and Units

Mechanical quantities frequently used in the sciences and engineering include parameters such as

energy	force	area	velocity	acceleration
work	power	volume	pressure	momentum

Several of these parameters are defined in Appendix C. All of them can be expressed in soil physics in terms of four dimensional quantities:

length	mass	time	temperature

To communicate without ambiguity, a consistent terminology and a unified system of measurement should be used.

The dimension of a physical quantity is composed of a *numerical value* (a number) and its *unit*. This can be represented as

physical quantity = numerical value and unit

The magnitude, or numerical value, conveys the relative extent of the dimension, while the units define the basis on which the comparison is made. Since neither the physical quantity nor the symbol used to denote it implies the choice of the unit, a coherent system of units must be constructed from a set of *base units* from which all *derived units* are obtained by multiplication or division without introducing numerical factors other than powers of 10.

Example
The bulk density of soil is defined as oven-dry mass/total soil volume and can have the following units: g/cm^3, Mg/m^3, kg/m^3, lb/ft^3.

Dimensional Quantities

The building blocks on which soil physical properties are based are the dimensional quantities. Dimensions are our basic concepts of measurements and can be divided into fundamental dimensions and derived dimensions. The units are the means of expressing the dimensions. The four fundamental dimensional quantities and the units that we are mostly concerned with in this course are length, mass, time, and temperature. A discussion of these quantities follows.

Length

The meter is a measure of distance. A meter is the distance that light travels in 1/299,793 seconds. The multiples and submultiples of the meter are

kilometer = km = 1000 m
centimeter = cm = 0.01 m = 10^{-2} m
millimeter = mm = 0.001 m = 10^{-3} m
micron = μm = 10^{-4} cm = 10^{-6} m
Ångstrom = Å = 10^{-8} cm = 10^{-10} m

The derived dimensions of length are area and volume, with units of m^2 and m^3, respectively. Key relationships for area are

1 hectare (ha) = 10^4 m^2 and   1 acre = 4046 m^2

and for volume,

1000 liters = 1 m^3

The process of measuring length, area, and volume is called mensuration.

Example

Determine the surface area and volume of each of the following shapes: cube, sphere, and right circular cylinder. Assume that the height is h, the length of a side is L, and, if needed, the radius is L/2.
Answer:

Shape	Surface Area	Volume	Application
Cube	$6L^2$	L^3	soil ped
Sphere	πL^2	$\pi L^3/6$	seed
Cylinder	πLh	$\pi h(L/2)^2$	root

Mass

The base unit for mass is the kilogram; 1 kg is the mass of a Pt-Ir metal cylinder kept near Paris, France. Mass is a measure of inertia. Inertia is the resistance to change in the amount and direction of motion. Simply, mass is a measure of the quantity of matter in an entity.

We do not measure mass directly, however; we measure its relationship to its weight. Weight is the force of attraction from the center of the earth. This relationship is derived as follows.

Newton's law of universal gravitation can be expressed as

$$F = \frac{GM_1M_2}{d^2} \qquad [1]$$

where F is the force (N), G is the universal gravitational constant, M is the mass of a given body (kg), and d is the distance between the bodies (m). The student should insert units for F, M, and d and solve for G to determine its units. Possible units include

$G = 6.67 \times 10^{-11}$ N·m^2/kg^2
$= 6.67 \times 10^{-8}$ dynes·cm^2/g^2
$= 6.67 \times 10^{-8}$ cm^3/s^2·g

This shows that the units of G depend on the units chosen for M, d, and F.

For a small body of mass m at the surface of the earth, the force of gravitation is

$F = mg =$ weight   [2]

Here

units of acceleration $= g = GM/d^2 = 9.8$ m/s^2

where

M = mass of the earth (6×10^{24} kg)
d = radius of the earth (6.4×10^{6} m)

The masses of bodies on the surface of the earth can be compared by computing the ratio

$F_1/F_2 = m_1g/m_2g = m_1/m_2$ [3]

This shows that the weights of two bodies at the same point on the surface of the earth are in strict proportion to their masses. Therefore, masses can be compared by weighing.

Example

Calculate the average bulk density of the earth and interpret your answer. Assume that the earth has a mass of 6×10^{27} g, a radius of 6.4×10^{8} cm, and a spherical shape.

$$\text{bulk density} = \rho_b = \frac{6 \times 10^{27}}{(4/3)\pi (6.4 \times 10^8)^3}$$
$$= 5.5 \text{ g/cm}^3$$
$$= 5500 \text{ kg/m}^3$$
$$= 5.5 \text{ Mg/m}^3$$

From these calculations, we conclude the following:

1. The average bulk density of the earth is 5.5 times as dense as water, the base measure.

2. The bulk density at the soil surface ranges from 1000 to 1800 kg/m^3. Therefore, in order to bring the average bulk density of the earth up to our calculated value, this must mean that the earth has a very dense core.

3. This bulk density value also has implications for porosity calculations. The depth of water in the earth is about 3 km on the average and as deep as 8 km in some places.

The density of the earth's crust varies with the gravitational acceleration constant; that is, g varies slightly with position on the earth's surface (as well as with the distance above the earth's surface). We can avoid these terrestrial variations when calculating the weight of a body by comparing the weight of the body with a matching known weight on a beam type of balance. This technique involves direct comparison of the unknown quantity with a known quantity. Another technique for measuring weight is to place the object on a scale and observe the linear deflection caused by the compression of a carefully calibrated spring. This is an indirect comparison, because we are measuring the predictable effect of the force W on the length of the spring.

Time

The fundamental invariable unit of time is the second, which is 1/86,400 of the mean solar day. We measure time in two different ways. On the one hand, we may want to know the time of day or the day of the year so we can properly sequence events; that is, we want to know an instant of time. Time in this case is an entity. On the other hand, we may need to measure the duration of a process, that is, a lapse of time. In this case, time is a phenomenon.

Temperature

Temperature may be measured in Kelvin (K), the measure of how hot or cold a body is with respect to a standard object. It is a measure of the thermal state of a body considered in reference to its ability to transfer heat to other bodies.

Absolute temperature in a gas or liquid is proportional to the average translational kinetic energy per molecule due to random motions (Miller 1959). It is a measure of the thermal energy. Temperature is the condition of an entity that determines its potential to transfer heat to another entity or to be heated by another entity.

The three temperature scales frequently used in the United States are shown in Figure 2.1. These relative temperature scales do not have a common zero, and the unit of measure differs.

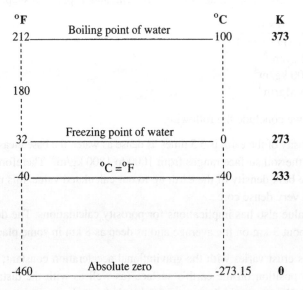

Figure 2.1. Temperature scales frequently used in this textbook.

The important temperature relations between the scales are written as follows:

$$°C = (5/9)(°F - 32) \qquad [4]$$

$$°F = [(9/5)(°C)] + 32 \qquad [5]$$

$$K = °C + 273.15 = [(5/9)(°F - 32)] + 273 \qquad [6]$$

We must be able to convert from one scale to another because no temperature scale is universally accepted in the United States. When we cite the temperature of a soil, we are stating the number of units of the temperature scale that occurs as measured from the reference point.

We cannot measure temperature directly. However, there are measurable physical properties that vary directly as temperature varies. Among these are the volume of a fluid, the length of a rod, the electrical resistance of a wire, or the color of a lamp filament. Any of these properties can be used to construct a thermometer. As an example, temperature can be measured by a mercury thermometer, in which the volume of mercury changes as it is heated or cooled. The height of a column of mercury in a glass tube increases or decreases as the mass of mercury in the base of the thermometer expands or contracts. Marking of the height of the mercury column after it stabilizes at a particular level corresponds to an indirect measurement of temperature.

Example

If the degrees centigrade, referred to now as Celsius, are 0, 10, 25, 30, and 40, determine the corresponding degrees Fahrenheit. Draw a graph of the relationship between the two temperature scales and determine the slope and the intercept of the line.

Answer: 32, 50, 75, 86, and 104 degrees; slope 9/5; and intercept 32.

Example

If the soil temperature is 0°F, what is the soil temperature in degrees Celsius?

Answer: $\approx -17.7°C$.

These example problems show that the temperature scales have their zero values at different points and that the unit degree of Celsius is the same as the unit degree of Kelvin but is 1.8 times greater than the unit degree of Fahrenheit.

Systems of Units

The MKS (meter-kilogram-second) system is an absolute system; that is, the definition of these units does not depend on local variations in the earth's gravitational field. The International System of Units (see below) is essentially the MKS system.

The cgs (centimeter-gram-second) system is also an absolute system.

The fps (foot-pound-second) system depends on gravity since the pound is a unit of force, or weight.

The International System of Units (Système International d'Unites, abbreviated internationally as SI) is an absolute system. Base and supplementary SI units are given in Table 2.1.

Table 2.1. Base and Supplementary SI Units and Their Symbols

Quantity	Unit	Symbol
Amount of substance	mole	mol
Electric current	ampere	A
Length	meter	m
Luminous intensity	candela	cd
Mass	kilogram	kg
Thermodynamic temperature	Kelvin	K
Time	second	s
Plane angle	radian	rad
Solid angle	steradian	sr

Table 2.2 Examples of Derived SI Units and Their Symbols

Quantity	Name	Symbol	SI Base Units	Derived Units
Area	square meter	—	m^2	
Volume	cubic meter	—	m^3	
Velocity	meter per second	—	m/s	
Acceleration	meter per square second	—	m/s^2	
Surface tension	newton per meter	—	N/m	
Density	kilogram per cubic meter	—	kg/m^3	
Force	newton	N	$kg \cdot m/s^2$	
Pressure	pascal	Pa	$kg/m \cdot s^2$	N/m^2
Energy (work)	joule	J	$kg \cdot m^2/s^2$	$N \cdot m$
Power	watt	W	$kg \cdot m^2/s^3$	J/s
Heat flux density	watt per square meter	—	kg/s^3	W/m^2
Specific heat capacity	joule per kilogram kelvin	—	$m^2/s^2 \cdot K$	$J/kg \cdot K$

Source: Adapted from Campbell 1977, p. 5.

Throughout this book we will mostly use units consistent with the SI system. Some of these units may seem unfamiliar at first, but the increased efficiency that results from using a consistent set of units easily offsets the small inconvenience associated with learning the new values.

Examples of Derived SI Units and Their Symbols

Many of the derived units in the SI system are given special names honoring physicists who have made a substantial impact on the science (Table 2.2). The units have also been given corresponding symbols. For example, the unit of force has the symbols $kg \cdot m/s^2$ and has been given the name newton and the symbol N. Similarly, the unit of energy is the newton-meter (N·m), with the short name joule and symbol J. The unit of power is the watt, defined as 1 joule per second, and its symbol is W.

Table 2.3. Accepted SI Prefixes and Symbols for Multiples and Submultiples of Units in the SI System

Multiple	Prefix	Symbol	Submultiple	Prefix	Symbol
10^{24}	yotta	Y	10^{-1}	deci	d
10^{21}	zetta	Z	10^{-2}	centi	c
10^{18}	exa	E	10^{-3}	milli	m
10^{15}	peta	P	10^{-6}	micro	μ
10^{12}	tera	T	10^{-9}	nano	n
10^{9}	giga	G	10^{-15}	femto	f
10^{6}	mega	M	10^{-12}	pico	p
10^{3}	kilo	k	10^{-18}	atto	a
10^{2}	hecto	h	10^{-21}	zepto	z
10^{1}	deka	da	10^{-24}	yocto	y

Sources: Adapted from Campbell 1977, p. 5; Bitton 1998, p. 275.

Accepted SI Prefixes and Symbols for Multiples and Submultiples of Units

One of the best features of the SI system is that units and their multiples and submultiples are related by standard factors designated by the prefixes (Table 2.3). Prefixes are not preferred for use in denominators (except kg). Do not use double prefixes (e.g., use nanometer but not millimicrometer). When a compound unit is formed by multiplication of two or more other units, its symbol consists of the symbols for the separate units joined by a centered dot. In this textbook, we use the convention whereby all units after the slash (/) are in the denominator. Hyphens should not be used in symbols for compound units. Positive and negative exponents may be used with the symbols for units. In the SI system, the symbols are not abbreviations. A symbol such as dm has to be taken as a whole. When prefixes are used with symbols raised to a power, the prefix is also raised to the same power. We should interpret cm^3 as $(10^{-2} m)^3$ and not as $10^{-2}(m^3)$.

The SI units required by the American Society of Agronomy (American Society of Agronomy et al. 1988) in its published manuscripts are presented in Table 2.4.

Constants and Conversions among Physical Quantities

Physical constants are frequently required in equations governing physical systems. In addition, sometimes there is a need to convert from one system of units to another. Tables 2.5 and 2.6 provide information on several useful constants and system conversions.

Volume and Mass Relationships in Soils

Review of Soil Materials

The soil is a very complex system. Its behavior depends on the magnitude and interactions among the three phases: solid, liquid, and gas. Recall that one of our definitions of

Table 2.4. Frequently Used Quantities in the Soil and Crop Sciences

Quantity	Application	Unit	Symbol
Length	soil depth	meter	m
	plant height		
Area	pot area	square centimeter	cm^2
	leaf area	square meter	m^2
	land area	hectare	ha
	specific surface area of soil	square meter per kilogram	m^2/kg
Volume	lab	liter	L
	field	cubic meter	m^3
Water content	soil	kilogram water per kilogram dry soil	kg/kg
		cubic meter water per cubic meter soil	m^3/m^3
Density	soil bulk density	megagram per cubic meter	Mg/m^3
Electrical conductivity	salt tolerance	decisiemen per meter	dS/m
Evapotranspiration rate	plants	cubic meter per square meter second	$m^3/m^2{\cdot}s$ or m/s
Elongation rate	plants	meter per second	m/s
		millimeter per second	mm/s
		meter per day	m/d
Ion transport	ion uptake	mole per kilogram (of dry plant tissue) per second	$mol/kg{\cdot}s$
		mole of charge per kilogram (of dry plant tissue) per second	$mol\ (p^+)/kg{\cdot}s$ $mol\ (e^-)/kg{\cdot}s$
Resistance	stomatal	second per meter	s/m
Transpiration rate	water mass flux density	milligram per square meter second	$mg/m^2{\cdot}s$
Yield	grain or forage yield	megagram per hectare	Mg/ha
	mass of plant or plant part	gram per square meter	g/m^2
		gram (gram per plant or per plant part)	g (g/plant; g/kernel)
Specific heat	storage	joule per kilogram kelvin	$J/kg{\cdot}K$
Flux density	heat flow	watt per square meter	W/m^2
	gas diffusion	gram per square meter per second (P)	$g/m^2/s$
		mole per square meter per second (A)	$mol/m^2/s$
	water flow	kilogram per square meter second (P)	$kg/m^2{\cdot}s$
		cubic meter per square meter second (A)	$m^3/m^2{\cdot}s$ or m/s
Thermal conductivity	heat flow	watt per meter kelvin	$W/m{\cdot}K$
Gas diffusivity	gas diffusion	square meter per second	m^2/s
Hydraulic conductivity	water flow	kilogram second per cubic meter (P)	$kg{\cdot}s/m^3$
		cubic meter second per kilogram (A)	$m^3{\cdot}s/kg$
		meter per second (A)	m/s

Table 2.5. Some Useful Constants

Constants	Dimension
Atmospheric pressure	1.01325×10^5 Pa
Avogadro's number	6.0221367×10^{23} atoms/mol
Boltzmann constant	1.380658×10^{-23} J/K
Gas constant	8.31451 J/mol·K
Gravitational acceleration	9.80665 m/s^2
Molar volume of an ideal gas at 1 atm and 25°C	24.465 L/mol
Planck constant	6.626×10^{-34} J·s
Stefan-Boltzmann constant	5.67×10^{-8} J/m^2·s·K^4
Zero on the Celsius scale	273.15 K

Source: Adapted from Bitton 1998, p. 275.

Table 2.6. SI System Conversions of Useful Parameters

Length
meter (m) = 1000 mm = 100 cm = 10 dm = 39.37 in = 3.281 ft = 1.094 yd = 6.21×10^{-4} mi
kilometer (km) = 10^6 mm = 10^5 cm = 10^3 m = 39,370 in = 3281 ft = 1093.6 yd = 0.621 mi

Area
square meter (m^2) = 10^4 cm^2 = 10^{-6} km^2 = 10^{-4} ha = 1550 in^2 = 10.76 ft^2 = 1.196 yd^2 = 3.861×10^{-7}
    mi^2 = 2.47×10^{-4} ac
hectare (ha) = 10^8 cm^2 = 10^4 m^2 = 0.01 km^2 = 1.55×10^7 in^2 = 10.76×10^4 ft^2 = 1.196×10^4 yd^2 =
    3.861×10^{-3} mi^2 = 2.471 ac

Volume
cubic meter (m^3) = 10^6 cm^3 = 1000 L = 61,023 in^3 = 35.314 ft^3 = 264.17 U.S. gal = 220.08 imp. gal =
    8.107×10^{-4} ac-ft
liter (L) = 1000 cm^3 = 0.001 m^3 = 61.023 in^3 = 0.0353 ft^3 = 0.26417 U.S. gal = 0.22008 imp. gal =
    8.1×10^{-7} ac-ft

Flow
cubic meter per second (m^3/s) = 86,400 m^3/d = 1000 L/s = 3.051×10^6 ft/d = 70.045 ac-ft/d = 15,850
    gal/min = 22.82×10^6 U.S. gal/d = 22.824 Mgal/d
liter per second (L/s) = 0.001 m^3/s = 0.0353 ft^3/s = 3051.2 ft^3/d = 0.070 ac-ft/d = 15.85 gal/min = 22,824
    U.S. gal/d = 2.28×10^{-2} Mgal/d

soil physics involved the impact of pressure and energy on the three phases of the soil. The conceptual unit volume of soil is shown in Figure 2.2.

Solid Phase

Under field conditions, the solid phase occupies from 30 to 60% of the total soil volume. Ideally, the solid phase occupies approximately 50% of the soil by volume. Except for the organic-matter content, the solid phase is relatively stable over time. Based on the unit volume of soil, the general weight and volume relationships of the inorganic and organic material are

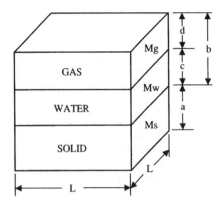

Figure 2.2. Three phases of a unit volume of soil. The symbols Mg, Mw, and Ms represent the masses of the gaseous, aqueous, and solid phases, respectively. The symbol L represents the length of one side of the unit volume of soil. The thicknesses of the three phases are represented by lowercase letters.

	Weight (%)	Volume (%)
Inorganic	98	49
Organic	2	2

Some of the solid material consists of crystalline inorganic particles, and some consists of amorphous gels that may coat the crystals and modify their behavior. The adhering amorphous material may consist of iron and aluminum oxides or of organic matter that attaches itself to soil particles and joins them together. The inorganic portion consists of particles having various shapes, sizes, and chemical compositions. The organic fraction consists of live active organisms and plant and animal residues in different stages of decomposition. Thus, the amount and kind of surfaces of the solid phase in soils and the interactions with the solution (liquid) and gaseous phases vary tremendously.

Liquid Phase

The liquid phase is dynamic rather than stable because its volume usually varies between 0 and approximately 50% in soil. It is the soil water that fills part or all of the open spaces between the solid phase. These open spaces are called *pores*.

The liquid phase varies in amount (water quantity), chemical composition (water quality), and the freedom with which it moves. Ionic composition in the soil solution includes inorganic ions, such as Na, K, Ca, Mg, Cl, NO_3, SO_4, HCO_3, and various organic ions and molecules.

Gaseous Phase

The gaseous phase is also dynamic and not stable. It occupies that part of the soil pore space not occupied by the liquid phase. Its amount, composition, and mobility may vary with time and position. The composition of the soil gas phase is mostly N_2, O_2, water vapor, and CO_2, with traces of other gases.

The amount of each of the three phases of the soil can be quantified by calculations of the mass and volume. A large proportion of applied soil physics is based upon the mass and volume relationships of the three phases, the proportion of a given phase in a unit volume of soil, and the rate of change of the mass and volume in space and time. The physical and chemical relationships among the solid, liquid, and gaseous phases are affected not only by the properties of each phase but also by temperature and pressure.

Volume/Mass Relationships of Soil Physical Quantities

Many soil physical properties are based on the mass and volume relationships shown in Figure 2.2. Here, L = length of the edge of the cube, a = depth of the solid, b = depth of the pores, sometimes called voids, c = depth of water, d = depth of gas, and Ms, Mw, Mg = mass of solids, water, and gas, respectively. Also, we can write

Mt = Ms + Mw + Mg = total mass of soil   [7]

Vt = Vs + Vw + Vg = total volume of soil   [8]

where Vs, Vw, and Vg are the volume of solids, water, and gas, respectively. In the SI system, the units of mass are kilograms (kg), and the units of volume are cubic meters (m^3).

The relationship between mass and volume for any material in a phase is

mass = volume $\times$ density   [9]

Example

An individual spherical soil particle has a diameter of 2 mm. What is its mass?

mass = volume $\times$ density
volume = $(4/3)\pi r^3 = (4/3)\pi(0.001)^3 = 4.189 \times 10^{-9}$ m^3
density = 2650 kg/m^3

Therefore, assuming the volume of the soil particle can be approximated by a sphere, the mass of the particle of soil is 1.11×10^{-2} g.

Several physical quantities have been derived to characterize the soil system. These quantities frequently use the mass and length units to define the characteristics of the solid, liquid, and gaseous phases of soils.

Water Density

Water density, ρ_w, can be calculated as

ρ_w = Mw/Vw = Mw/cL^2 = kg water/m^3 water   [10]

For most calculations, we will assume that the density of water is 1000 kg/m^3.

Dry Bulk Density

Dry bulk density, ρ_b, can be calculated as

ρ_b = Ms/Vt = Ms/Vt = Ms/L^3 [11]

The dry bulk density of most mineral soils varies from 1000 to 1800 kg/m^3. It may also vary due to tillage, compaction, swelling, aggregation, freezing and thawing, etc., and is not considered a static property at or near the soil surface. It is, however, reasonably constant in the lower portion of the soil profile.

Organic soils have lower bulk densities than mineral soils due to the lower densities of organic materials. Bulk densities ranging from 800 to 1000 kg/m^3 are common for these soils.

Particle Density

The particle density, ρ_p, can be calculated as

$$\rho_p = Ms/Vs = Ms/aL^2 = \text{kg solids/m}^3 \text{ solids} \qquad [12]$$

Particle densities of mineral soils vary between 2600 and 2700 kg/m^3 and for most mineral soils average 2650 kg/m^3. Particle density is considered to be a static property of a soil having a given mineralogical composition.

Total Porosity

The symbol for total porosity is f:

$$f = bL^2/Vt = (Vw + Vg)/Vt = \text{m}^3 \text{ voids/m}^3 \text{ soil} \qquad [13]$$

Total porosity is an index of the relative volume of pores in the soil. For mineral soils, its value generally is between 0.3 and 0.6 m^3/m^3. For coarse-textured soils, values of f tend to be less than for fine-textured soils, even though the average sizes of the pores are larger in the coarse-textured soils. Total porosity tends to decrease with depth in the profile due to compaction. Total porosity gives no information about the pore size distribution.

The total porosity of a core of soil is determined experimentally by saturating and weighing the core, then oven-drying and reweighing the same core. The difference in weight between saturated and oven-dry cores of soil represents a volume of water equal to the volume of the pore space in the soil.

Aeration Porosity

The symbol for aeration porosity is f_a:

$$f_a = Vg/Vt = d\ L^2/L^3 = \text{m}^3 \text{ air/m}^3 \text{ soil} \qquad [14]$$

This quantity is also known as the air-filled porosity. It represents the volume fraction of the soil filled with air and varies indirectly with volumetric soil water content.

Void Ratio

The symbol for void ratio is e:

$$e = (Vg + Vw)/Vs = \text{m}^3 \text{ voids/m}^3 \text{ solids} \qquad [15]$$

Void ratio expresses the relationship between the volumes occupied by solids and by voids. Therefore, the volume of voids in a soil volume is the sum of the volumes of the liquid and gaseous phases. Void ratio also is an index of the relative volume of soil pores but relates to the volume of solids rather than to the total volume of soil. It is used where the soil is undergoing compaction, shrinking, or swelling and mostly in soil engineering and mechanics. Note that a change in pore volume changes only the numerator. Values of e vary between 0.3 and 2.0 m^3/m^3. Compacted soils tend to have void ratios less than 1.0.

Water Content

The water content, θ, of a soil is a dimensionless ratio of two masses or two volumes or is given as a mass per unit volume. When either of the two dimensionless ratios is multiplied by 100, the values become percentages, and the basis (mass or volume) should be stated.

Soil water content by weight, θ_w, is calculated as

$$\theta_w = Mw/Ms = kg\ water/kg\ solids \qquad [16]$$

$$\% \text{ water (by weight)} = \theta_w \times 100$$

$$= \frac{\text{wt. (wet soil)} - \text{wt. (dry soil)}}{\text{wt (dry soil)}} \times 100$$

$$= \frac{\text{wt. (water lost)}}{\text{wt. (dry soil)}} \times 100$$

In order to show how the container weight drops out, we write

$$\frac{(\text{container wt.} + \text{wet-soil wt.}) - (\text{container wt.} + \text{dry-soil wt.})}{(\text{container wt.} + \text{dry-soil wt.}) - \text{container wt.}}$$

Example

Suppose the following conditions were operating:

container weight = 37.85 g
container weight + wet-soil weight = 45.38 g
container weight + dry-soil weight = 44.12 g

Calculate the percent water by weight.

$$\% \text{ water} = \frac{45.38\ g - 44.12\ g}{44.12\ g - 37.85\ g} \times 100$$

$$= \frac{1.26}{6.27} \times 100$$

$$= 20.1\%$$

For most example problems in this textbook, we will use θ_w rather than percent water to express the gravimetric soil water content. In nearly all cases for mineral soils, values of θ_w are less than 1.

It is frequently necessary to make a conversion from water content on a dry-weight basis to water content on a wet-weight basis. The equation for this conversion is

$$\theta_{ww} = \frac{\theta_{dw}}{1+\theta_{dw}}$$ [17]

where θ_{dw} and θ_{ww} represent the soil water content on a dry-weight and wet-weight basis, respectively.

Soil water content by volume, θ_v, is calculated as

$$\theta_v = Vw/Vt = m^3 \text{ water}/m^3 \text{ soil}$$ [18]

Soil water content by volume is usually determined by filling a cylinder of known volume with soil in its natural state of packing. This operation is carried out by first placing the cylinder in a cutting-head assembly, then forcing the cutting head into the soil, extracting the core, and then drying the core in a forced-draft oven at 105°C for 12 hours. Another method is to measure the volume of a small clod. This is done by coating the soil with paraffin, immersing the clod in water, and measuring the amount of water displaced. Usually, the soil water content by volume is determined experimentally or is calculated from the equation

$$\theta_v = \theta_w(\rho_b/\rho_w)$$ [19]

The water volume ratio, θ_r, is calculated as

$$\theta_r = Vw/Vs = m^3 \text{ water}/m^3 \text{ solids}$$ [20]

This ratio is sometimes known as the liquid ratio. It has the same form as the void ratio but without Vg in the numerator and is primarily used when the soil volume changes.

Equivalent Depth of Water per Depth of Soil

The quantity D_{eq} is

$$\text{depth of water} = D_{eq} = Vw/L^2 = cL^2/L^2 = \theta_v\Delta z$$ [21]

where Δz is the depth interval. The SI units of D_{eq} are m^3 of water per m^2 of soil surface. This equation shows that D_{eq} can be calculated by the product of the volumetric water content and the depth interval Δz and has units of volume of water per area of soil. The θ_v is an average θ_v over the depth interval. When θ_v varies in a soil profile, most often we do not know the mathematical form of $\theta_v(z)$. We then use numerical techniques such as regression or compartmentalization to approximate the relationship between θ_v and z (Appendix B). For example, numerically, we can write the equation for D_{eq} for a profile using calculus as follows:

$$D_{eq} = \int_0^z \theta_v(z)dz \approx \sum_{i=0}^{i=n} (\theta_{vi}\Delta z_i)$$ [22]

where n is the number of layers.

A graph of θ_v versus z for a given profile water content is shown in Figure 2.3. Over time, more than one curve may exist, and then the differences in water depth between the two curves can be determined. This graphing technique can be used to determine the amount of water in the profile at any given time and the amount of water lost from the profile between two sampling times.

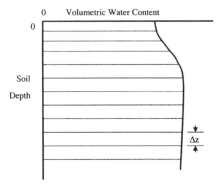

Figure 2.3. Volumetric soil water content distribution in the profile. The equivalent depth of water is the area to the left of the curve.

Relative Saturation Ratio

The relative saturation ratio, θ_s, is calculated as

$$\theta_s = Vw/(Vw + Vg) = m^3 \text{ water}/m^3 \text{ voids} \tag{23}$$

Values of θ_s express the volume of water present in the soil relative to the volume of voids. The ratio ranges from 0 to 1.0 m^3/m^3, with the highest value of θ_s at saturation. A soil can maintain a θ_s of 1.0 m^3/m^3 even though its water content is changed if the soil experiences compression or expansion, since these indicate a decrease or increase in void space. This parameter is sometimes known as the degree of saturation.

Relationships among Soil Physical Quantities

From the basic mass-volume definitions given above, relationships between several of the physical quantities have been derived. The derivation of several of these quantities are given below.

The relation between void ratio and porosity is

$$e = f/(1 - f) \quad \text{or} \quad f = e/(1 + e) \tag{24}$$

$$
\begin{aligned}
e &= [(Vg + Vw)/Vs](Vt/Vt) \\
&= [(Vg + Vw)/Vt](Vt/Vs) \\
&= f\{Vt/[Vt - (Vw + Vg)]\} = f/[1 - (Vw + Vg)/Vt] \\
&= f/(1 - f)
\end{aligned}
$$

The relation between volumetric water content, saturation ratio, and total porosity is

$$\theta_v = \theta_s f \tag{25}$$

The relation between total porosity, bulk density, and particle density is

$$f = 1 - (\rho_b/\rho_p) \tag{26}$$

$$
\begin{aligned}
f &= (Vw + Vg)/Vt = (Vt - Vs)/Vt \\
&= [1 - (Vs/Vt)]\{1 - [(Vs/Vt)(Ms/Ms)]\} \\
&= 1 - [(Ms/Vt)(Ms/Vs)] \\
&= 1 - (\rho_b/\rho_p)
\end{aligned}
$$

The relation between volumetric water content, water content by weight, bulk density, and water density is

$$\theta_v = \theta_w(\rho_b/\rho_w) \tag{27}$$

$$\begin{aligned} \theta_v &= (Vw/Vt)(Mw/Mw)(Ms/Ms) \\ &= (Mw/Ms)(Ms/Vt)(Vw/Mw) \\ &= \theta_w(\rho_b/\rho_w) \end{aligned}$$

The relation between aeration porosity, total porosity, volumetric water content, and saturation ratio is

$$f_a = f - \theta_v = f(1 - \theta_s) \tag{28}$$

$$\begin{aligned} f_a &= Va/Vt = [(Va + Vw) - Vw]/Vt \\ &= [(Va + Vw)/Vt] - (Vw/Vt) \\ &= f - \theta_v \end{aligned}$$

The relation between water volume ratio, volumetric water content, and void ratio is

$$\theta_r = \theta_v(1 + e) \tag{29}$$

$$\begin{aligned} \theta_r &= Vw/Vs = (Vw/Vs)(Vt/Vt) \\ &= (Vw/Vt)[(Vs + Vw + Vg)/Vs] \\ &= \theta_v\{1 + [(Vw + Vg)/Vs]\} \\ &= \theta_v(1 + e) \end{aligned}$$

Experimental Determination of Soil Physical Quantities

The field weight of soil can be determined from the mass of the phases as follows:

Mt = Ms + Mw + Mg
Mt = Ms + Mw

because Mg is negligible.

Proof: Using equation [9] we write the following:

$$\begin{aligned} Mt &= (\text{volume of solids} \times \rho_s) + (\text{volume of water} \times \rho_w) + (\text{volume of gas} \times \rho_g) \\ &= (Vs \times 2650) + (Vw \times 1000) + (Vg \times 1.3) \end{aligned}$$

Since Vg is seldom greater than 0.5 m^3/m^3 and the density of the gas is low compared to the density of water and solids, the magnitude of Mg is negligible.

The oven-dry weight of soil, W_{od}, is

$$W_{od} = Ms$$

Mw becomes 0 as a result of evaporation during the drying process, and Mg is negligible because of the low density of gases.

To determine the mass of water in a soil sample,

$$\begin{aligned} Mw &= (\text{field weight}) - (\text{oven-dry weight}) \\ &= W - W_{od} \\ &= (Ms + Mw) - Ms \\ &= Mw \end{aligned}$$

Calculations of Soil Physical Quantities

1. A soil having a bulk density of 1.2 Mg/m^3 and a particle density of 2.65 Mg/m^3 weighed 0.1 kg when sampled and 80 g after oven drying. Find values of f, f_a, θ_w, and θ_v. Calculate the equivalent depth (in meters) of water in each meter of soil when sampled.

f = 1 − (1200/2650) = 0.547 m^3/m^3
θ_w = (0.10 − 0.080)/0.080 = 0.25 kg/kg
θ_v = 0.25 × (1200/1000) = 0.30 m^3/m^3
f_a = f − θ_v = 0.547 − 0.30 = 0.247 m^3/m^3
D_{eq} = $\theta_v \Delta z$ = 0.30 m^3/m^3 × 1 m = 0.30 m of water

2. The water contents of the Ap and Bt horizons of a soil profile were determined to be 0.15 and 0.25 kg/kg, respectively. If the bulk densities of these two horizons were 1.3 and 1.4 Mg/m^3, respectively, determine the volumetric water contents, porosities, and saturation ratios of the two horizons.

For the Ap horizon:

θ_v = 0.15 × (1.3/1.0) = 0.195 m^3/m^3
f = 1 − (1.3/2.65) = 0.509 m^3/m^3
θ_s = 0.195/0.509 = 0.383 m^3/m^3

For the Bt horizon:

θ_v = 0.25 × (1.4/1.0) = 0.35 m^3/m^3
f = 1 − (1.4/2.65) = 0.472 m^3/m^3
θ_s = 0.35/0.472 = 0.742 m^3/m^3

3. If the thicknesses of the Ap and Bt horizons in problem 2 are 15 and 30 cm, respectively, what is the equivalent depth of water in each horizon? What is the equivalent depth of water in the profile (assume that the profile is composed only of the two horizons)? What is the average water content of the profile?

For the Ap horizon:

D_{eq} = 0.195 × 15 = 2.93 cm = 0.0293 m

For the Bt horizon:

D_{eq} = 0.35 × 30 = 10.50 cm = 0.105 m
profile = D_{Ap} + D_{Bt} = 2.93 + 10.50 = 13.43 cm = 0.1343 m
profile mean θ_v = 13.43 cm/45 cm = 0.298 m^3/m^3

4. Chong et al. (1991) gave the following (with some modifications) sample problem. Assume that the effective root zone for soybeans in a field was 45 cm. A soil sample was taken from the field 48 h after the field was uniformly saturated and allowed to drain. The following information was known or was obtained from the soil sample:

soil sample volume = 700 cm^3
wet soil weight = 1150 g

dry soil weight = 875 g

particle density of soil = 2.60 g/cm^3

permanent wilting point = 0.13 cm^3/cm^3

Based on the information given, determine the values of the quantities listed below for the soil sample. The answers are given in parentheses.

1. Water content by weight (0.314 kg/kg)
2. Bulk density (1.25 Mg/m^3)
3. Water content by volume (0.393 m^3/m^3)
4. Volume of water (275 cm^3)
5. Total porosity (0.519 m^3/m^3)
6. Volume of soil solids (336.7 cm^3)
7. Volume of soil pores (363.3 cm^3)
8. The sum of macro- and mesoporosity after 48 h of drainage (0.126 m^3/m^3)
9. Microporosity (0.393 m^3/m^3)
10. Equivalent depth of water (in cm) of soil water at field capacity in effective root zone (17.7 cm)
11. Plant available water content after 48 h of drainage by volume (0.263 m^3/m^3)
12. Available water content in centimeters of water after 48 h of drainage (11.8 cm)
13. If the water content in the root zone must be maintained at 30% by volume, how many centimeters of water must be removed from the profile? (4.2 cm)
14. If the initial water content is 20% by volume, how many centimeters of water must be added to the soil so that the water content in the soil can be brought to saturation? (14.36 cm)
15. If the initial water content is 23% by volume and 5 cm of water are uniformly added to the root zone, the water content in the root zone will be increased to what percentage by volume? (34.1%)
16. What is the maximum retentive capacity of this soil? (0.519 m^3/m^3)
17. What is the permanent wilting point in percentage by weight? (10.4%)
18. If a soil sample is taken 24 h after the field was saturated and the sample had a water content by weight of 37.5%, what is the aeration porosity of the soil at this time? (0.05 m^3/m^3)

Cited References

American Society of Agronomy, Crop Science Society of America, and Soil Science Society of America. 1988. Conventions and style. Chapter 6 in *Publications Handbook and Style Manual*. Madison.

Bitton, G. 1998. *Formula Handbook for Environmental Engineers and Scientists*, p. 275. Wiley-Interscience. New York.

Campbell, G. S. 1977. *An Introduction to Environmental Biophysics*. Springer-Verlag. New York.

Chong, S. K., B. P. Klubek, and E. C. Varsa. 1991. Teaching the concepts of soil physical and static water properties in introductory soil science. *Journal of Agronomic Education* 20:153–156.

Miller, F. 1959. *College Physics*. Harcourt, Brace, and World. New York.

Additional References

Gajda, W. J., and W. E. Biles. 1978. *Engineering Modeling and Computation.* Houghton Mifflin Co. New York.

Hillel, D. 1982. *Introduction to Soil Physics.* Academic Press. New York.

Himmelblau, D. M. 1982. *Basic Principles and Calculations in Chemical Engineering.* Prentice-Hall. Englewood Cliffs, NJ.

Klute, A. 1986. *Methods of Soil Analysis*, 2nd Ed, Part 1, Physical and Mineralogical Methods. Agronomy monograph No. 9. Madison, WI: Soil Science Society of America.

Rose, D. A. 1979. Soil water: Quantities, units, and symbols. *Journal of Soil Science* 30:1–15.

Problems

1. Convert the following:

    a. 10 gal/h to L/s

    b. 1.1 mi to m

    c. 3 ac to m^2

    d. 32 ft/s^2 to m/s^2

    e. 3 yd to m

2. Fill in the blanks:

    a. _____ grams of prevention is worth _____ kilograms of cure.

    b. A miss is as good as _____ kilometers.

    c. Big sale on _____-liter hats.

    d. The lengths of football, basketball, and soccer fields are _____, _____, and _____, in meters, respectively.

3. In the SI system of units, the weight of a 225-pound man standing on the earth's surface is approximately _____.

4. The molecular diffusivity can be related to the volumetric soil water content via the empirical equation $D = A \exp(B\theta_v)$, where D is the molecular diffusivity (m^2/s), θ_v is the volumetric soil water content (m^3/m^3), and A and B are constants. What are the units of A and B?

5. Complete the following table with the proper equivalent temperatures.

°C	°F	K
−40.0		
	77.0	
		69.8

6. Define and give the SI units of the following parameters:

    a. temperature

    b. pressure

    c. velocity

    d. acceleration

    e. work

    f. power

    g. density

    h. energy

    i. conductance

7. Explain the difference between absolute pressure and relative pressure.

8. Calculate the bulk density of a soil sample that has a wet weight of 15.2 g, a dry weight of 14.5 g, and a volume of 12.1 cm^3.

9. A soil sample has a bulk density of 1.12 Mg/m^3 and a particle density of 2.62 Mg/m^3. Calculate the total porosity of the soil.

10. Calculate the aeration porosity of a soil sample that has a bulk density of 1.35 Mg/m^3, a particle density of 2.65 Mg/m^3, and a gravimetric soil water content of 0.18 kg/kg.

11. Calculate the percent solids of a soil sample that has a bulk density of 1.13 Mg/m^3 and a particle density of 2.65 Mg/m^3.

12. How much wet soil at a water content of 0.20 kg/kg would be required to give 50 g of oven-dry soil?

13. If a soil has a bulk density of 1.25 Mg/m^3 and a water content of 0.20 cm^3/cm^3, how much wet soil would be required to have 50 g of dry soil?

14. How many centimeters of rainfall would be required to bring a soil from a water content of 0.08 kg/kg to 0.25 cm^3/cm^3 at a depth of 15 cm if the bulk density of the soil was 1.35 Mg/m^3?

15. Calculate the amount of water (in liters per hectare) that a soil will hold in the top 100 cm assuming that the average water content is (a) 0.25 cm^3/cm^3 and (b) 0.08 cm^3/cm^3. Assuming that these water contents are the upper and lower limits of extractable water, how many liters of water in the soil are available for plant use (assuming no losses from drainage and evaporation from the soil surface)? How many days will this water last if the average evapotranspiration rate is 0.65 cm/d?

16. A silt loam soil has a bulk density of 1.43 Mg/m^3. How much mass is there in 1 hectare of plow layer?

17. If the particle density of the same soil as given in problem 16 is 2650 kg/m^3, what is the volume of voids in the plow layer? What percentage of the plow layer is void space?

18. In the B horizons of certain soils, accumulations of clay along with chemical cementing agents such as iron oxides create bulk densities often obtaining values of 1700–1800 kg/m^3. What is the volume of voids in a soil with a bulk density of 1800 kg/m^3? What is the percentage of pore space? Assume a particle density of 2650 kg/m^3.

19. A homogeneous soil has an initial soil water content of 0.15 m^3/m^3 and a saturated water content of 0.40 m^3/m^3. How deep will a 5 cm rain wet the profile? How much water is needed to wet the soil to the 30 cm depth?

20. A soil profile has the properties given in the table below. Fill in the missing data and determine the depth of penetration of a 7.5 cm rain. The columns for water needed and water left are the depths of water needed to fill the pores to field capacity in the depth interval and the cumulative depth of water left over, respectively.

Depth Interval (cm)	θ_w (kg/kg)	ρ_b (Mg/m³)	θ_v (m³/m³)	θ_{fc} (m³/m³)	Water Needed (cm)	Water Left (cm)
0–15	0.08	1350		0.35		
15–35	0.12	1420		0.34		
35–62	0.15	1480		0.32		
62–100	0.20	1520		0.32		

21. Soil water contents were determined in a wheat field during early spring and again at harvest. The volumetric water content distributions within the profile are given in the table below. Plot the water content distributions by depth. Determine the equivalent depth of water stored in the profile at both dates. Also, determine the net change in depth of water in the profile. What can you conclude from your results?

Soil Depth (cm)	Volumetric Water Content	
	April 1	June 16
5	0.249	0.062
15	0.255	0.105
30	0.258	0.140
45	0.272	0.161
60	0.275	0.188
75	0.313	0.235
90	0.313	0.274
105	0.313	0.313

22. A soil ped has dimensions of 0.001 m³ and a mass of 1.52 kg, of which water was 0.24 kg. Assume that the density of water is 1000 kg/m³ and the particle density is 2650 kg/m³. Determine the following parameters for this soil volume.

    a. soil water content by weight on a dry-weight basis

    b. soil water content by volume on a dry-weight basis

    c. depth of water

    d. bulk density

    e. total porosity

    f. relative saturation

    g. void ratio

    h. aeration porosity

    i. equivalent depth of water needed to bring the ped water content to 0.4 m³/m³

    j. volumetric soil water content on a wet-weight basis

23. Convert the following physical properties of liquid water at 20°C to their SI unit equivalent.

a. density: 0.99823 g/cm^3

b. gravimetric heat capacity: 0.999 cal/g·°C

c. latent heat of vaporization: 586 cal/g

d. surface tension: 72.7 g/s^2

e. thermal conductivity: 0.00144 cal/cm·s·°C

f. viscosity: 0.001 g/cm·s

24. Suppose that the runoff from a small watershed was 500 liters per second. Convert this rate to the following:

a. m^3/s

b. m^3/d

c. U.S. gallons per minute

CHAPTER THREE

Soil Texture

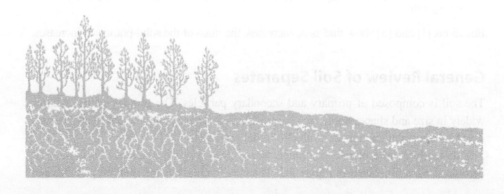

Introduction

The solid phase of the soil consists of mineral and organic matter in various amounts, shapes, sizes, and mineralogical compositions. The particle size distribution of a soil determines to a large extent its physical, chemical, and microbiological behavior. Since the distribution of particle sizes varies little with time during our lifetime, it has been extensively used to classify soils.

In this chapter, we explore the physical nature of the solid phase and some of its influences on soil physical behavior. Techniques for the determination of the sizes of soil particles are presented along with practical implications on soil behavior.

Amount of Solid Phase

The amount of the solid phase is determined by the volume fraction, volume, and mass of soil solids. The volume fraction of the solid phase, S, is calculated from the ratio of the bulk density and particle density:

$$S = \rho_b/\rho_p = m^3 \text{ solids/m}^3 \text{ soil} \tag{1}$$

where ρ_b and ρ_p are the bulk and particle densities (kg/m^3), respectively. When expressed on a percentage basis, S is multiplied by 100.

The total volume of soil solids, Vs, in a given soil volume is calculated from

$$Vs = S \times Vt = m^3 \text{ solids} \tag{2}$$

where Vt is the total soil volume (m^3).

The total mass of the solid phase, Ms, in a given volume of soil is calculated from

$$Ms = \rho_b \times Vt = S \times Vt \times \rho_p = kg \text{ solids} \tag{3}$$

Example

Calculate the mass of oven-dry soil having a bulk density of 1400 kg/m^3, depth of 15 cm, and an area of 1 ha. What is the influence of bulk density on the mass of soil in this area?

According to equation [3],

$$\frac{1400 \text{ kg / m}^3}{2650 \text{ kg / m}^3} \times 2650 \text{ kg / m}^3 \times (10,000 \text{ m}^2 \text{ / ha}) \times (0.15\text{m} \times \text{ha}) = 2.1 \times 10^6 \text{ kg}$$

Equations [1] and [3] show that as ρ_b increases, the mass of the solid-phase soil increases.

General Review of Soil Separates

The soil is composed of primary and secondary particles. The primary particles differ widely in size and shape. Some are coarse enough to be seen with the naked eye, whereas others are small enough to exhibit colloidal properties.

Soils rarely consist of one size range. For a given soil, the frequency distribution of particle sizes is illustrated in Figure 3.1. The clay fraction has the smallest particles and

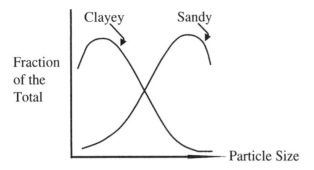

Figure 3.1. Soil fractions by particle size for a clayey soil and a sandy soil.

includes the clay minerals that have formed as secondary products from the weathering of rocks or have come directly from a parent material or transported deposits. The nonclay fraction consists of other minerals, rock fragments, and sometimes secondary concretions.

The term *soil texture* is an expression of the predominant size, or size range, of the particles. Quantitatively, soil texture refers to the relative proportions of the various sizes of particles in a given soil. Soil texture is considered a permanent, natural attribute of the soil; therefore, it is a static soil physical property.

The traditional method of characterizing particle sizes in soil is to divide the particles into size ranges known as *separates.* A soil separate consists of mineral particles between designated maximum and minimum diameters.

There is no natural classification of particle size; the limits are based on the contribution that particles of different sizes make to the physical and chemical properties of soil. As a result, several classification systems have been developed around the world. The two systems of most interest to us are given in Tables 3.1 and 3.2.

There are several other particle size classification systems (see Steinhardt 1979, p. 351; Gee and Bauder 1986).

Table 3.1. The International Soil Science Society
Classification System of Soil Particles

Separate	Particle Diameter	
	mm	μm
Coarse sand	2.0–0.2	2000–200
Fine sand	0.2–0.02	200–20
Silt	0.02–0.002	20–2
Clay	<0.002	<2

Table 3.2. The U.S. Department of Agriculture
            Classification System of Soil Particles

| | Particle Diameter | |
Separate	mm	μm
Very coarse sand	2.0–1.0	2000–1000
Coarse sand	1.0–0.5	1000–500
Medium sand	0.5–0.25	500–250
Fine sand	0.25–0.10	250–100
Very fine sand	0.10–0.05	100–50
Silt	0.05–0.002	50–2
Clay	<0.002	<2

Often the silt fraction is subdivided into three additional separates:

Separate	Diameter (μm)
Coarse silt	50-20
Medium silt	20-5
Fine silt	5-2

The particle diameter is defined in terms of a sphere that would pass through a screen with square openings equal to the diameter. Nonspherical particles are considered to have an equivalent diameter near the average of their maximum and minimum dimensions for determining their size classification.

Particles larger than 2 mm in diameter are excluded from soil texture determinations. Stones, gravels (2 mm to 7.5 cm), and cobbles (>7.5 cm) influence the use and management of land but they make little or no contribution to the basic soil properties, such as water retention and the capacity to store and release plant nutrients.

Textural Classes of Soil Particles

Most soils and sediments are mixtures of particles of various sizes. To make it easier to understand, use, and describe these mixtures, classes have been defined according to their relative proportion of sand, silt, and clay. The overall textural designation (class) is determined on the basis of mass ratios of these separates. Soils with different proportions of sand, silt, and clay are given different designations. The USDA textural triangle has 12 classes, which were defined for their practical value in agriculture (Figure 3.2).

The numerical values on the three scales are angled to indicate the slopes of the lines to which they apply. The intersection of lines gives the coordinates and textural class. When the coordinate occurs exactly on the line between two textural classes, it is customary to use the class associated with the finer particle size.

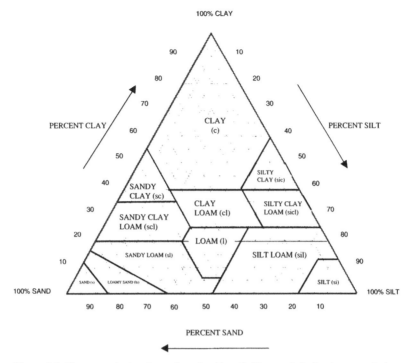

Figure 3.2. The textural triangle used to classify soil. The symbols for the textural classes are given in parentheses.

Some examples of soil textures are given in the following table:

| Soil | Sand | Mass Percentages | | Textural Class |
		Silt	Clay	
1	35	50	15	sil-l
2	18	25	57	c
3	25	50	25	sil-l
4	52	15	33	scl

Textural classes are named after the predominant size fraction occurring in them plus the word "loam" whenever all three major size fractions occur in sizable proportions. *Loam* is an old English word formally applied to crumbly soils, rich in humus. The definitions of these classes were developed from long experience and special research to establish boundaries between classes so that they have maximum general use for soil definitions and interpretations.

A glance at the textural triangle indicates the importance of specific surface, that is, surface area per unit mass or volume. Specific surface is an indicator of how much of the particle surface is available for the sorption of solute ions and molecules. It takes more than 80% silt to call a soil a "silt"; more than 85% sand to call a soil a "sand"; but only

Table 3.3 General Characteristics of Soil Textures

Characteristics	Sand	Loam	Silt loam	Clay
Feel	Gritty	Gritty	Silky	Cloddy or plastic
Identification	Loose	Cohesive	Shows fingerprint	Gives shiny streak
Internal dainage	Excessive	Good	Fair	Fair–poor
Plant available water	Low	Medium	High	High
Drawbar pull	Light	Light	Medium	Heavy
Tillability	Easy	Easy	Medium	Difficult
Runoff potential	Low	Low–medium	High	Medium–high
Water transportability	Low	Medium	High	High
Wind erodibility	High	Medium	Low	Low

Source: Adapted from Kohnke 1968.

40% clay is required to call a soil a "clay." In general, the smaller the particle, the greater the specific surface.

A knowledge of the texture of a soil is of obvious practical importance. It is a guide to the value of the land (Table 3.3). Land use capability and methods of soil management are largely determined by soil texture. Soil texture affects the movement and retention of water, solutes, heat, and air. The consistency and particle size distribution are only slightly influenced by tillage unless tillage is drastic. The differences in physical behavior resulting from differences in particle size affect the way soils are managed.

The use of the words "light" and "heavy" to describe soils originated from the difference in the amount of work needed to cultivate soils, not from the density of the particles.

Soils dominated by a single particle size tend to be less suitable for crop production than loams. The soil characteristics that optimize rice production may not be the same as those that optimize soybean production. Thus, soil texture and other soil physical, chemical, and biological properties have a large role in determining the best land use. Generally, the best soils for arable crops are those that contain 10–20% clay, around 3% organic matter, and the rest divided equally between sand and silt. The loamy soils tend to have the best combination of physical and chemical properties for agricultural uses.

A general grouping of the 12 basic classes of soil texture is as follows:

Coarse-textured soils:
    Sands and loamy sands
    Sandy loams, including fine sandy loams

The sands may be characterized as very fine, fine, coarse, and very coarse.

Medium-textured soils:
    Loamy soils: loam, silt loam, silt
    Moderately heavy soils: clay loam, sandy clay loam, and silty clay loam
Fine-textured soils:
    Clay and silty clays
    Sandy clays

Table 3.4. Definitions of the Soil Textural Classes in the Textural Triangle

---

Sands: Soil material that contains 85% or more of sand; percentage of silt, plus 1.5 times the percentage of clay, does not exceed 15.

Coarse sand: 25% or more very coarse and coarse sand and less than 50% any other one grade of sand.

Sand: 25% or more very coarse, coarse, and medium sand and less than 50% fine or very fine sand.

Fine sand: 50% or more fine sand or less than 25% very coarse, coarse, and medium sand and less than 50% very fine sand.

Very fine sand: 50% or more very fine sand.

Loamy sands: Soil material that contains at the upper limit 85–90% sand, and the percentage of silt plus 1.5 times the percentage of clay is not less than 15. At the lower limit, it contains not less than 70–85% sand, and the percentage of silt plus twice the percentage of clay does not exceed 30.

Loamy coarse sand: 25% or more very coarse and coarse sand and less than 50% fine or very fine sand.

Loamy sand: 25% or more very coarse, coarse, and medium sand and less than 50% fine or very fine sand.

Loamy fine sand: 50% or more fine sand or less than 50% very fine sand.

Loamy very fine sand: 50% or more very fine sand.

Sandy loams: Soil material that (*a*) contains 20% clay or less and 52% or more sand and the percentage of silt plus twice the percentage of clay exceeds 30 or (*b*) contains less than 7% clay, less than 50% silt, and between 43 and 52% sand.

Coarse sandy loam: 25% or more very coarse and coarse sand and less than 50% any other one grade of sand.

Sandy loam: 30% or more very coarse, coarse, and medium sand but less than 25% very coarse sand and less than 30% very fine or fine sand.

Fine sandy loam: 30% or more fine sand and less than 30% very fine sand or between 15 and 30% very coarse, coarse, and medium sand.

Very fine sandy loam: 30% or more very fine sand or more than 40% fine and very fine sand at least half of which is very fine sand; and less than 15% very coarse, coarse, and medium sand.

Loam: Soil material that contains 7–27% clay, 28–50% silt, and less than 52% sand.

Silt loam: Soil material that contains (*a*) 50% or more silt and 12–27% clay or (*b*) 50–80% silt and less than 12% clay.

Silt: Soil material that contains 80% or more silt and less than 12% clay.

Sandy clay loam: Soil material that contains 20–35% clay, less than 28% silt, and 45% or more sand.

Clay loam: Soil material that contains 27–40% clay and 20–45% sand.

Silty clay loam: Soil material that contains 27–40% clay and less than 20% sand.

Sandy clay: Soil material that contains 35% or more clay and 45% or more sand.

Silty clay: Soil material that contains 40% or more clay and 40% or more silt.

Clay: Soil material that contains 40% or more clay, less than 45% sand, and less than 40% silt.

Specific definitions of the soil textural classes are presented in Table 3.4.

In soil survey reports, percentages are based on oven-dry mineral soil particles passing through a 2 mm sieve. The percentages of the individual fractions are calculated from organic-matter-free, oven-dry soil particles less than 2 mm (or 2000 μm or 2×10^{-3} m) in diameter. Gravel and stones are reported separately (Soil Survey Division Staff 1993).

Particle Composition

The minerals of which the particles are composed reflect the nature of the parent materials and the degree of weathering that the soil has undergone. The minerals of the parent materials vary greatly in their stability and hence in their occurrence in soil. In particular, quartz (SiO_2) usually dominates the nonclay fraction because of its resistance to weathering and its abundance in certain parent materials such as granite, sandstone, and surface deposits. The principal minerals of the clay fraction are secondary products of weathering referred to collectively as clay minerals. The clay particles separate when soil is dispersed and, in a suspension, result in the long-lasting cloudiness. The particle size of the clay mineral fraction ranges downward to about 5×10^{-7} m. The clay fraction is the predominant site of phenomena such as swelling, plasticity, and cation exchange (Marshall et al. 1996).

Summation Curves and Uniformity Index

The distribution of particle sizes in a soil can be represented by computing summation curves. This is done as follows.

1. Compute values of percentage composition, %p:

%p = (weight of fraction/total weight of washed soil) $\times$ 100   [4]

2. Sum values of %p over size ranges beginning with the percentage of particles, by weight, smaller than the upper size limit.

Thus, %p represents the percentage of oven-dry soil mass with particles smaller than diameter d. Examples of two horizons of a Peridge soil are given in Table 3.5 and are shown in Figure 3.3.

Table 3.5. Particle Size Distribution, Percentage Composition, %p, and Accumulative Percentage Composition, Σp, of the Ap and B21t Horizons of the Peridge Soil

Diameter (μm)	Separate	p (%) Ap	p (%) B21t	Σp (%) Ap	Σp (%) B21t
2000–1000	very coarse sand	0.6	0.3	100.0	100.0
1000–500	coarse sand	1.3	0.5	99.4	99.7
500–250	medium sand	2.4	1.1	98.1	99.2
250–100	fine sand	9.7	5.2	95.7	98.1
100–50	very fine sand	11.3	7.0	86.0	92.9
50–20	coarse silt	35.4	28.9	74.7	85.9
20–5	medium silt	23.3	21.5	39.3	57.0
5–2	fine silt	5.1	6.7	16.0	35.5
<2	clay	10.9	28.8	10.9	28.8

Note: p = (weight of fraction) $\times$ (100/weight of washed soil); Σp = percentage of particles, by weight, smaller than the upper size limit.

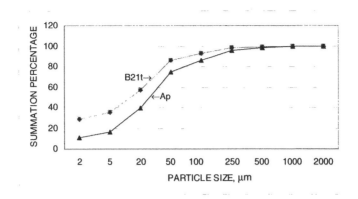

Figure 3.3. Summation curves for two horizons of a Peridge soil.

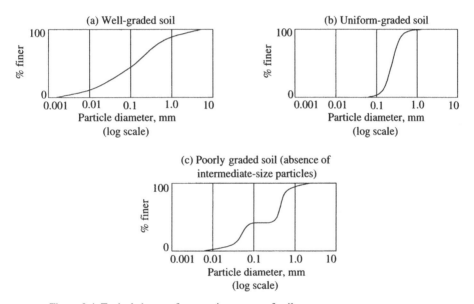

Figure 3.4. Typical shapes of summation curves of soil.

Summation curves are prepared by plotting %p versus the logarithm of the effective diameter. From the summation curves, the percentage of particles in a given size fraction can be determined by taking the difference in %p at the upper and lower size limits.

The shapes of the summation curves also give information on the grading pattern (Figure 3.4). For example, a well-graded soil has a flattened and smooth summation curve (Figure 3.4a), a uniformly graded soil has an intermediate summation curve (Figure 3.4b), and a poorly graded soil has a steplike summation curve (Figure 3.4c).

Another method used to represent soil particle size distribution is the uniformity index. The uniformity index, Iu, is mostly used for coarse-textured soils in engineering. It is defined as

$$Iu = D_{60}/D_{10}$$ [5]

where D_{60} and D_{10} are the diameters of the soil particle at summation percentages 60 and 10 respectively. For example, the uniformity index of the Ap horizon of the Peridge soil is

  Iu = 0.043/0.002 = 21.5

For the B21t horizon of this same soil, the uniformity index is

  Iu = 0.025/???

which cannot be calculated because the summation value at D_{10} has too much clay.

  Soils with the same particle size have Iu values of 1.0. Some well-graded sandy soils have Iu values greater than 1000.

Determination of Particle Size Distribution

The determination of the relative distribution of the size groups of soil particles is called a mechanical analysis and is one of the most common soil physical analyses. Success of the analysis depends on two factors:

  1. Preparation of the sample to completely disperse all aggregates into their individual primary particles without breaking up the particles themselves.
  2. Accurate fractionation of the sample into the various separates.

A flowchart of the mechanical analysis of soils is shown in Figure 3.5.

  The differing textural separates are classified into various sized groups on the basis of their *equivalent diameters*. The equivalent diameter of larger particles that are separated by sieving is the diameter of a sphere that will pass through a given sized opening. For the smaller particles that are separated by sedimentation techniques, equivalent diameter refers to the diameter of a sphere that has the same density and velocity of settling in a liquid medium. These methods are discussed below.

Preparation of the Sample

The main objectives of sample preparation are to obtain maximum dispersion and to maintain this dispersion while performing the particle size analysis. Therefore, secondary

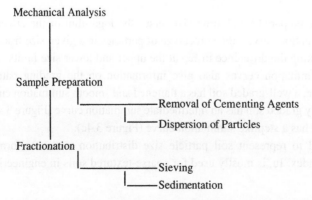

Figure 3.5. Flowchart of the procedures for a mechanical analysis of soil.

soil particles, or aggregates, that are formed by the union of primary particles into units of varying stability must be broken apart. This can be accomplished by

1. removal of binding (cementing) agents such as organic matter and Fe and Al oxides,
2. mechanical rehydration of clay particles, and
3. physical and chemical dispersion of the particles.

Aggregates of soil particles are usually held together by a binding agent, which may be

1. clay or other colloidal material,
2. organic matter,
3. Fe and Al oxides,
4. silicates, or
5. carbonates of Ca and Mg.

The dispersion process must remove these binding agents or at least render them ineffective. Dispersion may be achieved by mechanical and chemical means. Mechanical methods used to achieve dispersion include

1. shaking in water,
2. stirring,
3. rolling with a wooden roller,
4. agitating with air jets in a specially constructed apparatus, and
5. boiling with water.

Chemical methods may also be used to remove the binding agents and to raise the electrokinetic potential so that the colloids remain dispersed. Examples of chemicals used include

1. hydrogen peroxide, H_2O_2, which is used to destroy organic binding agents,
2. Sodium acetate, which is used to remove the carbonate cements, and
3. Na-dithionate in citrate solution (chelating agent), which makes the iron cements soluble.

Raising the electrokinetic potential until particles repel each other will keep colloidal particles in suspension. This can be done effectively by replacing the adsorbed polyvalent cations (particularly divalent and trivalent cations) with Na or Li. This has the effect of increasing the hydration of the particles, thus causing them to repel each other. Some of the chemicals that can be used are

1. sodium hexametaphosphate, $Na_2(PO_3)_2$,
2. sodium hydroxide, $Na(OH)$,
3. sodium oxalate, $Na_2C_2O_4$, and
4. sodium carbonate, Na_2CO_3.

Fractionation of the Sample

There are two primary methods of determining the size distribution of the soil particles after they have been dispersed.

Table 3.6. Common Sieve Types and Mesh Openings

Sieve Size Designation	U.S. Standard		Tyler Standard	
	inches	mm	inches	mm
#4	0.187	4.76	0.185	4.7
#8	0.0937	2.38	0.093	2.362
#10	0.0661	1.68	0.65	1.651
#20	0.0331	0.84	0.0328	0.833
#40	0.0106	0.42	—	—
#60	0.0098	0.25	0.0097	0.246
#100	0.0059	0.149	0.0058	0.147
#200	0.0029	0.074	0.0029	0.074
#270	0.0021	0.053	0.0021	0.053
#400	0.0015	0.037	0.0015	0.038

Source: Data from McCarthy 1993.

Sieving

Sieving is the preferred method for separating the coarser fraction of soils (sand fraction). Usually a "nest" of sieves placed one above the other is used to fractionate the sands. The top sieve has 2 mm holes; below it are sieves with successively smaller holes (e.g., 1 mm, 0.5 mm, 0.25 mm, 0.10 mm, and 0.05 mm). A soil sample is fractioned into the various sand separates by placing it on the top sieve and shaking the nest until each sand particle is caught on a sieve with holes too small for it to pass through. The silt and clay fractions pass through. The sand separates are weighed to determine the amount of each present. The relation between sieves in common use in the United States is presented in Table 3.6.

Sieving may be done with dry or wet soil. When the particles are dry, most of the clay should be removed before sieving. When they are wet, special precautions have to be taken to overcome the surface tension below the finest screen. After wet sieving, sand should be dried and sieved again because part of the silt fraction will be caught on the 0.05 mm screen.

Sedimentation

Sedimentation is the most widely used technique for determining the size distribution of dispersed soil particles. In sedimentation techniques, the settling rates of particles in a viscous fluid are measured, and these rates are then used to determine particle sizes.

Derivation of the Sedimentation Equation

The principle behind the sedimentation of various sizes of soil particles is known as Stokes' law of sedimentation (1851). This law can be simply derived as follows.

Suppose one has a spherical particle that is falling in a fluid such as water. This particle will encounter a frictional resistance that is proportional to the product of its radius,

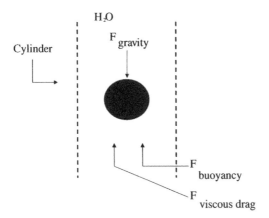

Figure 3.6. Schematic of the forces acting on a soil particle during sedimentation.

velocity of fall, and viscosity of the fluid. This retarding force is known as viscous drag, D, and was shown by Stokes to be

$$D = (2\pi r) \times \eta \times 3v = 6\pi r \eta v \qquad [6]$$

where r is the radius of the particle (m), η is the viscosity of the fluid (kg/m·s), and v is the velocity of the particle (m/s). As soon as the acceleration, acting for a finite time, results in the particle gaining a finite velocity, the viscous drag exerts the retarding force and the acceleration soon decreases. As long as there is any acceleration at all, the velocity continues to increase. The net force on the soil particle is

$$F_{net} = ma \qquad [7]$$

where m is the mass (kg) and a is the acceleration (m/s^2). The forces acting on the particles are shown in Figure 3.6 and include those associated with gravity, buoyancy, and viscous drag.

We take the sign convention that up is in the positive direction; therefore, the net force is

$$F_{buoyancy} + F_{drag} - F_{gravity} = ma = B + D - G \qquad [8]$$

This is a force balance equation where the difference between the upward and downward forces equals the net force. The forces B and D are due to buoyancy and viscous drag, respectively, and the force of gravity, G, is mass times the gravitational acceleration. The SI unit of force is the newton (N).

Eventually, a point is reached at which the sum of the resisting forces equals the constant downward force. The particle then continues to fall without acceleration but at a constant velocity known as the *terminal velocity,* or settling velocity. For sedimenting soil particles, the terminal velocity is achieved quite rapidly, and the resistance is primarily due to the viscous forces.

Since the acceleration is zero, we can equate the vertical upward and downward forces and obtain

B + D = G

$F_{gravity}$ = volume × density of particle × gravity
     = mass × gravity = $(4/3)\pi r^3 \rho_p g$
     = $m^3 × kg/m^3 × m/s^2$

$F_{buoyancy}$ = weight of the liquid displaced
        = volume × density of liquid × gravity
        = $(4/3)\pi r^3 \rho_l g$

$F_{viscous\ drag}$ = $6\pi r \eta v$

where r is the particle radius (m), ρ_p is the particle density (kg/m^3), and ρ_l is the liquid density (kg/m^3). Therefore, we write

$(4/3)\pi r^3 \rho_l g + 6\pi r \eta v = (4/3)\pi r^3 g \rho_p$

Solving for the terminal velocity gives

$v = (4/3)\pi r^3 g(\rho_p - \rho_l)/6\pi r \eta$

where v is the terminal velocity (m/s), g is the gravitational acceleration (m/s^2), and η is the viscosity (mPa·s, or g/m·s). Rearranging, we obtain the sedimentation equation

$$v = 2gr^2(\rho_p - \rho_l)/9\eta \qquad [9]$$

Substituting the diameter for the radius gives the more useful form of the sedimentation equation

$$v = gd^2(\rho_p - \rho_l)/18\eta \qquad [10]$$

Equation [10] predicts that the velocity of fall of particles with the same density, in a given liquid, will increase with the square of the diameter of the particle.

We usually correct to 20°C in water and assume ρ_p = 2.65 Mg/m^3. Substituting these values into equation [10] gives

$$v = 8983d^2 \qquad [11]$$

where d is the particle diameter in centimeters and v is in centimeters per second, or

$$v = 89.83 × 10^4 d^2 \qquad [12]$$

where d has units of meters and v is in meters per second.

The viscosity of water is strongly dependent on temperature and has units of poises or millipascal-seconds. The magnitude of η is a function of the liquid and its temperature; values for water are given in Table 3.7.

We can check the SI units of the sedimentation equation [9] as follows:

$$m/s = \frac{m/s^2 × m^2 × kg/m^3}{kg/m·s}$$

$$= \frac{kg/s^2}{kg/m·s}$$

$$= m/s$$

Table 3.7 Values of the Viscosity of Water, η, at Various Temperatures

Temperature (°C)	(g/m·s)	mPa·s
0	1.787	1.787
10	1.308	1.3037
20	1.002	1.0019
30	0.798	0.798
40	0.653	0.654
50	0.547	0.547

Source: Adapted from Marshall et al. 1996.

Also, remember the conversion

$$Pa\cdot s = N\cdot s/m^2 = kg\cdot m\cdot s/s^2\cdot m^2 = kg/m\cdot s$$

Example
Calculate the rate of fall in water at 20°C of a silt particle with a particle diameter of 0.02 mm (i.e., 20×10^{-6} m). Assume the density of the particle is 2.65 Mg/m^3 and the density of water is 1.00 Mg/m^3.

$$v = \frac{(20\times10^{-6}\ m)^2\ (9.8\ m/s^2)\ (2650-1000\ kg/m^3)}{18\times1\times10^{-3}\ Pa\cdot s}$$
$$= 3.59\times10^{-4}\ m/s$$

Example
Calculate the terminal velocity corresponding to the upper size limit of the clay fraction, 2 μm, if the temperature is 20°C and the particle density is 2650 kg/m^3.

$$v = \frac{9.8\ (2\times10^{-6})^2\ (2650-1000)}{18\times1\times10^{-3}}$$
$$= 3.6\times10^{-6}\ m/s$$

Comparison of the answers in this problem and in the previous one indicates that the terminal velocity of the clay was 100 times lower than that of the medium-sized silt particle having a diameter 10 times larger. Examination of equation [10] shows that this result should be expected.

Assuming that the terminal velocity (L/t) is attained almost instantly, we can calculate the time needed for the particle to fall through a distance L in the cylinder to be

$$t = 18L\eta/gd^2(\rho_p - \rho_l) \qquad [13]$$

The diameter of a particle can also be calculated from the sedimentation equation as

$$d = [18L\eta/tg(\rho_p - \rho_l)]^{0.5} \qquad [14]$$

and with some rearrangement, we obtain

$$d = \{[30\eta/980(\rho_p - \rho_w)] (L/t)\}^{0.5} \tag{15}$$

where d is the particle diameter (mm), η is the coefficient of viscosity, L is the distance from the surface of the suspension to the depth at which the suspension density is measured (cm), t is the time interval (min), and ρ_p and ρ_w are the densities of the particle and water, respectively.

Since all terms on the right-hand side of equation [15] are constant except t, the sedimentation equation can be rewritten as

$$d = \theta/t^{0.5} \tag{16}$$

where θ represents the constant terms. Values of θ as a function of the hydrometer reading have been tabulated for temperatures of 30°C by Day (1965) and of 20 and 25°C by Green (1981).

The use of Stokes's law for measurement of particle sizes is dependent on certain simplifying assumptions:

1. The particles must be large in comparison to liquid molecules so that Brownian movement will not affect the fall. In water, this occurs for all soil particles with diameters greater than 1×10^{-6} m.

2. The volume of the liquid must be great in comparison to the size of the particles. The fall of the particles must not be affected by the proximity of the wall of the vessel or by adjacent particles; the particles must fall independently of each other.

3. Particles must be rigid, smooth, and spherical. This requirement is difficult to fulfill with soil particles. It is highly probable that the particles are not completely smooth and are not spherical but are irregularly shaped and include a large number of plate-shaped particles in the clay fractions. Since variously shaped particles fall with different velocities, the term *equivalent diameter* is used to overcome this difficulty in Stokes's law. Equivalent diameter is the diameter of a sphere of the same material that would fall with the same velocity as the particle in question. This approximation is easily fulfilled with soil because of the water hulls around the falling particles.

4. There must be no slipping between the particle and the liquid so that the liquid offers only resistance. All particles have the same density.

5. The velocity of fall (terminal velocity) must not exceed a certain critical value so that there is no turbulent flow and resistance of the fall of the particle is due only to the viscosity of the liquid. There is no turbulent flow if $v < \eta/\rho_p r$. If the fluid is water at 25°C and the falling particle has a density of 2.7 Mg/m^3, the maximum radius for nonturbulent flow is 0.4 mm. Therefore, the sedimentation equation is not applicable to soil particles having diameters larger than very fine sand (i.e., roughly 80 µm).

Stokes' law of sedimentation has certain limitations:

1. Shape of the particles: We must assign an "equivalent diameter" to the particles, which is an approximation. Rod-shaped particles are not separated as accurately as disk-shaped ones.

2. Temperature: A constant temperature must be maintained to prevent convection currents. The rate of fall varies inversely with the viscosity of the medium. Therefore, convection currents would prevent the uniform settling of the particles.

3. Density of the particles: This may account for the largest error in the calculations. The particle density depends on the mineralogical and chemical constitution as well as degree of hydration. A pycnometer can be used to measure ρ_p. Particle densities for a few soil constituents are 1.37 Mg/m^3 for humus, 2.50 for kaolin, and 5.20 for hematite. Usually, we substitute an average ρ_p of 2.6–2.7 Mg/m^3 for mineral soils, which is no better than an educated guess.

Techniques of Mechanical Analysis Based on Stokes' Law

There are two primary methods used to quantify particle sizes of soils by sedimentation. These methods are presented more fully by Gee and Bauder (1986).

Pipette Sampling Method

This method is considered to be the most accurate in determining particle size distributions and is a direct sampling of the density of the solution. Using a pipette, samples of the suspension (usually 20 cm^3) are withdrawn at a given depth after various periods have elapsed after initiation of sedimentation.

At a depth L below the surface of the suspension and at time t, all particles whose terminal velocity v is greater than L/t will have passed below this level—for example, silt passes through but clay remains. All smaller particles will still be descending through the layer above this depth, and the coarser particles have been eliminated.

We assume that (1) there is an initial uniform spatial distribution in the water column, and (2) the particles descend a negligible distance before obtaining their terminal velocity. This allows us to also assume that the number of particles entering a layer of infinitesimal thickness at this depth equals the number leaving the layer in any given period. Thus, the concentration of particles of such a size that they have terminal velocity v is zero above the value of L given by vt, and constant below it. Since L is fixed, v is determined by the time lapse, t.

The amount of solid in the sample is found by evaporating the water by oven-drying, then weighing the residue. The ratio of the weight, W, of particles present in that volume (20 mL) at time t, divided by the initial weight, W_0, of particles, is equal to P/100, where P is the percentage of particles, by weight, smaller than d; that is,

$$W/W_0 = C/C_0 = P/100 \qquad [17]$$

where C and C_0 are the oven-dry masses of particles in the residue and initial sample.

Example

Determine the sedimentation time for a soil particle having a diameter of 2 µm and a particle density of 2.60 Mg/m^3. Assume that the sampling was made at the 0.10 m depth and at a temperature of 20°C.

$v = L/t$

$t = 18\eta L/gd^2(\rho_p - \rho_w)$

$\quad = 18\eta L/g(\rho_p - \rho_w)d^2 = 18 \times 0.001 \times 0.10/9.8 \times 1600 \times (2 \times 10^{-6})^2$

$\quad = 2.87 \times 10^4$ s

$\quad = 8$ h

Hydrometer Method

Although somewhat less accurate than the pipette sampling method, another widely used method of determining the amount of soil particles in suspension is to measure the density of the suspension with a special hydrometer. The hydrometer method also depends upon Stokes's law.

The depth to the hydrometer's center of buoyancy varies with the density of the suspension and also with the particle size distribution. The concentration, C, of the soil in the suspension, in g/L, can be calculated from the following equation:

$$\rho_s = \rho_w + C[1 - (\rho_w/\rho_p)] \tag{18}$$

where ρ_s is the density of the suspension, ρ_w is the density of water, and ρ_p is the particle density of the suspension. The stem of the hydrometer is read in density units (g/L on a scale to 60) or in percentages ($100C/C_0$) when the initial concentration, C_0, is equal to 0.04 Mg/m^3 and ρ_p is 2.65 Mg/m^3. The sampling time is arbitrary, but a geometric progression of time intervals provides a spread of data. For example, at 20°C with the Bouyoucos hydrometer, sampling times of 40 seconds and 8 hours result in concentrations of silt + clay and clay, respectively.

The hydrometer method is quicker but less reliable than the pipette method. Some precautions should be kept in mind:

1. It is a calibrated method.
2. It gives an average reading of the suspension over a portion of the size distribution curve.
3. Settling of the soil particles on the shoulders of the hydrometer introduces uncertainty in the results, although the severity of this effect has been reduced in the new designs.

Specific Surface

The distribution of mineral particles within a soil is a parameter that is invariant in ordinary time and under normal conditions. There is also a basic relation between the particle size and the specific surface or colloidal activity of the soil. Many of the physical and chemical properties of the soil are associated with surface activity. Large particles present a significantly smaller area for the retention of water, mineral elements, and organics than do small particles. Since soils are heterogeneous, the contributions of each soil separate must be evaluated.

Relation of Surface Area to Particle Size

The large amount of surface per unit mass is a characteristic property of all dispersed systems. The extent of the surface of a dispersed system is usually expressed in terms of *specific surface,* which has the following two definitions:

$$s_m = \text{surface area/mass (m}^2\text{/kg)} \qquad [19]$$

$$s_v = \text{surface area/volume (m}^2\text{/m}^3) \qquad [20]$$

These two parameters are related by

$$s_m \rho_p = s_v \qquad [21]$$

The specific surface of a cube is inversely proportional to its size. Its specific surface increases as the size decreases by a factor of 1/(length of a side). For example, the volume of a 1 cm cube is

$$\text{volume} = L^3 = 0.01 \times 0.01 \times 0.01 = 10^{-6} \text{ m}^3$$

The surface area of this cube is

$$\text{surface area} = 6 \text{ faces} \times (0.01 \text{ m/face})^2 = 6 \times 10^{-4} \text{ m}^2$$

If the length of a side is l, the specific surface by volume of the cube is

$$s_v = 6l^2/l^3 = 6/l$$

demonstrating that as l decreases, s_v and s_m increase:

$$s_v = 6 \times 10^{-4}/1 \times 10^{-6} \text{ m}^3 = 6 \times 10^2 \text{ m}^{-1}$$

Now, let's cut this cube into smaller cubes whose sides are 1/100 as long as the original cube.

 For each cube:
$$\text{volume} = (0.0001)^3 = 1 \times 10^{-12} \text{ m}^3$$
$$\text{surface area} = 6(1 \times 10^{-4}) = 6 \times 10^{-8} \text{ m}^2$$
$$s_v = 6 \times 10^{-8}/1 \times 10^{-12} = 6 \times 10^4 \text{ m}^{-1}$$

However, since there are 10^6 of these smaller cubes in the original cube, the total surface area is equal to

$$10^6 \times 6.0 \times 10^{-8} = 0.06 \text{ m}^2$$

Then, the specific surface area per unit volume is

$$s_v = 0.06/10^{-6} = 6 \times 10^4 \text{ m}^{-1}$$

Thus, it can be seen that the specific surface increases directly as the size of the cube decreases. In soil, we are primarily interested in the total surface area of a given mass or volume of soil and not only the surface area of individual particles.

Relation between Specific Surface and Particle Shape

The shape of the soil particles is also an important property from a physical point of view. The amount of surface area per unit mass or volume varies with the shape of the particles. This can be determined by considering the following shapes:

CUBE

$\qquad s_v = 6d^2/d^3 = 6/d$ [22]

SPHERE

$\qquad s_v = 4\pi r^2/(4/3)\pi r^3$
$\qquad\qquad = 3/r$
$\qquad\qquad = 6/d$

Note that the expressions for s_v of the cube and sphere are similar if the magnitude of d is the same as L for both shapes.

Knowledge of the particle size distribution can allow us to approximate the specific surface by the summation equation. Specific surface on a mass basis is

$$s_m = (6/\rho_p)\sum_{i=1}^{n}[C_i\,(d_i^2/d_i^3)] = (6/\rho_p)\sum_{i=1}^{n}(C_i/d_i)$$ [23]

where C_i is the mass fraction of the particles, d_i is the average diameter (m), ρ_p is the average particle density (kg/m^3), Σ is the summation symbol, and n is the number of soil separates. This analysis assumes that all soil particles have either a cubic or a spherical shape.

Example

Approximate the specific surface of a sand composed of the following array of particles sizes.

Diameter interval:	2000–1000	1000–500	500–250	250–150	100–50
Average diameter (μm):	1500	750	375	175	75
% by mass:	5.0	23.5	53.6	15.4	2.5

The calculations of specific surface on a mass basis are made as follows:

$s_m = (6/2650)\Sigma(C_i/d_i) = (6/2650)\,[(0.05/1.5 \times 10^{-3})$
$\quad + (0.235/7.5 \times 10^{-4}) + (0.536/3.76 \times 10^{-4})$
$\quad + (0.154/1.75 \times 10^{-4}) + (0.025/7.5 \times 10^{-5})]$
$s_m = (6/2650)\,(2989) = 6.77\ m^2/kg$

Note the following percentage contributions by each sand separate.

Separate	% by Mass	% by s_m	$\%s_m/\%$ Mass
Very coarse sand	5	1.11	0.22
Coarse sand	23.5	10.48	0.45
Medium sand	53.6	47.82	0.89
Fine sand	15.4	28.44	1.85
Very fine sand	2.5	11.15	4.46

This table shows that the smallest-diameter fractions contributed the largest percentage of specific surface for the soil.

Plate Shaped

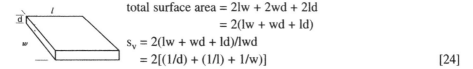

$$\text{total surface area} = 2lw + 2wd + 2ld$$
$$= 2(lw + wd + ld)$$
$$s_v = 2(lw + wd + ld)/lwd$$
$$= 2[(1/d) + (1/l) + 1/w)] \qquad [24]$$

Frequently, the width w and the length l have the same dimensions. Then, $w = l = a$; and we write $s_v = 2/d + 4/a$.

Example

Assume that a clay platelet has a thickness of 10 Å. Calculate s_v:

d = thickness = 10 Å = 10^{-9} m
w = width = 10^{-7} to 10^{-6} m

Therefore,

$s_v = 2/10^{-9} + 4/10^{-7} = 2/10^{-9}$

Thus, $s_v \approx 2/d$. Note that $2/10^{-9} >> 4/10^{-7}$.

edge area = 10^{-9} m $\times$ 10^{-7} m = 10^{-16} m^2
plane area = $(10^{-7})^2$ m^2 = 10^{-14} m^2

Thus, the area of the edges is negligible compared to that of the planar sides. However, the specific surface varies approximately as the inverse of the edge; that is,

$s_v \approx 2/d = 2/10^{-9}$ versus $4/10^{-7}$

Plate-shaped Disk

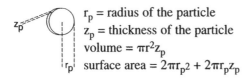

r_p = radius of the particle
z_p = thickness of the particle
volume = $\pi r^2 z_p$
surface area = $2\pi r_p^2 + 2\pi r_p z_p$

Thus,

$$s_v = 2\pi r_p(r_p + z_p)/\pi r_p^2 z_p = 2/z_p + 2/r_p \qquad [25]$$

The amount of contact per unit surface also changes with shape. Plate-shaped particles can be arranged into more intimate contact than spherical particles. Each sphere can make six-point contacts when placed in a close-packed arrangement. A disk, or plate, will make intimate contact with about 3/4 of its flat surface area when closely packed with other disks.

Spherical particles do not provide favorable conditions of surface contact for great cohesion. Plate-shaped particles cause high cohesion when stacked on each other. Also, laminar particles can slide over each other under an applied force, whereas spheres will permit a deformation only by breaking down or rolling apart.

The results of comparisons of various particle shapes show that a sphere has the smallest surface area per unit volume, and a cube and a sphere of the same volume can have the same specific surface. If a sphere is deformed into a rod or into a plate, the surface area increases, with the platelike particles exhibiting the greatest surface area.

Relation of Specific Surface to Surface Activity

Clay particles are generally platy and thus contribute even more to the overall specific surface of the soil than is indicated by their small particle diameters. In addition to the external surfaces, certain clay crystals exhibit internal surface areas, such as those that form when the open lattice of montmorillonite expands on imbibing water. Thus, the total specific surface of a soil consists of both external and internal surfaces and depends on the type and total amount of clay.

Many of the attributes of the soil relate to the interfacial surface phenomena. Therefore, almost every soil physical and chemical property has the same relationship to particle diameter. Examples of surface phenomena include retention of water at high pressures, cation exchange capacity, swelling, adsorption, plasticity, and strength (Petersen et al. 1996).

The general relationship between specific surface and particle size is given in Figure 3.7, which shows a curvilinear decrease in specific activity with an increase in particle size.

The general relationship between surface activity and specific surface is given in Figure 3.8, which shows a curvilinear increase in surface activity with an increase in specific surface.

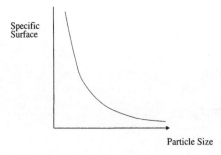

Figure 3.7. The general relationship between specific surface and particle size.

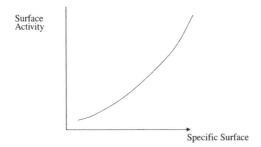

Figure 3.8. The general relationship between surface activity and specific surface.

Additional Calculations of Total Surface Area

The mass of solids can be used to estimate the total surface area available for surface reactions such as water absorption, cation exchange, sorption of organic molecules, etc.

Example

Assume that a silt loam soil with a mass of 1 kg is composed of 25% sand, 60% silt, and 15% clay and that the specific surface of these soil components is 0.5, 2.5, and 25.0 ha/kg, respectively. What is the total surface area (in ha) of the individual soil separates and of the soil?

Component	fraction	s_m (ha/kg)	Percent of total
Sand	0.25	0.125	2.3
Silt	0.60	1.50	27.9
Clay	0.15	3.75	69.8
Total		5.375	

The specific surface of 5.375 ha/kg of soil solids is equivalent to a surface area of 53,750 m^2/kg. It also shows that the clay fraction contributes almost 70% of the surface area of this soil.

Example

Assume that the soil given above has a particle density of 2600 kg/m^3. Calculate the specific surface on a volume basis.

Answer: 53,750 m^2/kg $\times$ 2600 kg/m^3 = 1.398 $\times$ 10^8 m^2/m^3.

Example

Assume that the same soil as in the above example has a cation exchange capacity of 25 cmolc (centimole charge)/kg of soil. How many sites are potentially available for exchange and what is the surface charge density?

We know that 1 cmolc/kg = 6.023 $\times$ 10^{21} sites. Therefore,

$$\text{number of sites} = 25 \text{ cmolc/kg} \times 6.023 \times 10^{21} \text{ sites}$$
$$= 1.506 \times 10^{23} \text{ sites/kg}$$

The surface charge density is the average number of sites per unit surface area. It is calculated from the following ratio:

$$\text{surface charge density} = 1.506 \times 10^{23} \text{ sites}/53{,}750 \text{ m}^2$$
$$= 2.802 \times 10^{18} \text{ sites/m}^2$$

In soils, the surface charge density varies with size and mineralogy of the particles.

Methods of Determining Specific Surface

Experimental determination of specific surface of soil usually involves the sorption of molecules such as ethylene glycol, glycerol, or ethylene glycol monoethyl ether by soil surfaces (Petersen et al. 1996). Usually a monolayer of the molecule is sorbed by a dried sample of soil. After equilibrium is obtained, the weight sorbed is used to calculate specific surface.

The Soil Texture Profile

Soil formation processes normally result in a distinct set of horizons in each soil. Arkansas has an interesting history of soil development and affords numerous applications of the sedimentation equation. For example, in the Mississippi Delta region of Arkansas, soils generally were formed from water (alluvial) and/or wind (eolian) transport. These soils have textures that are dependent on the velocity of the fluid flow, which is a function of the distance from the source.

Sedimentation Processes

Many of the soils in the Mississippi Delta region are derived from parent materials deposited mainly during Holocene geological time. The alluvium is a mixture of materials washed from many kinds of soils, rocks, and unconsolidated sediments from 24 states ranging from Montana to Pennsylvania, deposited by the Mississippi and Ohio Rivers, and, in part, reworked by local tributaries of the Mississippi River. The wide ranges in texture of the alluvium result from differences in the sites of deposition. When a river overflowed and spread over the floodplain, the coarse sediments were deposited first; therefore sandy and loamy sediments were deposited in bands along the channel, resulting in low ridges known as natural levees. Soils such as Beulah, Crevasse, Dubbs, and Robinsonville formed on the higher parts of these ridges. Finer sediments that have particle diameters greater than clay were deposited on the lower parts of natural levees as the floodwaters spread and velocity decreased. Soils such as Yancopin, Dundee, and Mhoon formed in these sediments. Where the water was left standing as shallow lakes or backswamps, the clays and finer silts settled. Soils such as Alligator, Bowdre, Earle, Forestdale, Sharkey, and Tunica formed in these sediments of small particle diameters.

This simple pattern of sediment distribution is not always found along the Mississippi River, because over the centuries, the river channel has meandered back and forth across the floodplain. Sometimes the river channel cut out all or part of the natural levees, and at other times sandy or loamy sediments were deposited on top of slack-water clay, or slack-water clay on top of sandy or loamy sediments. The normal pattern of sediment distribution from a single channel has been severely truncated in many places, and more recent alluviums have been superimposed. Soils such as Bowdre, Earle, and Tunica formed in these materials. Bowdre soils were formed in thin beds of clayey sediments over coarse sediments, and Tunica soils were formed in somewhat thicker beds of clayey sediments over coarser sediments.

Zone of Eluviation

In other areas where older soils are found, the changes required to form a soil are humus formation coupled with leaching and eluviation. The actual materials that move from the A and E horizons to the B horizon may be clay, humus, or certain calcium, iron, and aluminum compounds. As a result, the textures and thicknesses of the various horizons in a soil profile usually differ from one soil series to another. Therefore, each soil series has a characteristic texture profile. Since texture has a great influence on the physical, chemical, and biological properties of soil horizons, the development of the texture profile must be considered.

The process of clay particles slowly moving from the A and E horizons and depositing in the B horizon by percolating water is called eluviation. Thus, the A and E horizons are referred to as the *zone of eluviation* because they have lost the smaller-sized particles. Most old soils classified as Ultisols and Alfisols have higher clay contents in their B horizons than in their A horizons, and the difference in clay content increases as the soils become older. Therefore, the majority of the world's soils have a clay-enriched B horizon, which is indicated by a lowercase t as in Bt. When the Bt horizon has at least 20% more clay than the horizons above, it qualifies as an Argillic horizon. The B horizon is referred to as the *zone of illuviation.*

Soil Depth

Deep soils provide more adequate root zones and have greater capacities to store water, oxygen, and plant nutrients than do shallow soils. Therefore, deep soils tend to be more productive than shallower soils, and the differences are most often observed during drought stress. Generally, plants can endure a longer drought when they grow in soils having a higher available water-holding capacity.

Another aspect of soil depth is the thickness of the various horizons. The properties that distinguish one horizon from another influence the suitability of the horizons for various uses. The A horizon is generally the most favorable for plant growth, whereas a B horizon high in clay might be most useful for reducing the percolation of water through the profile, thereby making an impervious core in the dam when a pond is constructed or minimizing the drainage losses of water for rice production.

Examples of Texture Profiles

Table 3.8 presents the particle size distributions and organic-matter contents of several soil profiles commonly found in eastern Arkansas. These data were supplied by Dr. E. M. Rutledge, soil classifier, University of Arkansas.

Table 3.8 Profiles of Selected Soils in Arkansas

Depth		Particle Size Analysis (%)			Organic
Interval (cm)	Horizon	Sand	Silt	Clay	Matter (%)
Sharkey clay (Mississippi County)					
0–10	Ap	9.6	50.8	39.6	1.6
10–20	Bw	1.1	36.5	62.4	0.7
20–43	Bg1	1.6	28.1	70.3	0.6
43–58	Bg2	1.7	33.3	65.0	0.3
58–76	Bg3	4.5	35.7	59.8	0.3
76–94	Bg4	5.7	47.3	47.0	0.3
94–175	2Cg	4.2	60.5	35.3	0.2
Memphis silt loam (Lee County)					
0–15	Ap	0.8	77.4	21.8	1.4
15–25	Bt1	0.8	76.3	22.9	1.4
25–71	Bt2	0.3	69.1	30.6	0.6
71–117	Bt3	0.5	75.7	23.8	0.3
117–152	Bt4	0.5	79.9	19.6	0.3
152–203	C	0.5	84.1	15.4	0.3
Crevasse loamy fine sand (Johnson County)					
0–23	Ap	85.8	11.7	2.5	0.2
23–71	C1	80.4	15.8	3.8	0.1
71–114	C2	77.6	17.4	5.0	0.1
114–183	C3	94.6	4.1	1.3	0.1
DeWitt silt loam (Arkansas County)					
0–10	Ap1	3.2	78.0	18.8	1.6
10–18	Ap2	3.4	76.1	20.5	1.4
18–37	Eg	7.0	72.2	20.8	0.7
37–75	Btg	1.4	36.6	62.0	12.2
75–98	Bt1	1.6	43.6	54.8	1.1
98–125	Bt2	2.6	51.5	45.9	0.7
125–158	B′tg1	3.0	52.9	44.1	0.5
158–195	B′tg2	3.0	43.6	53.4	0.4
195–240	B′tg3	2.6	33.5	63.9	0.4
Calloway silt loam (Prairie County)					
0–18	Ap	9.5	82.1	8.4	0.8
18–41	EB	7.6	72.4	20.0	0.1
41–64	Btx	7.2	71.6	21.2	0.1
64–79	Btxg	8.4	73.8	17.8	0.1
79–127+	BC	10.4	57.5	32.1	0.1

Implications of Soil Texture Profiles

The distribution of particle size in the soil profile has much to do with the transport of air, water, and chemicals and with the distribution of plant roots. Sandy soils usually cause little restriction on movement through soil, whereas clayey materials often slow or even prevent movement of air, water, and roots. Another property related to soil texture is the amount of water that can be stored in the soil. Clayey soils tend to retain much more water than sandy soils.

The permeability and water storage aspects of texture combine to cause soils developing in materials with a high clay content to have shallower profiles than those developing in coarser-textured materials under comparable climatic and topographic conditions. This is because higher permeability allows the physicochemical effects of weathering to proceed deeper, assuming the same initial depth to bedrock.

A strongly differentiated profile normally restricts soil management and crop production. The A horizon of such a soil may have lost much of its chemical fertility because mineral weathering has become too slow to provide an adequate supply of plant nutrients. Water passes slowly through the B horizon because it has a high clay content. Root penetration is limited by the strength and the poor aeration of the B horizon. Most of the root development therefore occurs in the A horizon. Plant growth may suffer from too much water during wet seasons and too little water after a week or two of drought. A strongly differentiated soil can also be a problem for engineering uses because the A horizon becomes saturated easily. Saturation reduces its weight-bearing strength. Therefore, wet soils may fail to support the weight of a building.

Concluding Remarks

The textural composition of the solid phase affects the behavior and use of soil. Particle size distribution is used to classify and to predict the relative ease of transport of water, air, and solutes. Specific surface is used to estimate sorption of water and of inorganic and organic ions and molecules. An understanding of the spatial distribution of these physical properties enables soil scientists to infer their impact on soil-plant-water relationships, solute and heat transport, and geotechnical engineering properties.

Cited References

Day, P. R. 1965. Particle fractionation and particle-size analysis. In C. A. Black et al. (eds.), *Methods of Soil Analysis,* part 1, *Physical and Mineralogical Properties Including Statistics of Measurement and Sampling,* pp. 545–567. American Society of Agronomy Monograph 9. Madison.

Gee, G. W., and J. W. Bauder. 1986. Particle-size analysis. In A. Klute et al. (eds.), *Methods of Soil Analysis,* part 1, *Physical and Mineralogical Methods,* pp. 383–411. 2d ed. Agronomy Monograph 9. Soil Science Society of America. Madison.

Green, A. J. 1981. Particle-size analysis. In J. A. McKeague (ed.), *Manual on Soil Sampling and Methods of Analysis.* Canadian Society of Soil Science. Ottawa.

Kohnke, H. 1968. *Applied Soil Physics.* McGraw-Hill Book Co. New York.

Marshall, T. J., J. W. Holmes, and C. W. Holmes. 1996. *Soil Physics.* Cambridge University Press. New York.

McCarthy, D. F. 1993. *Essentials of Soil Mechanics and Foundations.* 4th ed. Regents/Prentice-Hall. Englewood Cliffs, NJ.

Petersen, L. W., P. Moldrup, O. H. Jacobsen, and D. E. Rolston. 1996. Relations between specific surface area and soil physical and chemical properties. *Soil Science* 161:9–21.

Soil Survey Division Staff. 1993. *Soil Survey Manual.* USDA Agriculture Handbook No. 18. U.S. Government Printing Office. Washington, DC.

Steinhardt, G. C. 1979. Particle size distribution. In R. W. Fairbridge and C. W. Finkl Jr. (eds.), *The Encyclopedia of Soil Science,* Part 1. Dowden, Hutchinson, and Ross. Stroudsburg, PA.

Additional References

Baver, L. D., W. H. Gardner, and W. R. Gardner. 1972. *Soil Physics.* 4th ed. John Wiley and Sons. New York.

Hillel, D. 1982. *Introduction to Soil Physics.* Academic Press. New York.

Koorevaar, P., G. Menelik, and C. Dirksen. 1983. *Elements of Soil Physics.* Elsevier. New York.

Problems

1. Assume a sampling depth of 5 cm and aqueous suspension temperature of 20°C. Calculate the amount of time that must elapse before sampling for soil particles 20 m or less in diameter.

2. Compare the specific surfaces of 1 cm^3 of spherical particles that have the following diameters: 0.1 cm, 0.002 cm, and 0.0002 cm. Explain your results.

3. If the B horizon contains 30% clay, 40% silt, and 30% sand by weight and has a bulk density of 1.5 Mg/m^3, calculate the average number of particles of each size separate in each 1 cm^3 of soil. Hint: Assume spherical particles and that each separate has an average particle diameter.

4. Estimate the specific surface (m^2/kg) of a soil composed of 15% coarse sand, 25% fine sand, 40% silt, and 20% clay. Use the average particle diameters of the respective separates in your calculations and a particle density of 2.65 Mg/m^3.

5. Suppose that a 25 cm^3 pipette is used to sample a soil suspension at the 25 cm depth in a 1.1 L cylinder for the clay fraction. If the temperature of the aqueous suspension is 20°C, and the oven-dry weights of sand and clay are 18 and 28 g of soil, respectively, determine the following for a 60 g sample of soil:

    a. mass percentages of sand, silt, and clay

    b. textural classification

    c. amount of time necessary to sample the suspension for clay.

6. Use the particle size distributions given below to classify the soils and to draw the particle size distribution curves (i.e., summation curves).

	Sand					Silt			
Soil	Very coarse	Coarse	Medium	Fine	Very fine	Coarse	Medium	Fine	Clay
1	0.2	0.8	0.4	1.2	4.8	33.8	36.7	6.5	15.6
2	0.1	0.2	0.5	1.8	6.5	10.5	12.7	14.0	53.7

Calculate the uniformity index, Iu, for each soil. Then compare and interpret the uniformity indexes of the two soils.

7. Assume that a cube of quartz 1 cm on a side is weathered into cubic particles 2 m on a side. How many particles will be obtained from the parent material?

8. Assume that a cube of quartz 1 cm on a side is weathered into spherical particles 2 m in diameter. How many particles would result? If the particles were packed so that each sphere touched six other spheres so that the maximum possible void space were created, approximately what volume of voids would result?

9. A soil was found to have a bulk density of 1.3 g/cm^3, a clay content of 30%, and a surface area of 300 m^2/g of clay. What is the total surface area (m^2) available for surface reactions in the surface 15 cm of 1 ha?

10. If the B horizon has a bulk density of 1.8 g/cm^3, how many particles would 1 cubic cm of soil contain if it consisted of 60% clay, 30% silt, and 10% sand? Assume the average particle density is 2.65 g/cm^3 and that the average particle diameters are 0.10, 0.01, and 0.0001 mm for the sand, silt, and clay fractions, respectively.

11. Draw a graph of the soil profile distributions of sand, silt, clay, and organic matter of each soil series given in Table 3.8. Interpret these distributions in terms of water retention and the transport of water, inorganic fertilizers, and pesticides.

12. Thought-provoking questions
    a. What are the soil physical characteristics that are best for rice production?
    b. What are the soil physical characteristics that are best for cotton production?
    c. What are the soil physical characteristics that are best for soybean production?
    d. What are the soil physical characteristics that are best for corn production?
    e. Assume that a farmer has two 20-ha fields and plans to grow rice and soybeans in a 1:1 rotation. If one field has a DeWitt silt loam and the other has a Memphis silt loam, compare how the different physical properties of these soils (see Table 3.8) would affect the expected growth of these two crops.

CHAPTER FOUR

Soil Structure

Introduction

The individual primary particles are not usually randomly arranged in field soils but are combined to form secondary soil particles called aggregates or peds. An *aggregate* is a group of primary soil particles that cohere to each other more strongly than to other surrounding soil particles (Kemper and Rosenau 1986). A *ped* is a unit of soil structure formed by natural processes, in contrast to a *clod,* which is formed artificially (Soil Science Society of America 1997). Therefore, two soils with the same texture may have distinctly different soil physical properties and behaviors because the particles are arranged in different ways. The primary particles are bound together into aggregates by colloidal substances such as organic matter, clay, metallic oxides, and carbonates. Over the years, tillage due to cropping, harvesting, and irrigation has resulted in deterioration of the structure in many soils, primarily due to losses of organic matter and increases in compaction.

In this chapter, we define soil structure and examine the components, genesis, characteristics, and manifestations of soil structure, along with the impact of structure on water and solute transport. We also include a discussion of organic-matter dynamics. The emphasis is placed on developing a greater understanding of the role of soil aggregation in soil physical behavior, particularly in the transport of water, heat, air, and solutes.

Definition of Soil Structure

Soil structure is defined as the size, shape, and arrangement of aggregates and pores of the soil. It is sometimes called soil architecture.

The nature, size, and distribution of aggregates and the pore space between these natural units of soil play a very important role in determining soil physical behavior. In particular, soil physical conditions and characteristics such as water movement, heat transfer, aeration, strength, and erodibility are strongly influenced by structure. Soil structure is not a plant growth factor, but it influences practically all factors that affect plants grown in soil, primarily through soil aeration, compaction, water relations, and temperature. Consequently, poor soil structure may be an indirect factor limiting plant growth. Management operations such as cultivation, plowing, irrigating, draining, cropping, liming, and manuring directly influence soil structure.

Components of Soil Structure

The arrangement of soil particles implies that the solid particles and pore spaces are interrelated. In a given volume of soil, the greater the volume fraction of the solid phase, the lower the volume fraction of pores. The solid phase is represented by the soil particles, which are grouped in various ways and have various shapes and sizes. These natural soil units are called peds. Peds can be distinguished from clods, which are caused by disturbance such as plowing when the soil is wet. The surfaces of peds persist through cycles of wetting and drying in place. Within and between the peds are the pores, which contain varying volumes of water and air.

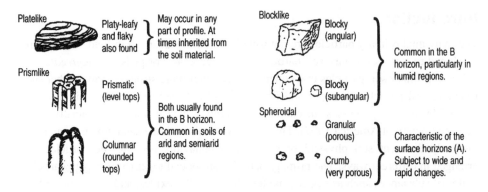

Figure 4.1. Types of soil structural units (peds) and their accompanying subtypes.

Systems for identifying soil structure have evolved because of the need for classifying soils based on their characteristics in their natural state. These classification systems are based on the shape and arrangement (type), the size (class), and the distinctness and durability (grade) of soil particles and pores. We will examine the qualitative nature of both the solid phase and the pores with respect to their physical characteristics.

Soil Peds

The dominant shape of the ped determines the type of structural units. Type refers to the shape of the ped. Four types of soil structural units, along with their subtypes, are described below and are shown in Figure 4.1. In general, the shapes of the peds are similar to these shapes of the individual soil particles discussed in the previous chapter.

Platelike

This structural type consists of aggregates arranged in relatively thin horizontal plates, leaflets, or lenses. The term *platy* is often used to describe this type of structure. Platy structure is sometimes found in the surface layers of virgin soils but may characterize the lower horizons as well.

Most structural features are usually a product of soil-forming forces. The platy types are often inherited from soil parent materials, especially those deposited by water or ice. Platy structure can also be created by compaction of silty and clayey soils by heavy machinery. Platy structure makes discontinuous pores that reduce the redistribution of water and air and the penetration of roots.

Prismlike

This type of soil structure is characterized by vertically oriented aggregates (prisms or pillars) that vary in height among different soils and may reach a diameter of 15 cm or more. They commonly occur in B horizons in arid and semiarid regions and, when well developed, are a very striking feature of the profile. They also occur in some poorly drained soils in humid regions and often are associated with the presence of smectitic clays.

The subtype of prismlike structure is determined by the shape of the prism tops. When the tops of the prisms are rounded, the term *columnar* is used. Columnar structure is especially common in subsoils high in sodium (i.e., natric horizons). When the tops of the prisms are angular and relatively flat horizontally, the structural pattern is designated as *prismatic*. Prismatic structure is moderately permeable; columnar structures are slowly permeable.

Blocklike

In this type of soil structure, the aggregates have a blocklike shape, usually six-faced, with their three dimensions more or less equal. These peds range from about 1 to 10 cm in thickness.

When the edges of the blocks are sharp and the rectangular faces distinct, the subtype is designated as *angular blocky*. When some rounding has occurred, the aggregates are referred to as *subangular blocky*. These subtypes usually are confined to the B horizon, and their stage of development and other characteristics have much to do with soil drainage, aeration, and root penetration. Blocky structure has medium permeability.

Two or more of the structural types can occur in the same soil profile. For example, it is common to find the surface horizon having a granular structure with a subangular blocky or platy structure in the lower horizons.

Spheroidal

Aggregates that have a curved, irregular shape are classified as the spheroidal type. They usually lie loosely and are separated from each other. Relatively porous spheroidal aggregates are called *granules,* and their shape is granular. However, when the granules are especially porous, they are called *crumbs.*

Granular and crumb structures are characteristic of many mineral soils in the A horizon, particularly those relatively high in organic matter. They are especially prominent in soils in grassland and pasture and in soils that have active earthworm populations. The peds are small, usually between 1 and 10 mm in size and rounded in form. Granular and crumb structures are the only types of aggregation that are commonly influenced by practical methods of soil management.

Class and Grade

A full classification of soil structure requires descriptions of not only the *type* or shape of the structural aggregates but also the relative size and the degree of development or distinctness of the peds. The relative size of the peds is subdivided into five *classes:* very fine (or very thin), fine (or thin), medium, coarse (or thick), and very coarse (or very thick). Note that the names of these classes correspond to the designations of the sand separates.

The development of the peds is known as the *grade,* which refers to how distinct and strong the peds are in undisturbed soil. The grade is subdivided into strong, moderate, and weak categories.

Structure is commonly described as a part of the basic morphological characterization of soils. The three elements of the classification system for soil structure are combined in the following order: (1) grade, (2) size, and (3) shape. Thus, a given soil horizon might be described as having "moderate, fine, subangular blocky structure."

Some soils and/or horizons lack structure and are referred to as structureless. In these soils or horizons, no structural units are observable in place, or when disturbed, soil fragments or single primary particles or both result. A soil fragment is a broken ped. Structureless soils may be either single grain or massive.

Soil Pores

The quantity, size, shape, and continuity of soil pores are also used to characterize soil structure. In particular, the total porosity, the size distribution of the individual pores, and the continuity of the pores are important characteristics. Soil pores influence the ability of soils to support plant, animal, and microbial life. Soil pores retain water, allow drainage, allow entry of O_2 and removal of CO_2, allow roots to penetrate, and are indirectly responsible for modifying the mechanical properties of soils so that cultivation can be carried out successfully.

Observation of soil pores in the field is confined to those pores visible without magnification or with a $10\times$ magnification using a hand lens. The visible pores are larger than 50 μm (0.5 mm). Luxmoore (1981) arbitrarily classified pore sizes into three groups (Table 4.1). Macropores, which are the largest pores, allow rapid drainage of water after heavy rainfall or irrigation. Once these pores are empty, drainage becomes slow. They are formed by roots, animals, insects, etc. The size and distribution of macropores usually bear no relation to the particle size distribution and the related micropore distribution. They generally favor high infiltration rates, good tilth, and adequate aeration for plant growth.

The visible pores are described morphologically by *quantity* per unit area, *size, shape,* and *vertical continuity*. The number of pores per unit area is divided into the following three classes: few (<1), common (1–5), and many (>5) pores per square centimeter. Pore size is divided into the following four classes: very fine (<0.5 mm), fine (0.5–2 mm), medium (2–5 mm), and coarse (>5 mm). Pore shapes are either vesicular (spherical or elliptical) or tubular (approximately cylindrical). Vertical continuity involves assessment

Table 4.1. Soil Pore Classification Scheme in Relation to Equivalent Pore Diameter and Equivalent Suction

Pore Name	Pore Size (μm)	Pore Function
Macropore	>1000	Allows rapid drainage of water after heavy rainfall or irrigation.
Mesopore	10–1000	Some water is available for plant use and drainage.
Micropore	<10	Corresponds to the soil matrix. Holds water tightly, some of which is available for plant use.

Source: After Luxmoore 1981.

of the average vertical distance by which the minimum pore diameter exceeds 0.5 mm when the soil is moderately moist or wet. Three classes are used: low—less than 1 cm; moderate—1 to 10 cm; and high—10 cm or more (Soil Survey Division Staff 1993).

The micropores closely correspond to the solid phase of the soil. These small pores dominate the total porosity of most fine-textured soils and retain water, some of which is available for plant use. Small pores (<1 μm equivalent diameter) can exclude bacteria, while slightly larger pores (1–2 m) allow bacteria but exclude protozoa and nematodes (Carter 1996).

A summary of Luxmoore's pore classification scheme in relation to equivalent pore diameter and equivalent soil water pressure is shown in Figure 4.2. There has been little or no consensus on the definition of pore sizes for the categories.

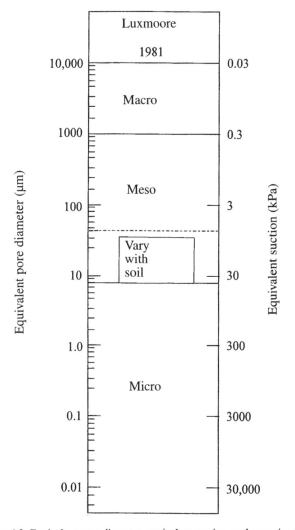

Figure 4.2. Equivalent pore diameter, equivalent suction, and pore size terminology. (Adapted from Luxmoore 1981.)

Genesis of Soil Structure

The size, quantity, and stability of aggregates recovered from soils reflect an environmental conditioning that includes factors that enhance the aggregation of soil particles and those that cause their disruption.

Physical, chemical, and biological processes are involved in the subsequent development of soil structure. These processes have been summarized by Rowell (1994).

Physical Processes

The physical processes important to soil structure formation include

1. drying and wetting, which cause shrinkage and swelling and the development of cracks and channels;
2. freezing, which creates spaces as ice is formed, and thawing.

Wetting and drying, especially with smectitic clay soils, cause swelling and shrinkage stresses that result in the weakening of the larger clods and the development of shrinkage cracks in clods and in fresh sediments. Clayey soils whose clods do not decompose upon weathering remain very cloddy and are usually difficult to manage. Generally, clayey soils tend to weather easily. The clay loams and silty clays are more difficult to manage. Furthermore, water flowing between separated surfaces can deposit clay, iron oxides, organic matter, and other materials carried in suspension to form skins or oriented clay. Surfaces modified by deposition of clay are called *cutans,* which are zones where naturally occurring aggregates can be separated from one another.

Freezing and thawing have a major impact on soil structure. Freezing affects aggregation by the expansion of water as it changes to ice within the profile. In coarse-textured soils, the water freezes in place, but in fine-textured soils, there is movement of water toward the freezing sites so that ice lenses are formed (Marshall et al. 1996). Rapid freezing, when little water movement can occur, results in the freezing of water within pores before it can move out, leading to considerable and usually excessive breakdown, which is later revealed on thawing. Under field conditions, however, freezing is usually not rapid, which allows considerable movement of water upward toward the lower soil temperatures and also toward any ice formed as the water potential in contact with it decreases. As the soil temperature falls, ice forms between the peds of a clay soil, with shrinkage of the unfrozen clay until, at a certain soil temperature, ice begins to form within the pores of the peds and shrinkage ceases. Compression and heaving of the soil occur as the ice lenses grow. Cracks enlarge by expansion on freezing, and because they take in more water when the ice has thawed, they open further on refreezing.

Chemical Processes

Soil chemical processes important to soil structure formation involve

1. clay and exchangeable ions,
2. inorganic cements,
3. organic compounds and cementing agents.

Generally, clay assists in aggregating soil by acting as a cementing agent and also by its ability to swell upon wetting and to shrink upon drying. Thin films of clay, called clay films, cover small aggregates of particles of certain soils and serve to hold these aggregates together.

Ions such as Ca, Mg, and K ions have a flocculating effect on clay, while H and Na ions have a dispersing effect. Flocculation is an electrokinetic phenomenon that can be created in aqueous suspensions. The flocculation of clay results in aggregates having particle sizes in the silt range. Na tends to disperse clay and to create a reaction in which several of the plant nutrients become unavailable, consequently depressing the growth of roots and microbes. Therefore, a poor soil structure results.

Sesquioxides form irreversible and slowly reversible colloids and help to form water-stable aggregates. This effect is particularly noticeable in Oxisols found in the Tropics. In spite of their high content of acid clay, these soils are generally well aggregated. The smallest peds are of fine-sand size. This results in pores large enough to allow rapid percolation of water. $CaCO_3$ precipitating out around soil particles also acts as a cement. Salts tend to enhance flocculation of clay, even in Na-saturated clays.

Organic compounds are effective in inducing aggregation and in stabilizing soil structure. They form irreversible or slowly reversible colloids that serve as cements. Organic compounds such as sugars are ineffective before they are changed to microbial tissues and decomposition products. Fats, waxes, lignins, proteins, resins, and certain other organic compounds have a direct stabilizing effect. Several organic compounds have been used as so-called soil conditioners. These are mostly long C-chain compounds that attach themselves to the exchange sites of the clay, and thus bind together many clay particles.

Biological Processes

Biological processes that affect soil structure include the following:

1. Roots extract water, resulting in the formation of spaces by shrinkage, and release organic materials; they leave behind organic residues and root channels when they die.

2. Soil animals move material, create burrows, and bring mineral and organic residues into close association.

3. Microorganisms decompose plant and animal residues, leaving humus, which is an important material binding particles together.

A plant root grows by forcing its tip, sometimes less than a millimeter in diameter, into soil pores of about the same size. The young root then swells, making the pore larger. This swelling involves the movement of primary and secondary soil particles, or aggregates, and results in an adjustment of the arrangement of soil particles near and within the rhizosphere. In addition, if the root extracts water from the rhizosphere, the soil may shrink as it loses water. Cracks created by shrinkage can last throughout the drier portion of the growing season.

The other biological agents that create structural pores, channels, and burrows are some of the larger members of the soil fauna. Earthworms are the most important group. They make their burrows through a considerable depth of soil and leave worm casts, which are fine ribbons of soil excreted by the worm and left in a small heap. In addition, the larvae of a number of insects, particularly beetles, make channels in the soil as they move about in search of food. Ants make numerous channels in the upper part of the profile.

These physical, chemical, and biological processes show that the formation of soil structure requires both physical rearrangement of groups of particles to form a loose aggregate and the stabilization of the new arrangement. The loose aggregates are weak and easily crushed. Stability is particularly associated with organic materials linking mineral particles together and with the clay minerals and sesquioxides. The binding between clay particles depends on the ions associated with the particle surfaces and in soil solution. Soil organisms provide the best cement when they decompose plant residues to produce polysaccharides that glue the peds together.

Characterization of Soil Structure

Researchers have long sought a physical measurement that would give an overall evaluation of the structure of soil. However, there are no theoretical or experimental methods developed so far to quantitatively express soil structure from measurements of aggregate size or stability. Therefore, the structure of soil is characterized indirectly by soil physical and morphological techniques and by inferences made about the structural suitability for the intended use. Several of these techniques are discussed below.

Soil Physical Techniques

Soil physical parameters used to characterize soil structure include bulk density, total porosity, pore size distribution (obtained from water retention data), analysis of aggregate size, and stability. These characteristics are considered to be static parameters. Dynamic physical characteristics of soil structure can be obtained by measuring hydraulic conductivity, infiltration rate, and air permeability. All of these soil physical characteristics are strongly affected by the size, distribution, and quantity of the solid particles and pore spaces.

Total Porosity and Bulk Density

As soils become more porous, they also become less dense. Thus, total porosity is determined from a measurement of the dry bulk density of a soil sampled so as to preserve its natural structure. The mathematical relationship between these parameters is

$$f = 1 - \rho_b/\rho_p \qquad [1]$$

where f is the total porosity (m^3/m^3), ρ_b is the bulk density (kg/m^3), and ρ_p is the particle density (kg/m^3). Total porosity in a soil varies depending on (1) the texture and organic-matter content, (2) the depth in the profile, and (3) soil management. In general, higher porosities are found in soils having higher organic and clay contents. Depth in the profile is important because soil structure–forming processes are most active near the soil surface and compaction often increases with depth. Soil management affects porosity because it affects organic-matter content over time and applies forces to soils that may loosen or compact the aggregates and/or particles.

Bulk density is frequently used to evaluate soil structure. The higher the ρ_b, the more compacted the soil and the lower the porosity. Bulk densities tend to be lowest at or near the surface and often increase with depth to some maximum value depending upon soil formation processes. Tillage pans, fragipans, and some genetic horizons can have significantly higher bulk densities than the horizons immediately above and below them.

Example
The A horizon of a soil had a dry bulk density of 1.3 Mg/m^3 in the crop row and 1.6 Mg/m^3 in the middle between the rows where tractor wheel tracks were observed. Calculate the effects of compaction on the volume of the solid phase and total porosity.

Position	Volume of Solid Phase (m^3/m^3)	Total Porosity (m^3/m^3)
Row	0.491	0.509
Middle	0.604	0.396

Pore Size Distribution

Pore size distribution is strongly influenced by aggregation of soil particles as well as by soil texture. The size distribution of soil pores is determined from the water retention curve (also known as the characteristic curve). The equivalent radius of the largest pore that will be filled with water is a function of the soil water pressure through the capillarity equation. Mathematically, this is expressed as

$$r = (2\sigma \cos \theta)/(\rho_w gh) \approx 0.15/h \qquad (h < 0) \qquad [2]$$

where r is the equivalent radius of the pore (cm), σ is the surface tension (kg/s^2), θ is the contact angle between the water and the pore wall (usually assumed to be zero), ρ_w is the density of water (kg/m^3), g is the gravitational acceleration (m/s^2), and h is the soil water suction (cm of water). This equation can be used to represent the minimum radii for a given soil water pressure range. The equation shows that pore size and water pressure are indirectly related and that the macropores can retain water only at low negative soil water pressures. Therefore, a water retention curve can be used to show the amount of pore space that has pores smaller than a given effective size.

Example
Show the influence of soil water suction on the equivalent pore radius at suctions of −10 and −100 cm:

$$r = 0.15/10 = 0.015 \text{ cm} \qquad \text{and} \qquad r = 0.15/100 = 0.0015 \text{ cm}$$

In applying equation [2], the following assumptions are made: (1) the soil is rigid and does not change its volume with changes in soil water content; (2) the pores are tubes of circular cross section, so that an effective, rather than actual, size is measured; and (3) a drying, rather than a wetting, soil is used because the contact angle is more likely to be zero.

Wilson and Luxmoore (1988) showed that the hydrologically effective porosity, f_e, associated with each size class of pores can be calculated from equation [2] and Poiseuille's equation by

$$f_e = (8\eta I_m)/(g\rho_w r^2) \tag{3}$$

where η is the viscosity of water (kg/m·s), I_m is the macropore flow (m/s), and r is the minimum radius of the size class. Values of I_m were calculated as the difference between infiltration rates at saturation and at −2 cm pressure. This calculation results in upper bound estimates of the macroporosity associated with the size classes.

Since the effective porosity equals the number of pores of a certain size class per unit area times the average area of a pore of that size class, the number of effective pores, N, per unit area for each size class is given by

$$N = f_e/(\pi r^2) \tag{4}$$

This method of estimating macro- and mesoporosity is based on the following assumptions: (1) laminar flow of water; (2) smooth, cylindrical pores; and (3) a pore radius determined by capillary theory. The method ignores the effects of pore length and dead-end pores on the flow rate.

Aggregate Analysis

One of the most widely used techniques for assessing soil structure is the analysis of the size distribution and water stability of aggregates. The aggregates are separated into groups of different sizes by means of a nest of sieves or a settling velocity using the sedimentation equation. Sieving may be done wet or dry. The size of the openings in each sieve that is selected to be included in the nest of sieves will depend on the desired groups of aggregates.

There is no universal agreement on the method to be used in the sieving of aggregates. However, sieving of dry aggregates indicates the size distribution in a soil sample without revealing information about the stability of the aggregates. Sieving in water indicates the distribution of aggregates that are stable after immersion, giving results that may be different from those obtained by dry sieving. Aggregate size probability distributions tend to be logarithmic normal and can be characterized by specifying the geometric mean diameter and the geometric standard deviation (see Appendix A).

Infiltration Rate

The time variations in the infiltration rate of water can be expressed mathematically by the empirical power equation

$$i = at^{-b} \tag{5}$$

where i is the infiltration rate (mm/h), that is, the volume of water infiltrating a unit soil surface area per unit time t (h); and the parameters a and b are statistically fitted to the data that reflect soil properties. The parameter a is approximately the initial infiltration rate (mm/h), and b is related to the stability of soil structure. A small value of b indicates a stable soil structure with little change in infiltration rate with time. Because of the shape of the distribution of the data, equation [5] can be applied only to short-time infiltration experiments.

Example
Calculate the infiltration rate after 5 minutes if the values of a and b are 20 mm/h and 0.12, respectively:

5 min/(60 min/h) = 0.0833 h
$i = 20 \times 0.0833^{-0.12} \cong 27.0$ mm/h

Show how your answer would change if the value of b was 0.012:

$i = 20 \times 0.0833^{-0.012} \cong 20.6$ mm/h

The condition of the soil surface—such as the degree and stability of the soil aggregates, texture, organic-matter content, and protection by crop residues and aboveground crop morphology—has significant influences on the infiltration rate of soil. In addition, soil compaction and the presence of tillage pans and restrictive permeable horizons negatively affect the infiltration rate. These effects are reflected in the magnitude of the model parameters a and b.

Soil Morphological Techniques

Soil profile descriptions typically include a field evaluation of soil structure that includes sizes, shapes, and degrees of development of pores and of natural aggregates such as the peds. Frequently, macropores formed by roots or soil animals are observed. Soil profile descriptions are mostly qualitative in nature, but major differences among structures in the same soil can be determined quite reproducibly by trained observers.

One morphological technique used for soil structure evaluations is to develop and examine soil thin sections. As shown in Figure 4.3, the lattice clays in the Eg/Bt horizon are generally platy and, when moved and deposited, tend to be oriented. The continuous orientation gives them a bright and strong color intensity. These features indicate transportation of clay along with soluble iron. Because clay translocation is accomplished through percolating water, the translocated clay tends to form coatings of oriented clay particles along the channels through which water and solutes move. The channels or crevices or voids formed by the cleavage faces of the peds are the pores or channels formed by biogenic activity. In Figure 4.4, the dark brown areas represent areas of iron and clay accumulation in the B2t horizon. The gray areas are areas of clay depletion.

Manifestations of Soil Structure

Soil structure exerts great influences on those soil physical properties that are important in plant growth and in the transport of mass and energy in soil (Hamblin 1985). A brief general discussion follows.

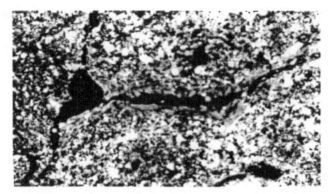

Figure 4.3. Ferriargillan lining of a channel in the Eg/Bt horizon. (This picture of a soil thin section was kindly provided by Jawad Khan.)

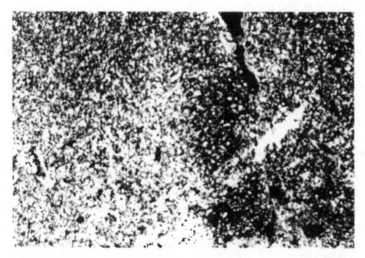

Figure 4.4. Areas of accumulation and depletion of clay in the B2t horizon. The darker areas show accumulation of iron and clay; the lighter areas show clay depletion. (This picture of a soil thin section was kindly provided by Jawad Khan.)

Soil Temperature

Soil temperature affects the rates of many physical, chemical, and biological processes in the soil. For example, organic-matter decomposition with resultant release of nutrients increases with increasing soil temperature over a wide range of temperatures commonly found in the field. Fluctuations in soil temperature are influenced by both the heat capacity and the thermal conductivity of soil. These are in turn directly affected by the soil bulk density, mineralogical composition, and soil water content.

Soil Aeration

Oxygen is consumed and carbon dioxide is produced by plant roots and microorganisms in the soil at a rate that appears to be characteristic of the soil-plant system. This process results in increased concentrations of CO_2 and decreased concentrations of O_2 in the pores. As a result, concentration gradients are created, and diffusion transports O_2 into the profile and CO_2 from the profile. The rate of diffusion is directly proportional to the aeration porosity. When the concentration of CO_2 gets too high and the O_2 gets too low, plant growth is retarded. Therefore, it is desirable for a soil to have a moderate number of large, continuous pores that are air filled so that aeration is adequate for plant needs.

Soil management practices that improve soil structure also help improve soil aeration. These include conversion to pasture, plowing at the appropriate soil water content, incorporation of crop residues into the soil after each crop, and application of animal manures.

Soil Water

The physical processes of infiltration and permeability are closely related to the pore size distribution and the stability of soil structural units. Initial infiltration rates may be high when the soil is dry. However, infiltration rates decline rapidly when the soil becomes saturated and water is not redistributed to lower parts of the profile as rapidly as it is received from rainfall. This may result in significant runoff.

The soil physical environment is determined by processes that transfer water, salts, air, and heat and by mechanical stresses with the soil profile (Cresswell et al. 1992). For each process, there exists a well-established theory. The transfer relations are based on conservation of mass or energy, and the rate equations are based on relating the flux density of an entity to the space gradient of potential of the entity. The transfer coefficients and the functions necessary to define these processes are measurable, macroscopic, and average properties of the soil material and its structure. The transfer coefficients also depend on the soil water content.

Research studies have shown that infiltration is influenced much more by the effects of soil organic-matter content on water-stable aggregation, initial soil water at the surface, and management than by texture and topography. Numerous studies have indicated that a single application of crop residue, straw, or hay or moderate to large applications of animal manure turned under in the spring do not greatly increase infiltration during that crop year. The increases in infiltration appear to be the result of gradual distribution of organic materials throughout the plow layer by successive annual additions and plowings.

Other studies have shown that infiltration of water into soil is directly related to macroporosity. Runoff is much less when soil contains a large number of macropores. Macropores are important for root growth and solute and water movement. Increases in water flux density can be attributed to root channels, whereas decreases in water flux density can be attributed to cropping and different tillage practices and wheel traffic.

Plant Growth

Soil structure affects plant growth through its influence on soil heat, temperature, air, water, and mechanical impedance to roots. Adverse effects of soil compaction on crop growth have been recognized for many years (Unger and Kaspar 1994). Plant roots differ in their ability to penetrate compacted soils, and this determines how effectively the plants use soil water and nutrient supplies. To alleviate soil compaction conditions, many studies involving tillage practices such as deep plowing and chiseling have been conducted. According to Unger and Kaspar (1994), these studies often gave inconsistent results because the soil conditions causing the problems and the soil environment resulting from use of the problem-alleviating practices were not measured or fully understood.

The influence of soil structural characteristics on crop productivity is normally defined by comparing the actual yield with the potential yield as influenced by the prevailing climate and genetic factors.

Effects of Cultivation on Soil Organic-Matter Dynamics

Tillage operations are used to provide more favorable soil conditions for crop growth and development. In general, soils are tilled for three reasons: (1) to change soil structure, (2) to manage crop residues, and (3) to control weeds. Moderate tillage facilitates root growth by loosening surface and subsurface soil and improves aeration and water infiltration of the soil profile. Frequent tillage operations using either moldboard plows, disk, or chisel cultivators or a combination of disk-chisel or disk cultivators have been used in crop production for many years. Tillage fractures the soil into its natural aggregates and into clods and fragments whose sizes vary with the tillage operation and initial conditions. The cumulative effect of frequent tillage operations and cropping is an undesirable change in soil physical, chemical, and biological properties. There is now a tendency to reduce the number and intensity of tillage operations in cropping. Various schemes are being tried and promoted and are known as no till, zero till, conservation tillage, minimum tillage, etc. All of these involve fewer operations across the field with tillage implements.

Comparison of virgin prairie soils with cultivated soils has shown that grassland/pasture soils have greater stability of aggregates, hydraulic conductivity, water infiltration rate, organic C and total N concentrations, biological organisms, and water retained, along with lower bulk density, than cultivated fields. As a result, extensive tillage of the soil for long periods of time may have detrimental effects on soil properties, crop establishment, and crop yield. One of the most important soil properties affecting soil structure and aggregation is organic matter. Soil organic matter is essential to productive soils. It promotes desirable soil physical and chemical properties and serves as a base for a diverse population of soil organisms.

Mathematical Models for Soil Organic-Matter Dynamics

Several mathematical models have been used to describe the relationships between soil organic-matter content and time of cropping. For example, Scott and Wood (1989) compared two groups of mathematical models to predict the temporal changes in organic-matter content of a Typic Albaqualf developed under prairie vegetation. The first group of models, which assume that the rate of loss of organic matter is independent of time, included the linear, first-order exponential, power, and modified exponential model. The best fit of this group of models to field data was obtained from the rate equation

$$dN/dt = A - KN \qquad [6]$$

where N is the organic-matter content (g/kg), t is the time of cropping (yr), A is the constant yearly application rate of residue (g/kg·y), and K is the sum of the transfer coefficients of organic matter into the fast and slow fractions (per year), or the decay rate constant. This model assumes that the rate of addition of organic matter is constant over time and the rate of loss is proportional to the amount of organic matter present. Equation [6] also indicates that at equilibrium (i.e., dN/dt = 0), soil organic-matter content is directly proportional to A and inversely proportional to K (i.e., N = A/K) and will increase to new equilibrium levels as A increases. Scott and Wood (1989) found that A was 0.96 g/kg·y and K was 0.066/y for the Typic Albaqualf that had been continuously cropped in rice, soybeans, and winter wheat for 30 years. Thus, the rate of loss of organic matter was greater than the addition of organic matter from cropping, and as cropping proceeded, the difference in the rate of addition and the rate of loss decreased.

The two-variable rate loss models examined included the logistic and the double first-order equations. Both models fit the experimental data well. The logistic rate model was

$$dN/dt = rN[1 - (N/M)] \qquad [7]$$

and resulted in values of 12.0 g/kg and –0.037/y for M and r, respectively. The parameter M represents the equilibrium organic-matter content after continuous, long-term cropping of rice, soybeans, and wheat. The parameter r represents the relative depletion rate. The double first-order rate model was

$$dN/dt = N_1 r_1 e^{-r_1 t} + N_2 r_2 e^{-r_2 t} \qquad [8]$$

with values of 0.053 and 0.01/y for r_1 and r_2, respectively, and 16.3 and 18.3 (g/kg) for N_1 and N_2, respectively. Application of this model to the field organic-matter contents showed that the rate of decomposition in the 0–5 cm depth interval was more than five times more rapid during the first phase than during the second phase.

In the same study, they found that the total organic carbon content in the 0–15 cm depth interval was significantly linearly related to bulk density, total porosity, and water retained at 10 kPa. The slope of the line indicated that a loss of 10 g/kg of organic matter resulted in an increase in bulk density of 0.143 Mg/m^3 and a loss of 0.05 m^3/m^3 of water retained at 10 kPa. In a later study at the same site, Scott et al. (1991) found that organic-matter content was also linearly related to the logarithm of the saturated hydraulic conductivity. No significant relationships were found between total carbon content and water retained at 1500 kPa (Figure 4.5).

These results are in agreement with those of Hudson (1994), who found that within three textural groups, as organic matter increased, the volume of water held at field capacity increased at a much greater rate than that held at the permanent wilting point. As a result, highly significant positive correlations were found between organic-matter content and available water content for sand, silt loam, and silty clay loam. In all textural groups, as organic matter increased from 0.5 to 3% by weight, available water content of the soil more than doubled.

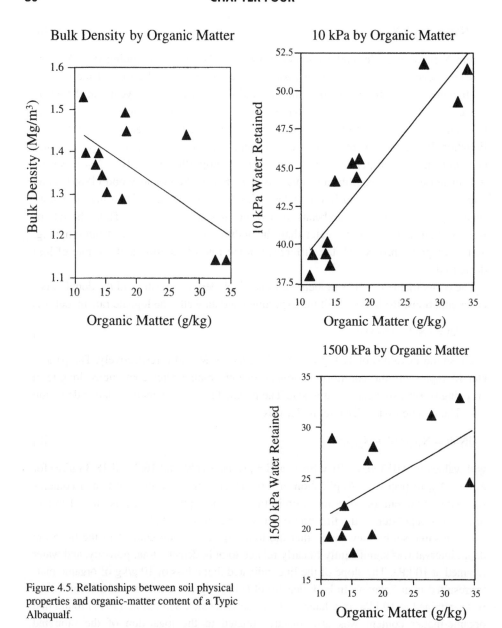

Figure 4.5. Relationships between soil physical properties and organic-matter content of a Typic Albaqualf.

Rapid and Slow Fractions

The previous discussions about the accumulation and loss of organic matter focused on the total amount of organic matter in the soil. Perhaps a more useful approach to understanding the dynamics of organic matter is to recognize that soils contain different fractions, or pools, of organic C that vary in their susceptibility to microbial decomposition. Probably the most frequently used method is to divide the observed decomposition into two fractions: rapid and slow.

The rapid fraction of soil organic matter consists of materials with low C/N ratios and short half-lives. These plant components include the living biomass, some of the fine-particulate detritus, most of the polysaccharides and other nonhumic substances, and some easily decomposable fulvic acids.

The slow fraction of soil organic matter consists of very stable materials remaining in the soil for hundreds or even thousands of years. This fraction includes most of the humus physically protected in clay-humus complexes, most of the humins, and much of the humic acids.

Models having two to four pools, or fractions, have been developed over the years. These models usually assume that all C and N transformations are first-order reactions under the assumption that the concentration of the material involved rather than the bio-logical capacity is rate limiting in decomposition. The first-order decomposition rate equations have the form

$$dC_s/dt = -kC_s \qquad\qquad [9]$$

where k is the first-order decomposition rate constant (per day), C_s is the amount of sub-strate (kg/ha), and t is the time (days). In addition, the dynamic changes in organic C can be quantified by calculations of turnover times (tt) and half-lives ($t_{0.5}$) according to

$$tt = 1/k \qquad\qquad [10]$$

$$t_{0.5} = 0.693tt \qquad\qquad [11]$$

Balesdent et al. (1988) found for an Alfisol developed under prairie vegetation in Missouri that the loss of organic matter from cropping was primarily from the rapid frac-tion. Mechanical analysis of the soil showed a heterogeneity with regard to organic-matter stability. The clay fraction and concretions in the soil contained the most-stable organic matter. They concluded that the initial rapid loss of prairie C occurred from the rapid pool, which had a short half-life of 10–15 years and complete turnover in 30–40 years. The extent of this rapid loss was dependent on the type of cultivation. The slow, or stable, fraction constituted about 50% of the current level and had a complete turnover of 600 years or more. The major replacement of prairie C by cultivated crop C occurred during the initial period and was characterized by marked microbial mineralization of prairie organic matter, with the easily mineralizable component disappearing first. After a period of a decade or two, continued turnover predominantly involved organic matter of crop origin.

The temporal changes in the rapid and slow fractions of organic matter in the surface of the Typic Albaqualf after the initial cultivation of the virgin prairie are shown in Figure 4.6. These curves are from the data of Scott and Wood (1989) using the double first-order model and show that the active fraction of the total organic matter decays rapidly due to cropping, whereas the slow fraction is relatively resistant to change. Turnover times were 18.9 and 100 years for the rapid and slow fractions, respectively. In the prairie, the rapid fraction represented about 47% of the total organic matter. After 30 years of cropping to rice, soybeans, and winter wheat, the organic matter declined to 15.2 g/kg, with a simulated value of 16.9 g/kg. The rapid fraction declined from 16.3 in the prairie to 3.3 g/kg, or 17.8% of the simulated organic-matter content. During this

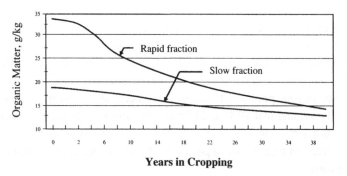

Typic Albaqualf

Seidenstricker Farm

Figure 4.6. Temporal variation of organic-matter content in a Typic Albaqualf in Arkansas.

same time interval of 30 years, the slow fraction declined from 18.3 g/kg in the prairie to 13.6 g/kg in the cropped field. This represented a reduction of about 80 and 25% for the rapid and slow fractions, respectively.

Scientists familiar with these findings have concluded that these fractions represent organic material with different stabilities with regard to decomposition and hypothesize that the greater stability of the slow fraction is caused by physical protection of soil organic matter and by differences in microbial biomass. This protection of part of the organic matter against microbial attack is lowered by processes that disrupt the soil, such as tillage. Tillage leads to an increase in mineralization of both C and N. The stabilizing effect is assumed to be the result of sorption of the organic materials by clay and silt particles and by the aggregates. This renders the organic material less accessible to microbial decomposition. This explains why a small loss in the rapid fraction of organic matter can significantly affect changes in soil physical properties such as bulk density, saturated hydraulic conductivity, and aggregate stability.

Managing Soil Structure

Changes in soil management often result in relatively small effects on soil organic-matter content but have pronounced effects on soil physical properties and aspects of soil productivity attributed to soil organic matter. Land use practices that sustain soil productivity must include combinations of practices in which structural degradation over time caused by one practice is balanced by regeneration due to other practices. The extent of degradation or regeneration is determined by the rate of change in soil structure and the duration of the practice.

Structural form is the structural characteristic that determines the availability of water and oxygen in soil and the resistance of the soil to penetration by roots. Stability and resiliency relate to the dynamics of soil structural form, that is, its variation over time. The introduction of new land use practices seldom results in step changes in structural form, and therefore, stability, resiliency, and vulnerability must be defined in terms of rates of change in soil structural form.

Resiliency is characterized by the maximum recovery in structural form when it is possible and the rate at which recovery occurs. The resiliency potential is defined as the maximum recovery in structural form that a soil in a given state can experience.

Kay et al. (1993) hypothesized that the rate of change of a given characteristic of structural form can be related to three functions that are additive: a function related to temporal stresses arising from the land use practice, a biological-related function, and a weather-related function. Each of the functions includes a response characteristic. The equation used to describe the temporal variation in soil structural form and stability is

$$\partial S_f/\partial t = f(R, F/A) + f(Lb, b) + f(Lw, w) \qquad [12]$$

where S_f is the structural form for a given land use practice.

The stress-related function includes two variables, the stability parameter, R, which is appropriate for a given characteristic of structural form and a given stress, and the stress or force per unit area, F/A. The form and magnitude of the stress are determined by the land use practice and are normally related to tillage and traffic. The stability is determined by soil properties and the influence of biological factors and weather on these properties.

The biological-related function describes the rate of change in a characteristic of structural form caused by soil flora and fauna such as roots and earthworms. It includes a parameter that describes the population/activity of organisms, b, and a resiliency parameter, Lb, that relates the rate of change in structural form to the population/activity of the organisms.

The weather-related function is composed of a parameter that describes the number/intensity of wetting-drying cycles or freezing-thawing cycles, w, and a resiliency parameter, Lw, that relates the rate of change in structural form to w.

Brady and Weil (1996) summarized six general principles useful in managing soil organic matter:

1. A continuous supply of organic materials must be added to the soil to maintain an appropriate level of soil organic matter, especially in the rapid fraction. Cover crops can provide protective cover and add organic material to the soil as green manure. It is almost always preferable to keep the soil vegetated than to keep it in bare fallow.

2. No attempt should be made to try to maintain higher soil organic matter levels than the soil-plant-climate control mechanisms dictate.

3. Because of the linkage between soil N and organic matter, adequate N inputs are requisite for adequate organic-matter levels. Accordingly, the inclusion of legumes in the crop rotation and the judicious use of N-containing fertilizers to enhance high soil productivity are two desirable practices. At the same time, steps must be taken to minimize the loss of N by leaching, erosion, or volatilization.

4. Maximum plant growth will increase the amount of organic matter added to soil from crop residues. Even if some plant parts are removed in the harvest, vigorously growing plant leaves, root, and top residues serve as major sources of organic matter for the soil. Moderate applications of lime and fertilizers should help remove chemical toxicities and deficiencies that might constrain plant growth.

5. Because tillage accelerates the loss of organic matter both by increased oxidation of soil organic matter and by erosion, it should be limited to that needed to control weeds and to maintain adequate soil aeration.

6. Perennial vegetation, especially natural ecosystems, should be encouraged and maintained wherever feasible.

Impact of Preferential Flow on Water and Solute Transport

The occurrence of solutes leaching to groundwater at a rate faster than expected has raised concerns. One factor that leads to rapid transport of a contaminant is preferential flow of water and solutes in well-structured soils. Field soils, especially those in pasture or forest, usually are structured in some way by containing large continuous macropores, such as drying cracks, earthworm channels, gopher holes, decayed root channels, or interpedal voids in naturally occurring aggregates (Quisenberry and Phillips 1976; Beven and Germann 1982). Because of preferential flow, uniform flow does not occur in well-structured field soils, and surface-applied agricultural chemicals such as fertilizers and pesticides may leach to groundwater reservoirs quicker than would be expected from uniform movement through the soil matrix. In essence, most of the soil matrix is bypassed. This process has been termed preferential flow, macropore flow, incomplete or partial mixing, short-circuiting, and non-Fickian transport.

Among various types of preferential flow, macropore flow is most common. The significance of macropores in flow through soils has been recognized for more than a century. However, until the early 1970s, it was largely ignored in soil physical theory used to develop models of water and solute transport. Thomas and Phillips (1979) concluded that some consequences of water flow in macropores are that (1) the value of a rain or irrigation to plants will generally not be as high as anticipated since some of the water may move beyond the root zone; (2) recharge of groundwater and springs can begin long before the soil reaches the upper drainage limit (i.e., field capacity); (3) some of the salts in the surface of a soil will be moved to a much greater depth by rain or irrigation than predicted by piston displacement; on the other hand, much of the salt will be bypassed and remain near the soil surface; and (4) because of this, it is not likely that water will carry a surge of contaminants to groundwater at some time that is predictable by Darcian theory.

The influence of macropores on solute flow in the field has been best demonstrated with water-soluble dyes. Dye stain patterns have given visual evidence of where the water has moved in the soil and that macropore flow dominates solute transport when the soil water matrix potentials are <2.5 kPa. The pattern of macropore flow in structured soil is controlled by soil porosity and pedality (structure), especially pore size distribution and pore continuity/connectivity. Root and earthworm channels, roots, cracks/fissures, slickensides, and interpedal pores appear to be the main macropore flow paths at high soil water potentials.

Field studies have also shown that macropores and mesopores constitute a small fraction of the total porosity but make the largest contribution to the rapid flow through the soil profile. Nonuniform flow patterns occur, which cause water and solutes to penetrate to depths many times deeper than what is expected from traditional infiltration models.

This leads to a spatial concentration of the water and solute transport and to deep penetration within the profile. The conventional assumption of uniform advancement of the wetting front is not valid in soils containing significant macropores. Marshall et al. (1996) suggest that to find whether macropores are significant, water should be applied (1) at a small soil water pressure and (2) at a small negative pressure. In the former case, the macropores conduct water, whereas in the latter case, those pores with an equivalent diameter greater than 1 mm remain empty when water is applied to the soil through a porous plate at a pressure such as –0.3 kPa.

Cited References

Balesdent, J., G. H. Wagner, and A. Mariotti. 1988. Soil organic matter turnover in long-term field experiments as revealed by carbon-13 natural abundance. *Soil Science Society of America Journal* 52:119–124.

Beven, K., and P. Germann. 1982. Macropores and water flow in soils. *Water Resources Research* 18:1311–1325.

Brady, N. C., and R. R. Weil. 1996. *The Nature and Properties of Soils.* 11th ed. Prentice Hall. Upper Saddle River, NJ.

Carter, M. R. 1996. Analysis of soil organic matter storage in agroecosystems. In M. R. Carter and B. A. Stewart (eds.), *Structure and Organic Matter Storage in Agricultural Soils,* pp. 3–11. Advances in Soil Science. Springer-Verlag. New York.

Cresswell, H. P., D. E. Smiles, and J. Williams. 1992. Soil structure, soil hydraulic properties and the soil water balance. *Australian Journal of Soil Research* 30:265–283.

Hamblin, A. P. 1985. The influence of soil structure on water movement, crop growth, and water uptake. *Advances in Agronomy* 38:95–158.

Hudson, B. D. 1994. Soil organic matter and available water capacity. *Journal of Soil Water Conservation* 49:189–194.

Kay, B. D., V. Rasiah, and E. Perfect. 1993. The structural resiliency of soils. In J. Caron and D. A. Angers (eds.), *Proceedings of the Second Eastern Canada Soil Structure Workshop,* pp. 73–86. Mont Sainte-Anne, Que.

Kemper, W. D., and R. C. Rosenau. 1986. Aggregate stability and size distribution. In A. Klute et al. (eds.), *Methods of Soil Analysis,* part 1, *Physical and Mineralogical Methods,* pp. 425–442. 2d ed. Agronomy Monograph 9. Soil Science Society of America. Madison.

Luxmoore, R. J. 1981. Micro-, meso-, and macroporosity of soil. *Soil Science Society of America Journal* 45:671.

Marshall, T. J., J. W. Holmes, and C. W. Holmes. 1996. Soil structure. In *Soil Physics,* pp. 199–228. Cambridge University Press. New York.

Mitchell, J. K. 1993. Soil structure: Its formation, stability, and relationships to soil properties. In *Fundamentals of Soil Behavior,* pp. 190–227. John Wiley and Sons. New York.

Quisenberry, V. L., and R. E. Phillips. 1976. Percolation of surface applied water in the field. *Soil Science Society of America Journal* 40:484–489.

Rowell, D. L. 1994. *Soil Science: Methods and Applications.* John Wiley and Sons. New York.

Scott, H. D., I. P. Handayani, and A. Mauromoustakos. 1991. Temporal variability of selected properties of two grand prairie soils as affected by cropping. In *Proceedings of the 1991 Southern Conservation and Tillage Conference,* pp. 79–83. Special Report 148. Arkansas Agricultural Experiment Station. Fayetteville.

Scott, H. D., and L. S. Wood. 1989. Impact of crop production on the physical status of a Typic Albaqualf. *Soil Science Society of America Journal* 53:1819–1825.

Soil Science Society of America. 1997. *Glossary of Soil Science Terms.* Soil Science Society of America. Madison.

Soil Survey Division Staff. 1993. *Soil Survey Manual.* USDA Agriculture Handbook No. 18. U.S. Government Printing Office. Washington, DC.

Thomas, G. W., and R. E. Phillips. 1979. Consequences of water movement in macropores. *Journal of Environmental Quality* 8:149–152.

Unger, P. W., and T. C. Kaspar. 1994. Soil compaction and root growth: A review. *Agronomy Journal* 86:759–766.

Wilson, G. V., and R. J. Luxmoore. 1988. Infiltration, macroporosity, and mesoporosity distributions on two forested watersheds. *Soil Science Society of America Journal* 52:329–335.

Additional References

Bouma, J. 1992. Effect of soil structure, tillage, and aggregation upon soil hydraulic properties. In R. J. Wagenet, P. Baveye, and B. A. Stewart (eds.), *Interacting Processes in Soil Science.* Lewis Publishers. Boca Raton, FL.

Kohnke, H. 1968. *Soil Physics.* McGraw-Hill Book Co. New York.

Luxmoore, R. J., P. M. Jardine, G. V. Wilson, J. R. Jones, and L. W. Zelazny. 1990. Physical and chemical controls of preferred path flow through a forested watershed. *Geoderma* 46:139–154.

Payne, D. 1988. Soil structure, tilth and mechanical behavior. In A. Wild (ed.), *Russell's Soil Conditions and Plant Growth,* pp. 378–411. John Wiley and Sons. New York.

Taylor, S. A., and G. L. Ashcroft. 1972. *Physical Edaphology: The Physics of Irrigated and Nonirrigated Soils.* W. H. Freeman and Co. San Francisco.

Verberne, E. L. J., J. Hassink, P. De Willigen, J. J. R. Groot, and J. A. Van Veen. 1990. Modelling organic matter dynamics in different soils. *Netherlands Journal of Agricultural Science* 38:221–238.

Problems

1. Draw diagrams of the following ped structures
    a. Moderate, fine, subangular blocky
    b. Weak, medium, platy; parting to moderate, coarse, angular blocky
    c. Weak, medium, angular blocky

2. Using capillary theory, calculate the equivalent radii of pores at –2, –5, and –14 cm of water pressure.

3. Calculate the effective porosity if the macropore flow rate is 2.67 × 10^{-5} m/s at a water pressure of –2 cm and at 20°C.

4. Calculate the number of effective macropores per square meter for the soil pores with equivalent radius given in the problem above.

5. One equation sometimes used to estimate the saturated flux density, q$_s$, of water moving through macropores is

$$q_s = kf_{ma}^2$$

where f_{ma} is the porosity of the macropores (m^3/m^3), and k is a constant (m/s). If the value of k is 0.5 m/s and f_{ma} is 0.02 m^3/m^3, calculate q_s.

6. A morphological description of a soil profile is given below. Describe the structure of the horizons of this soil. Note the description of the pores, boundaries between the horizons, and the presence of iron and manganese concretions in the B horizons.

Profile Description of a Fine Loamy, Siliceous, Thermic Plinthic Fragiudult

Horizon	Depth Interval (cm)	Description
A	0–7	Dark brown (10YR 3/3) fine sandy loam; weak, fine, subangular blocky structure; very friable; common, very fine roots; common, fine, discontinuous pores; few spots of red (2.5 YR 4/6) material; clear, smooth boundary.
BE	7–19	Strong brown (7.5YR 4/6) fine sandy loam; weak, medium, subangular blocky structure; friable; few very fine roots; common, fine, discontinuous pores; few 5 mm diameter vertical worm holes filled with brown to dark brown (10YR 4/3) material from the horizon above; gradual, smooth boundary.
B1t	19–33	Red (2.5YR 4/6) fine sandy loam; moderate, medium, subangular blocky structure; friable; very few faint clay films on faces of peds; few very fine roots; common, fine, continuous pores; common 5 mm diameter vertical worm holes filled with brown to dark brown (10YR 4/3) material from the horizon above; clear, wavy boundary.
B2t	33–47	Red (2.5YR 5/6) sandy clay loam; weak, medium, angular blocky structure; firm; few distinct clay films on faces of peds; few very fine roots; common, fine, continuous pores; few fine, moderately cemented, black (7.5YR 2/0) Fe-Mn concretions; few (1%) 1 cm rounded gravel; few vertical worm holes filled with brown to dark brown (10YR 4/3) material; gradual, smooth boundary.
B3t	47–80	Yellowish red (5YR 5/6) sandy clay loam; moderate, coarse, angular blocky structure; firm; common distinct clay films on faces of peds; few very fine roots; common, fine and common, medium continuous pores; few fine, moderately cemented, black (7.5YR 2/0) Fe-Me concretions; few (1%) 5–50 mm rounded gravel; clear, smooth boundary.
B4t	80–95	Yellowish red (5YR 5/6) sandy clay loam; weak, medium, angular blocky structure; firm; few distinct clay films on faces of peds; few very fine roots; common, fine, continuous pores; few medium, distinct, red (2.5YR 4/8) masses of Fe accumulation and few fine, distinct, yellowish brown (10YR 5/6) Fe concretions; clear, smooth boundary.

CHAPTER FIVE

Fate and Transport of Mass and Energy

Introduction

The fate and transport of mass and energy in the soil and atmosphere are two of the most important areas of study in soil physics. Transport of water, heat, oxygen, and solutes such as fertilizers, pesticides, heavy metals, and salts has practical significance for both

agriculture and the environment. In agriculture, for example, transport of water and heat energy strongly affects soil-plant-water relations and all aspects of crop production. In the environment, the soil is often used (and misused) as a medium to store and/or to degrade hazardous wastes. Transport of these wastes in and through soil is an important consideration in estimating the potential for pollution. For all these reasons and more, the description, measurement, prediction, and possibly management of mass and energy transport in soil are important concerns in soil physics.

With transport phenomena, we are primarily concerned with the prediction of the mass, temperature, and velocity variations (or distributions) within a medium such as soil. In order to obtain these profile distributions, we use two sets of equations: (1) conservation, or balance, equations, and (2) rate equations, or flux laws. This chapter presents the basics of constructing a balance equation for mass and energy in the soil, and the general form of the mass and energy transport rate equations. Both sets of equations can be used to make judgments about the potential of solutes to move within the soil profile and subsequently to pollute the environment. The approach taken is that the transport phenomena of mass and energy are analogous; that is, there are similarities in mathematical form and in physical mechanism. Therefore, an understanding of the balance and transport of one ion or molecule can readily lead to a greater understanding of the balance and transport of another.

Conservation of Mass and Energy

Conservation laws occupy a special place in science and engineering. No doubt you have heard statements such as "mass (or energy) is neither created nor destroyed," "the mass (or energy) of the universe is constant," "the mass (or energy) of any isolated system is constant." So important are these laws that to refute a conservation law, it would be sufficient to find just one example of a violation.

Why study mass and energy balances as a separate topic? You will find that mass balance calculations are almost invariably a prerequisite to all other calculations in the solution of both simple and complex problems in soil physics. Skills developed in analyzing mass and energy balances are easily transferred to other types of balances and other types of problems. In this section, we discuss the principle of conservation of matter and how it can be applied to scientific and engineering calculations. The aim is to help students acquire a generalized approach to problem solving. For some, this method of instruction may be entirely new. We begin our discussion by defining a system, then we derive the balance equation of a unit volume of soil.

Definition of a System

Inherent in the formulation of each of the balance equations is the concept of a *system* for which the balance calculations are made. In this textbook, a soil system is defined as any arbitrary portion or whole of the universe as specified by the scientist. The remainder of the universe is called the surroundings. Therefore, a system has distinct boundaries.

One of the best ways to visualize a soil system is to consider the system as a compartment

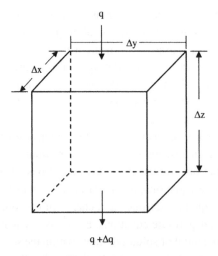

Figure 5.1. Schematic diagram of a unit volume element of soil.

into and out of which flow occurs, as well as where generation and/or extraction take place. A three-dimensional compartment idealizing a system can be represented as shown in Figure 5.1.

A balance equation can be applied to the system as a whole (macroscopic balance), to an increment (incremental balance), or to a differential element (differential balance). Sometimes, we are not concerned with the internal details within the system, only with the passage of material across the system boundaries.

An *open* system is one in which material is transferred across the system boundary, that is, enters the system, leaves the system, or both. A *closed* system is one in which there is no such transfer of mass and/or energy during the time interval of interest. Most agricultural systems are open.

There are two other ways to categorize open systems. A *steady-state* system occurs when there is no change in storage within the system. Therefore, for a steady-state system we can say, "What goes in must come out." The flow rate is constant and the accumulation is zero. In equation form, a steady-state system becomes

$$\text{rate in} - \text{rate out} = 0 \qquad\qquad [1]$$

Therefore, conditions are said to be steady or stationary if the concentration of all components or their spatial distribution in the system does not change with time.

A *transient-state* system occurs when there is storage and/or consumption within the system. Therefore, changes occur within the system during a specified time interval. Most transport systems in the field environment are transient-state systems. Therefore, a considerable amount of scientific interest has been placed on the development of the governing equations and analytical or numerical solutions of those equations that simulate transient-state transport of mass and energy.

Balance Equation

To account for the flow of mass or energy in and out of a system, the generalized law of the conservation of mass is expressed as a balance equation. A balance equation is nothing more than an accounting of flows and changes in inventory of mass or energy for a system. In words, the principle of the balance equation that is applicable to processes both with and without chemical reaction is

$$
\begin{array}{ccccccccc}
\text{accumulation} & & \text{inflow} & & \text{outflow} & & \text{generation} & & \text{consumption} \\
\text{within} & & \text{to} & & \text{from} & & \text{within} & & \text{within} \\
\text{the} & = & \text{the} & - & \text{the} & + & \text{the} & - & \text{the} \qquad [2] \\
\text{system} & & \text{system} & & \text{system} & & \text{system} & & \text{system}
\end{array}
$$

The inflow refers to all flow into a system across the system boundaries. The outflow refers to all flow leaving the system across the boundaries. Generation refers to all production within the system, consumption refers to all degradation within the system, and accumulation refers to the change in the total amount of mass or energy in the system. In equation [2], the generation and consumption terms refer to the gain or loss by chemical reaction, plant uptake, and/or microbial action. The accumulation term may be positive or negative.

Equation [2] can be expressed in either integral or differential form. In integral form, the SI units of each of the processes are amounts of mass (kg) or energy (J). In differential, or rate, form, the time interval can be any desired length, such as year, month, hour, or second, and the units of each of the processes are amounts per unit time (e.g., kg/s, J/s, or W).

Equation [2] reduces to equation [3] for cases in which there is no generation (or consumption) of mass or energy within the system:

accumulation = inflow – outflow   [3]

Equation [2] reduces further to equation [4] when there is no accumulation within the system:

inflow = outflow   [4]

If there is no flow in and out of the system, equation [2] reduces to the basic concept of the conservation of one species of matter within an enclosed isolated system:

accumulation = generation – consumption   [5]

The amount and distribution in soil of mass such as water and of heat energy dominate the control of the microclimate near the ground. Most of the phase conversions occur at or near the soil surface. Application of the principle of mass and energy conservation to the inflows and outflows in the soil profile enables a balance equation to be written for the soil and/or components in the soil.

Example: Part of a Mass Balance Calculation

Assume that a soil of unit area was sampled for pesticide residues and that there was no inflow or outflow. Calculate the accumulation of the pesticide in the soil based upon the following properties.

Depth Interval (cm)	Bulk Density (g/cm^3)	Concentration (μg/g)	Amount (μg)
0–5	1.24	1.7	10.54
5–10	1.32	1.9	12.54
10–15	1.35	2.3	15.53
15–20	1.40	1.8	12.60
20–25	1.45	1.0	7.25
25–30	1.47	0.1	0.74

The total amount of pesticide in the 0–30 cm depth interval is 59.20 μg/cm^2. This amount can be determined from

$$\text{total} = 5 \sum_{i=1}^{i=6} \rho_b C \quad \text{or} \quad \int_0^{30} \rho_b(z)C(z)dz = 59.2 \, \mu g / cm^2$$

where i is the depth interval and the volume of soil in each depth interval is 5 cm $\times$ 1 cm^2 = 5 cm^3.

Similar calculations of accumulation of mass within a soil profile at other sampling times can be used to quantify losses and gains in mass by inserting quantities into equation [2].

One important point to always keep in mind is that the basic material balance is a balance on mass, not on volume or moles. Therefore, there may not be such equality on the part of the total moles in and out of a system if a chemical reaction takes place within the system. What is true, however, is that the number of atoms of an element put into a system must equal the number of atoms of the same element leaving the system and/or stored within the system and/or consumed within the system.

Net accumulation is the change in mass (or energy) over a known space and/or time. Therefore, the total mass (or energy) is calculated at each location and/or time; the difference between these quantities is the net accumulation. We now consider two of the most important applications of the balance equation.

Water Balance

The water balance equation expresses the overall mass conservation for water in any given period. The water balance equation can be written by considering the possible fate of the precipitation received on a certain area of land in a certain time. The equation is concerned with a volume of soil contained by the imaginary surface that would be generated by a vertical line moving around the area of ground for which the precipitation is considered. The depth of the soil volume can be chosen arbitrarily, as will be seen from the equation. Although there will be a continuous distribution of water between the various terms, and a phase lag as water moves down into the profile, the following conservation, or balance, equation must be satisfied for any given period and volume of soil:

$$P + I \pm R - ET \pm D = \Delta W \qquad [6]$$

where P is the precipitation received in the area (of any size) for which the balance is being considered, I is the irrigation, R is the net surface runoff, ET is the evapotranspi-

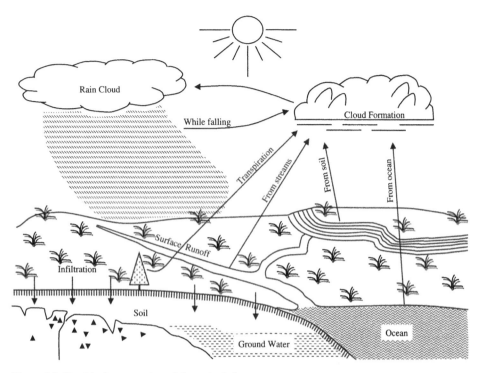

Figure 5.2. Graphical presentation of the water balance.

ration, D is the net drainage, and ΔW is the change in storage of water. In equation [6] the inputs are P and I and the outputs are R, ET, and D. Since the storage capacity of soils is limited, over a long period of time the outputs must equal inputs; that is, the system is at steady state. Over a short time period, however, changes in storage will be significant, so inputs and outputs will usually not be equal and the system will be at a transient state (Figure 5.2).

Three of the terms in equation [6] have been designated as general terms that express the sum of several physical processes. Those processes that add water to the volume of soil are positive, and those processes that subtract water from the soil are negative. For example, P can be used to describe additions of water from rainfall, hail, sleet, snow, etc. All of these terms are positive. The term R can be used to describe runoff but also runon. Runoff is a negative term because water is lost from the system, whereas runon is positive because water is gained. Lastly, D can be used to describe downward drainage but also the upward movement of water into the system. Downward drainage is negative, whereas upward flow of water is positive.

It is also important that the system boundaries (i.e., the volume of soil and the time interval) be defined. For example, if a water balance is to be calculated for soybeans, then the depth of the root zone as well as the area of the field should be stated. To calculate rates of processes, the duration between the measurements must be known.

Example: Water Balance Calculations

A daily water balance was determined for a soybean field during the growing season. Assuming that the profile initially had a 40-cm equivalent depth of stored water, compute on a daily basis the following: (1) amount of water in the profile, (2) changes in water content of the profile, and (3) evapotranspiration. Summarize the amounts of water gained or lost from the soil profile during a 5-day interval.

Suppose that we obtained the following data for the daily water balance (quantities are in centimeters).

Day	Rain	Runoff	Drain	Irr	Evap	Trans	ET	ΔW	Profile
					(cm)				
0									40.00
1	0	0	0	0	0.18	0.41	0.59	−0.59	39.41
2	0.66	0.13	0.05	0	0.04	0.43	0.47	+0.01	39.42
3	0.03	0	0	0	0.07	0.45	0.52	−0.49	38.93
4	0	0	0	0	0.04	0.41	0.45	−0.45	38.48
5	0	0	0.05	1.90	0.04	0.56	0.60	+1.25	39.73
SUM	0.69	0.13	0.10	1.90	0.37	2.26	2.63	−0.27	

Substituting the sums of the water balance categories into equation [6] results in

$$0.69 - (0.13 + 0.10) + 1.90 - 2.63 = -0.27 \text{ cm}$$
$$= 39.73 - 40.00 \text{ cm}$$

The data in this table illustrate the method used by many computer programs to calculate a soil water balance. Basically, equation [6] is used in a manner similar to the way one would balance a checkbook. In this particular example, a daily time step was used, but other time steps (e.g., monthly or annually) also could have been used. The sums at the bottom of the table give the overall water balance for this field over the 5-day period.

Energy Balance

Many of the important soil physical properties are directly influenced by the heat energy received at the surface. The energy balance equation is also a detailed statement of the law of conservation of energy (Figure 5.3). Mathematically, it is stated as

$$R_n = H + LE + M - G \qquad [7]$$

where R_n is the net radiation flux density (MJ/m^2·d), H is the energy flux density utilized in heating the air (sometimes called the sensible heat), LE is the energy utilized in evapotranspiration, L is the latent heat of vaporization (2.45 MJ/kg) and is the conversion factor needed to convert E to energy units, and E is the amount of water evaporated from soil and transpired by crops (usually called the evapotranspiration), M is the energy flux density utilized in miscellaneous energy terms such as photosynthesis and respiration, and G is the energy utilized in heating the soil, water, and vegetation (MJ/m^2·d).

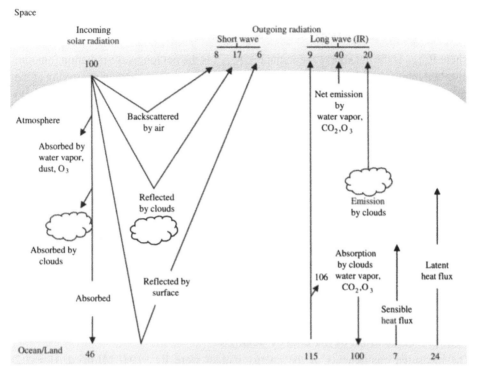

Figure 5.3. The distribution of 100 units of incoming solar radiation and shortwave, outgoing infrared radiation, globally and annually averaged. (Adapted from Moran and Morgan 1997.)

The sign convention normally used in micrometeorology is that all radiation arriving at the soil surface is given a positive sign and all radiation reflected from or emitted by the surface is given a negative sign. Evaporation from the soil surface is considered positive, and condensation is considered negative. In the flow of soil heat below the soil surface (either with G or H), the upward z direction is taken as positive, and downward z is taken as negative. The units of energy flux density are expressed as W/m^2 or $J/m^2 \cdot s$. Equation [7] is applicable on both micro- and macroscales and explains how energy is used to warm the soil and to evaporate water.

Solar radiation provides almost all of the energy received at the surface of the earth. Some of the radiant energy is reflected back to space. Wavelengths less than 4 μm are designated as shortwave radiation and result from hot objects such as the sun. The radiation emitted by "cold" objects such as soil and plants has wavelengths longer than 4 μm and is called longwave radiation.

The net radiation absorbed by any surface is the total (global plus longwave) arriving minus the reflected shortwave and the emitted longwave. Therefore, net radiation is the difference between total upward and total downward radiation fluxes and is a measure of the energy available at the soil surface. It is the fundamental quantity of net energy available at the earth's surface to drive the processes of evaporation, air and soil heating, photosynthesis, and respiration.

Net radiation can also be expressed mathematically as

$$R_n = R_{swbal} + R_{lwbal} = R_{sg}(1 - p) + R_{lwbal} \qquad [8]$$

where R_{swbal} is the net shortwave radiation, R_{lwbal} is the net longwave radiation (amount absorbed minus the amount emitted), R_{sg} is the global shortwave radiation arriving at the surface (direct beam plus scattered) measured with a pyranometer, and p is the albedo (fraction of global radiation reflected by the surface) measured with two pyranometers (one facing up and one facing down). The quantity $R_{sg}(1 - p)$ is the net shortwave radiation, R_{swbal}.

The albedo for most soil and plant conditions in the field varies from about 0.10 to 0.30 and for full canopy conditions averages about 0.23. Albedo tends to be lower for wet soil than for dry soil. Equation [8] relates the amount of energy absorbed by the soil surface but not how this energy is used by soil and plants.

When water does not limit transpiration, most of the energy in the form of R_n is utilized as energy in evapotranspiration, ET. Under humid conditions, the ratio LE/R_n is approximately 0.8. When bare soil is wet, most of the energy goes into evaporation. However, as the soil dries, water becomes less available for ET, and more of the energy goes into heating the soil.

Example: Calculation of the Energy Balance

The measured components of the energy budget were R_n = 9.0 MJ/m²·d, H = 1.2 MJ/m²·d, and G = –0.4 MJ/m²·d. Find the amount of energy used in evapotranspiration, LE, and calculate ET, the evapotranspiration rate.

$LE = R_n - H + G = 9.0 - 1.2 - 0.4 = 7.4$ MJ/m²·d
$LE/R_n = 7.4/9.0 = 0.82$
$ET = LE/L\rho_w = 7.4$ MJ/m²·d/(2.45 MJ/kg $\times$ 1000 kg water/m³)
$\qquad = 0.003$ m/d
$\qquad = 0.3$ cm/d

Derivation of the Continuity Equation for One-Dimensional Flow of Water in Soil without Extraction

Transport of mass and energy always obeys a balance equation, and mathematical derivations of the fundamental transport laws must reflect the balance, or conservation, laws. One of the most important uses of a balance equation is for the equation of continuity. The derivation of the equation of continuity for one-dimensional flow of water without extraction follows.

Assume that the volume element of soil can be represented as shown in Figure 5.1 and that there are no consumptive and generative processes operating. Then, in a given unit of time, the volume of water entering the volume element equals $q\Delta x\Delta y$, which is the flow rate (m³/s) per unit area. The parameter q is the flow rate of water per unit area and is known as the flux density, with units of m³/m²·s. The terms Δx, Δy, and Δz each represent the length of a side of a soil volume in the x, y, and z directions, respectively.

The volume of water expressed in cubic meters leaving the volume element equals $(q + \Delta q)\Delta x \Delta y$. Therefore, according to the conservation equation [2], the volume of water accumulated in the volume element during Δt equals

$$q\Delta x \Delta y \Delta t - (q + \Delta q)\Delta x \Delta y \Delta t = -\Delta q \Delta x \Delta y \Delta t$$

that is, inflow − outflow = net accumulation. This volume of water has accumulated in a volume $\Delta x \Delta y \Delta z$. The volumetric water content of the volume element is θ_v. Therefore,

$$\Delta \theta_v = -(\Delta q \Delta x \Delta y \Delta t)/\Delta x \Delta y \Delta z \qquad \text{or} \qquad \Delta \theta_v/\Delta t = -\Delta q/\Delta z$$

Taking the limit of the left-hand term as Δt approaches zero and the limit of the right-hand term as Δz approaches zero and recognizing that several independent variables are uncontrolled gives

$$\partial \theta_v/\partial t = -\partial q/\partial z \qquad\qquad [9]$$

Assuming that the time increment and distance element are infinitesimally small, equation [9] yields the conservation equation for one-dimensional flow of water in soil. A similar procedure can be used for the derivation of the energy balance equation. This will be done in a unit volume of soil in Chapter 6.

General Transport Mechanisms

Transport of mass and energy can occur by several physical mechanisms. In the case of mass, transport can occur by two mechanisms: diffusion and convection. In the case of heat (energy), three basic mechanisms are usually identified: conduction, convection, and radiation. Of these energy transport mechanisms, conduction and radiation can be considered pure in the sense that they can occur as the only propagating mechanisms. Convection, on the other hand, is a mixture of conduction and transport of a fluid, with radiation present in significant or insignificant amounts depending on the fluid present and the temperature level. We will begin with a discussion of transport of mass under steady-state conditions.

Concentration, Flux Density, and Influence of Soil Phase

Knowledge of the concentration of a solute is required to compute the rate of transport by diffusion and convection. As a result, considerable effort has been made to determine the concentration of solutes in soils, water, and air. The concentration of a solute in soil is defined as a measure of the amount of that solute (Q) in a specific volume (m^3) or mass (kg) of soil. Concentration of a solute can be defined in terms of molarity or normality or, most important in this textbook, mass per unit volume.

The frame of reference also affects calculations of the storage and movement of a solute through the concentration and transport mechanisms in the three phases of soil. A consistent system of units should be used throughout the calculations. For solutes that are soluble in water and have a nonnegligible vapor pressure, all three phases of soil can be represented, that is, the solid (or sorbed), aqueous, and gaseous phases.

The total solute concentration in soil can be defined with units of mass of solute per soil volume (kg/m³). Since the solute partitions into the three phases of soil, the total solute concentration C_t is

$$C_t = \rho_b C_s + \theta_v C_l + f_a C_g \qquad [10]$$

where C_s is the solid-phase solute concentration (kg/kg of solid), C_l is the solute concentration in water (kg/m³ of soil solution), and C_g is the solute concentration in the gaseous phase (kg/m³ of soil air). The parameters ρ_b, θ_v, and f_a are the bulk density, water content by volume, and aeration porosity, respectively. The units of C_t are kg/m³ of soil.

The mass flux, j, is defined as the amount of substance passing through a plane perpendicular to the direction z during a specified time interval, divided by the magnitude of the time interval. Mathematically, the mass flux, j, is

$$j = Q/t \qquad [11]$$

where Q is the amount of substance, which can be expressed as a mass, a volume, or a weight. The SI units of mass flux are kilograms or moles per second.

Quite often we also see the flux density reported. The flux density, J, is the flux divided by the cross-sectional area and mathematically is

$$J = Q/At \qquad [12]$$

where A is the cross-sectional area of the plane perpendicular to the direction of flow. The units of mass flux density are kilograms or moles per meter squared per second. The flux density is the preferred term in expressing the amount of mass transferred when the shapes of the boundaries are changing or are unknown.

The transport of the same solute in soil is the sum of the flux density in each phase. Mathematically, this is

$$J_t = J_s + J_w + J_a \qquad [13]$$

where J_t is the total flux density (in kilograms per square meter per second), and J_s, J_w, and J_a are the flux densities in the solid, aqueous, and gaseous phases, respectively. For almost all strongly sorbed solutes, transport in the solid or adsorbed phase is negligible, and for practical purposes, J_s can be assumed to be zero.

Mass Transport

First, we consider the movement of mass from one location in the soil to another location. Examples include the diffusion of oxygen and nutrients to plant roots and the convective movement of nitrates, chlorides, and pesticides within and below the soil profile.

Mass transport can occur by two physical mechanisms: molecular diffusion and convection. Molecular diffusion represents the mass transport that occurs as a result of concentration gradients; convective transport occurs by bulk-fluid motion. In soils, the fluids are water and air.

Table 5.1. Molecular Diffusivities as a
Function of Phase (in m^2/s)

Gas	1×10^{-5}
Liquid	1×10^{-9}
Solid	1×10^{-13}

Source: Fahien 1983, p.61.

Diffusion

Diffusive transport results from the movement of solute molecules from regions of higher concentration to regions of lower concentration. Mathematically, this transport mechanism can be represented by Fick's first law of diffusion as

$$j = -DA \, \partial C / \partial z \qquad [14]$$

where j is the solute flux and is the quantity of solute per unit time (kg/s), A is the cross-sectional area (m^2), C is the solute concentration (kg/m^3), z is the spatial coordinate (m), and D is the transport coefficient, whose value depends on the nature of the solute and that of the diffusing medium. The parameter D is also known as the diffusivity or molecular diffusion coefficient and has dimensions of length squared per unit time (m^2/s). The quantity $\partial C / \partial z$ is known as the concentration gradient (kg/m^4). As Fick's law indicates, diffusion results from concentration inequalities and occurs in the direction of decreasing concentration. Equation [14] is known as the simplified form and holds for ideal dilute binary solutions of constant density. It is a steady-state mass transport equation.

Typical values of the molecular diffusivity are given in Table 5.1. These data show that the phase of the soil has a great influence on the molecular diffusivity. Under conditions where the concentration gradient is 1, transport by diffusion is 10,000 times faster in the gaseous phase than in the aqueous phase, which in turn is 10,000 times faster than in the solid (adsorbed) phase.

The physical meaning of Fick's law can be summarized as follows:

1. Due to Brownian motion, there is a natural tendency for solutes to diffuse from a region of higher concentration to a region of lower concentration.

2. This tendency depends upon the nature of the soil system and the solute.

3. The larger the magnitude of the concentration gradient, the greater the driving force per unit distance and the greater the flow.

4. The flux and the concentration gradient have opposite signs.

The concentration gradient is the slope of the tangent line to the curve of C(z) at the point of interest. In calculus notation, this slope or tangent is called the derivative of the concentration function. When the concentration is a function of only one variable (e.g., distance), we can express the gradient, or slope, by means of the incremental operator $\Delta C / \Delta z$ or the differential operator dC/dz. The slope is defined as

$$\text{slope} = (C_2 - C_1)/(z_2 - z_1) = \Delta C / \Delta z \qquad [15]$$

Solute Concentration

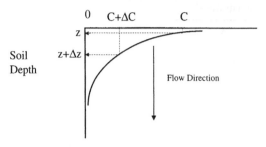

Figure 5.4. Concentration distribution curve of nitrate in soil.

Since the slope in the direction of flow is negative, the negative sign in equation [14] makes the diffusive flow positive in the positive z direction. Note that the sign of the slope is such that indices of C and z are subtracted in the same sequence.

Example

Suppose NH_4NO_3 is broadcast on a bare soil surface. After a given time interval, the concentration distribution of nitrate within the soil profile is measured and found to have the shape shown in Figure 5.4. What is the direction of flow?

Example

A soluble, nonvolatile pesticide, Kill All, was broadcast-applied to a given soil surface. After 2 weeks, the concentration of the pesticide at the 5 cm soil depth was 5 μg/L, while the concentration at the 10 cm depth was 15 μg/L. If the average soil water content was 0.25 cm³/cm³, the soil water flux was zero, and the diffusion coefficient for this pesticide in this soil was 4×10^{-4} cm²/h, calculate or determine the following:

   a. the concentration of Kill All in the soil at both depths
   b. the average concentration gradient of Kill All in this horizon
   c. the rate of diffusive movement of Kill All in μg/cm²·h
   d. the diffusion rate in kg/ha·day
   e. the direction of transport of this herbicide in the profile

Solution:

$$C_1 = \theta_v C_1 = (5 \text{ μg/L})(0.25 \text{ cm}^3/\text{cm}^3)(\text{L}/1000 \text{ cm}^3)$$
$$= 0.00125 \text{ μg/cm}^3 \text{ soil}$$
$$C_2 = \theta_v C_1 = (15 \text{ μg/L})(0.25 \text{ cm}^3/\text{cm}^3)(\text{L}/1000 \text{ cm}^3)$$
$$= 0.00375 \text{ μg/cm}^3 \text{ soil}$$
$$dC/dz \approx [(0.00375 \text{ μg/cm}^3) - (0.00125 \text{ μg/cm}^3)]/[(-10 \text{ cm} - (-5\text{cm})]$$
$$= -0.0005 \text{μg/cm}^4$$

Here, we assume that z is positive in the upward direction.

$$J_d = (4 \times 10^{-4} \text{ cm}^2/\text{h})(-0.0005 \text{ μg/cm}^4)$$
$$= 2 \times 10^{-7} \text{ μg/cm}^2 \cdot \text{h}$$

Since J is positive, the direction of flow is upward.

Convection

Convective transport is sometimes called advection and depends on the velocity of the moving fluid and the concentration of solute in the fluid. Mathematically, the flux density due to convection can be represented by

$$J_c = q_z C_l \qquad [16]$$

where q_z is the flux density of water (m^3 of water/m$^2 \cdot$s of soil) moving through the soil in the z direction, and C_l is the concentration of solute in the water (kg/m^3). The ratio of q_z and θ_v is known as the pore water velocity, v_z (m/s). One should be careful to have a consistent frame of reference with equation [16]; that is, the system or phase should be well defined. Examples of these systems include the solution phase, the aqueous phase, the gaseous phase, and the solid phase or the entire soil.

Example

After 2 days of a heavy rain, the average concentration of Kill All in the Ap horizon is 10 µg/L. The rate of downward water movement through the horizon has a pore water velocity of 0.15 cm^3/cm$^2 \cdot$h, and the water content is 0.40 cm^3/cm^3. What is the transport rate by convective flow of Kill All in kilograms per hectare per day? What is the direction of convective flow?

$$J_c = q_z C_l = (-0.15 \text{ cm}^3/\text{cm} \cdot \text{h})(10 \text{ µg/L})(\text{L}/1000 \text{ cm}^3)(0.40 \text{ cm}^3/\text{cm}^3)$$
$$(10^8 \text{ cm}^2/\text{ha})(24 \text{ h/d})(1 \text{ kg}/10^9 \text{ µg})$$
$$= -0.0036 \text{ kg/ha} \cdot \text{day}$$

Since the sign of q_z is negative, J_c is negative, and the direction of flow is downward.

In soil, equation [16] does not accurately represent the total convection since the flow pathways are tortuous (i.e., not straight), and the velocity of water is faster in the center of the pores than at the edges and is faster through the larger pores than through the smaller pores. This creates a condition in which the rate of movement of the solute around particles and water films may differ from the average transport rate. Therefore, an additional contribution of convective transport is created. This is called hydrodynamic dispersion, J_h. Mathematically, equation [16] can be written for soils as

$$J_c = q_z C_l + J_h \simeq q_z C_l - D_h \partial_z C_l \qquad [17]$$

where J_h accounts for the contribution to convective flow from hydrodynamic dispersion, which has a form that is mathematically identical to Fick's first law.

Heat (Energy) Transport

Heat transfer (or heat) is energy in transit due to a difference in temperature. Therefore, whenever a difference in temperature exists in the soil or between phases in the soil, there must be a transfer of heat. In this section, we will consider three heat transfer processes: conduction, convection, and radiation.

Table 5.2. Thermal Conductivity as a
            Function of Phase (in J/s·m·K)

Gases	0.001–0.1
Lquids	0.01–1.0
Solids	1.0–100

Source: Fahien 1983, p. 14.

Conduction

Conduction is the primary heat-energy transfer mechanism in soil. It also acts to transfer energy up and down the stems of plants.

Transport of heat or energy by conduction in a soil system is accomplished by the transmission of translational, rotational, and vibrational energy from molecule to molecule. Conduction may be viewed as the transfer of energy from the more energetic to the less energetic particles and molecules in a porous material due to the interactions between the particles and between the molecules. Considering only macroscopic effects, conduction of energy in one dimension can be described by the steady-state form of Fourier's law:

$$h = -kA(\partial T/\partial z) \tag{18}$$

where h is the heat flux in J/s or W, and $\partial T/\partial z$ is the temperature gradient in degrees per unit length. The ratio $\partial T/\partial z$ is the slope of the temperature-distance relationship. The temperature gradient, or slope of the temperature profile, is the tangent to the curve T(z) at the point of interest. Equation [18] indicates that energy transfer occurs in the direction of decreasing temperature.

The thermal conductivity, k, is the proportionality constant between the flux density, h/A, and the temperature gradient, $\partial T/\partial z$. It is defined by equation [18]. In the SI system, k is defined with units of $J/m·s·K = W/m·K$. For most agronomic uses in soils, the value of k can be considered to be independent of temperature.

Typical values of the thermal conductivity are presented in Table 5.2. These data show that the highest thermal conductivity occurs in the solid phase and the lowest in the gaseous phase. Under conditions where the temperature gradient is 1, conduction of heat is 100 times faster in the solid phase than in the solution phase, in which it is 10–100 times faster than in the gaseous phase.

The heat flux, h, has units of J/s or W. The heat flux density, H, is obtained by dividing the heat flux by the cross-sectional area. The units of H are W/m^2. It is the amount of heat conducted across a unit area per unit time and is the preferred term to use in describing heat transport in soil systems.

Example

Consider a 1 cm thick soil crust having a temperature at the surface of 35°C and at the bottom of 25°C. If the thermal conductivity through the crust is 1.1 W/m·K, calculate the flux density of heat conducted through the soil crust. In which direction is heat conducted? In incremental form, Fourier's equation is

$$H = -k(\Delta T/\Delta z) = -1.1[(308 - 298)/0.01] = -1100 \text{ W/m}^2$$

Since the highest temperature occurs at the soil surface, heat is conducted down through the crust.

Convection

When a fluid such as water is at rest or in motion with a surface at a temperature that is different from the fluid, energy flows in the direction of the lower temperature. The energy flux leaving or entering the surface can be calculated from Newton's law of cooling as

$$h = C_v Av(T_s - T_o) \qquad [19]$$

where T_s is the temperature of the soil in contact with the fluid (K), T_o is the temperature of the fluid far away from the surface so that no influence of the surface is evident (K), v is the velocity of the fluid (m/s), C_v is the volumetric heat capacity (J/m³·K), and h is the heat flux (J/s or W). As equation [19] indicates, with known values of v and of the temperature difference, the convective heat flux can be computed.

Convection of heat energy is extremely important in the atmosphere, where air is warmed or cooled as air masses come in contact. A warm air mass can give up heat to substances that are cooler. Convection of heat energy is of lesser importance in soils but may occur during infiltration and redistribution of water from rainfall and irrigation, which has a temperature that differs from the soil matrix.

Radiation

All forms of matter emit thermal radiation. The emission may be attributed to changes in the electron configurations of the constituent atoms or molecules. The energy of the radiation field is transported by electromagnetic waves (or photons). Transport of energy by radiation requires no carrier and thus can be transmitted through a vacuum. The energy travels as discrete packets called quanta or photons, whose energy content depends on their wavelengths or frequencies. If we are interested in the sum total over all frequencies, we can use the Stefan-Boltzmann law to obtain

$$h = \sigma A T^4 \qquad [20]$$

where σ is the Stefan-Boltzmann constant, a natural constant that depends on the units used. In the SI system, σ is equal to 5.675×10^{-8} W/m²·K⁴. Equation [20] gives the maximum energy flux that can leave a surface of area A at absolute temperature T.

A surface that transmits the maximum amount of radiant energy is called a blackbody or a perfect radiator, e_b. Soil surfaces, however, emit at a rate e_s, which is less than the rate e_b at a given absolute temperature. We compute the ratio $e = e_s/e_b$, which is called the emittance of the surface or the emissivity. The parameter e serves as a correction factor or indicator of the efficiency of natural sources and varies as a function of the wavelength of the radiation. Thus, equation [20] can be written as

$$h = e\sigma AT^4 \tag{21}$$

Equation [21] can be used to quantify the rate at which energy is emitted by a surface. Determination of the net rate at which radiation is exchanged is much more complicated.

Analogies between One-Dimensional Transport Phenomena

So far, we have used the same basic mathematical models to develop the steady-state flux laws for mass and energy in one dimension. These laws were written as

Fick's law: $j = -DA(\partial C/\partial z)$ or $J = -D(\partial C/\partial z)$
Fourier's law: $h = -kA(\partial T/\partial z)$ or $H = -k(\partial T/\partial z)$

In each case, the transport rate equations take the same fundamental form of

Flux density = (flow rate)/area = Q/At
$$= -\text{transport coefficient} \times \text{potential gradient} \tag{22}$$

where Q is the quantity, A is the cross-sectional area, t is the time, D and k are the molecular transport properties, and C and T are the potentials.

Although these equations are similar, they are not completely analogous because the transport coefficients have different units. For example, the molecular diffusivity has units of length squared per unit time, while the thermal conductivity has units of power per unit length per unit degree. In order to make these terms compatible, we define the thermal diffusivity as

$$\alpha = k/\rho_b C_w = k/C_v \tag{23}$$

where α is known as the thermal diffusivity in units of length squared per unit time (m^2/s), k is the thermal conductivity (J/m·s·K), ρ_b is the bulk density (kg/m^3), C_w is the gravimetric heat capacity (J/kg·K), and C_v is the volumetric heat capacity (J/m^3·K). Values of ρ_b and C_w normally are assumed to be constant within a horizon.

The various quantities in mass and energy transport are summarized in Table 5.3.

Concluding Remarks

Although much of the material in this chapter will be discussed in greater detail in the chapters to follow, the student should now have some understanding of mass and energy balances and the rate equations used to describe transport. The student should also be aware of the empirical, physical laws of mass and energy transfer and be able to specify the important processes operating in a given situation in the field. Also, there should be some appreciation of the significance of the conservation laws and the transport rate equations and their similarities. The rate equations and conservation laws are tools with which calculations are made of the amounts of mass and energy moved from one location to another and/or from one time to another time.

Table 5.3. Analogous Terms in the One-Dimensional Flux Laws

Property	Mass	SI Units	Energy	SI Units
Flux	j	kg/s	h	J/s
Flux density	J	kg/m²·s	H	J/m²·s
Transport property	D	m²/s	k	J/m·s·K
Potential gradient	$\partial C/\partial z$	kg/m⁴	$\partial T/\partial z$	K/m
Diffusivity	D	m²/s	$\alpha = k/\rho_b C_w$	m²/s
Concentration	C	kg/m³	$\rho_b C_w T$	J/m³
Gradient of concentration	$\partial C/\partial z$	kg/m⁴	$\partial(\rho_b C_w T)/\partial z$	J/m⁴

Notes: The concentration term for energy is derived from $\rho_b C_w T = (kg/m^3)(J/kg \cdot K)(K) = J/m^3 =$ energy/volume. This is analogous with concentration of mass/volume.

The flux laws can be written as flux density = –diffusivity × concentration gradient and also flux density = –driving force/resistance, where driving force = concentration difference = $C_2 - C_1$, resistance = distance/(transport coefficient × area), e.g., $\Delta z/DA$.

Cited References

Fahien, R. W. 1983. *Fundamentals of Transport Phenomena.* McGraw-Hill Book Co. New York.

Moran, J. M., and M. D. Morgan. 1997. *Meteorology: The Atmosphere and the Science of Weather.* Prentice-Hall. Upper Saddle River, NJ.

Additional References

Hanks, R. J. 1992. *Applied Soil Physics.* 2d ed. Springer-Verlag. New York.

Incropera, F. P., and D. P. DeWitt. 1985. *Fundamentals of Heat and Mass Transfer.* 2d ed. John Wiley and Sons. New York.

Koorevaar, P., G. Menelik, and C. Dirksen. 1983. *Elements of Soil Physics.* Elsevier. New York.

Rosenberg, N. J., B. L. Blad, and S. B. Verma. 1983. *Microclimate: The Biological Environment.* 2d ed. John Wiley and Sons. New York.

Wolf, H. 1983. *Heat Transfer.* Harper and Row. New York.

Problems

1. The global shortwave radiation was measured as 26.15 MJ/m² and the net longwave radiation was –5.02 MJ/m². Find the net radiation for a dry silt loam and for the same soil with a wet surface.

2. Derive the equation of conservation of mass for the transport of 2,4-D through a unit volume of soil. Assume that there is no generation or consumption in the soil system.

3. Write the conservation of mass equation that shows the flow with consumption of 2,4-D within the unit volume of soil.

4. Determine the monthly water balance for a rice field during a growing season. Assume that the rice was grown on a DeWitt silt loam and was planted on April 1 and harvested on September 15. Draw a cross-sectional view of the system showing water flow pathways.

5. Several rice growers in Arkansas have developed a water reuse system where the water lost by surface drainage from the rice field is reused by another rice or soybean field at a lower elevation or is pumped back to the inlet at the top of the rice field. Show mathematically how this reuse system would influence the water balance of the rice field.

6. Determine the seasonal balance for nitrogen in the rice field given in Problem 4. Assume that 112 kg/ha of nitrogen was applied, with one-half applied at planting and one-half at the midseason application during panicle initiation. Describe how you would determine these data experimentally.

7. A heat flux of 0.5 kW is conducted downward through a soil horizon of cross-sectional area 10 m^2 and thickness 15 cm. If the upper surface boundary temperature is 40°C and the thermal conductivity of the soil is 1.2 W/m·K, what is the lower surface boundary temperature?

8. The heat flux density through a soil horizon 15 cm thick, whose upper and lower boundary temperatures are 25°C and 18°C, respectively, has been determined to be 40 W/m^2. What is the thermal conductivity of the soil?

9. If the soil temperature is 20°C at the soil surface and 28°C at the 15 cm depth, in which direction is heat moving? Is the soil being heated or cooled?

10. The convective heat transfer coefficient between a soil surface at 40°C and ambient air at 20°C is 5 W/m^2·K. Calculate the heat flux density leaving the soil by convection.

11. Bromide as KBr was broadcast at 2.5 kg to a DeWitt silt loam having a plot size of 100 m^2. The soil profile was subsequently sampled twice in 10 cm depth intervals to 1 m and analyzed for bromide concentration. The concentration distributions in the profile were as follows:

Middle Soil Depth (cm)	Median Bromide Concentrations (mg/kg soil)		Bulk Density (kg/m^3)
	(1)	(2)	
5	39.8	18.1	1226
15	30.5	12.8	1427
25	14.7	11.0	1454
35	6.7	8.7	1462
45	2.7	6.0	1412
55	2.7	4.2	1425
65	2.0	3.6	1420
75	1.5	2.5	1443
85	1.3	1.9	1560
95	1.2	1.5	1630

Plot the concentration distribution curves. For each sampling time, calculate the mass of bromide recovered in the soil profile in grams and the fraction of the bromide remaining at each sampling. Also, calculate the net accumulation. Explain the results.

12. A soluble pesticide, Kill All, was broadcast-applied to a given soil surface. After 2 weeks, the concentration of the pesticide at the 5 cm soil depth was 5 µg/L while the concentration at the 10 cm depth was 15 µg/L. If the average soil water content was 0.25 cm^3/cm^3, the soil water flux was zero, and the diffusion coefficient for this pesticide in this soil was 4×10^{-4} cm^2/h, calculate or determine the following:

   a. The concentration of Kill All in the soil at each depth.

   b. The rate of diffusive movement of Kill All in µg/cm^2·h.

   c. The diffusion rate in kg/ha·day.

   d. The direction of transport of this herbicide in the profile.

Soil Temperature

Introduction

The temperature is one of the most important factors affecting the rates of physical, chemical, and biological processes in soil. For example, seeds will not germinate until the soil temperature has reached a certain critical value. Each plant needs a certain soil temperature before a normal growth rate can be expected. The biological process known as respiration is temperature dependent and has been described mathematically by

$$R_T = R_o Q_{10}^{(T-T_o)/10} \tag{1}$$

where R_T is the respiration rate at temperature T, R_0 is the rate at the reference temperature T_0, and Q_{10} is a factor relating the change in respiration rate for each 10°C change in temperature.

A model often used to describe the temperature dependence of reaction rates is the Arrhenius equation:

$$k = A \exp(-E/RT) \tag{2}$$

where k is the reaction rate constant, A is called the preexponential factor, E is the activation energy (J), R is the gas constant (8.314 J/mol·K), and T is the absolute temperature (K). Both E and A are empirical constants found from the slope and intercept, respectively, of a plot of log k versus 1/T.

The thermal properties of soils also influence the plant environment. Dry soils warm up quickly during the day and cool rapidly at night and cause the air temperature about the plant to do the same. Heat stored near the soil surface has a great effect on the amount of evaporation that occurs at the soil surface. Soil temperature is, therefore, a very important parameter when energy balance relationships are used to measure evapotranspiration (ET) rates.

In this chapter we explore the factors affecting the status and distribution of soil temperature and the transport of heat in soil profiles. The purpose is to gain a greater appreciation for the dynamics of the soil thermal environment.

Classes of Soil Temperature

The influence of soil temperature has been recognized by defining six classes at the family level in the taxonomy of soils (SSSA 1987). These classes are based on the mean annual soil temperature and are given below.

Pergelic: These soils have a mean annual temperature lower than 0°C. Permafrost is present.

Cryic: The mean annual soil temperature is higher than 0°C but lower than 8°C, the difference between mean summer and mean winter soil temperature is more than 5°C at a depth of 0.5 m, and summer temperatures are cold.

Frigid: This class of soils has warmer temperatures in the summer than those in the cryic regime, but its mean annual temperature is lower than 8°C, and the difference between mean winter and mean summer soil temperature is more than 5°C at a depth of 0.5 m. Isofrigid soils are the same except the summer and winter temperatures differ by less than 5°C.

Mesic: The mean annual soil temperature is 8°C or higher but lower than 15°C, and the difference between mean summer and mean winter soil temperature is greater than 5°C at 0.5 m. If the difference is less than 5°C, the temperature regime is isomesic.

Thermic: The mean soil temperature is 15°C or higher but lower than 22°C, and the difference between mean summer and mean winter soil temperature is more than 5°C at 0.5 m. If the difference is less than 5°C, the temperature regime is isothermic.

Hyperthermic: The mean soil temperature is 22°C or higher, and the difference between mean summer and mean winter soil temperature is greater than 5°C at 0.5 m. Isohyperthermic is the same except the summer and winter soil temperatures differ by less than 5°C.

Thermal Concepts

Several thermal parameters affect the temperature of the soil. These are defined below.

Heat

Heat is the kinetic energy of the random motion of the particles of which the soil is composed. Heat is a form of energy. The SI units of heat are those of energy (joules).

Temperature

Temperature is a measure of the thermal state considered in reference to its ability to transfer heat. It is a measure of the intensity of heat or the potential energy of heat and is used to refer to a particular level or degree of molecular activity. Temperature serves the same purpose as a driving force or energy potential in heat flow and as a pressure head in saturated water flow. Measurement of the thermal state is based on a relative scale (degrees Fahrenheit and Celsius) or on an absolute scale (Kelvin). Because the temperature scales do not have a common zero, mathematical relations have been developed to convert from one scale to another as follows:

Temperature Scales	Useful Formulas
Celsius (°C)	$°C = (5/9)(°F - 32)$
Kelvin (K)	$K = °C + 273.15$
Fahrenheit (°F)	$°F = [(9/5)°C] + 32$

Heat Capacity

Heat capacity is the amount of temperature change in the soil in response to the absorption or release of heat. The heat capacity governs how rapid the change in soil temperature will occur in response to an absorption or dissipation of heat. There are two types of heat capacities.

Gravimetric heat capacity, C_g, is the amount of heat required to raise the temperature of 1 kg of soil by 1 degree Kelvin. In the metric system, the units are cal/g·deg. However, in the SI system, the units are J/kg·K. The specific heat of materials is the ratio of the heat capacity of the material to that of water and is dimensionless.

Volumetric heat capacity, C_v, is the amount of heat required to raise the temperature of 1 m^3 of soil by 1 degree Kelvin. In the metric system, the units of C_v are cal/cm^3·deg. In the SI system, the units are J/m^3·K.

The gravimetric and volumetric heat capacities of soil are related by

$$C_g \rho_b = C_v \tag{3}$$

where ρ_b is the soil bulk density (kg/m^3).

Amount of Heat

The absolute heat content of soil cannot be calculated, but the changes in the amount of heat, Q_q, can be calculated. The amount of heat needed to change the temperature of a given volume or mass of soil from an initial state, T_1, to a final state, T_2, can be calculated on either the mass basis,

$$Q_q = C_g M \Delta T \tag{4}$$

or the volume basis,

$$Q_q = C_v V \Delta T \tag{5}$$

where Q_q is the amount of heat (J), C_g is the gravimetric heat capacity (J/kg·K), C_v is the volumetric heat capacity (J/m^3·K), M is the soil mass (kg), V is the soil volume (m^3), and ΔT is the change in soil temperature (K).

Heat Flux

Heat flux, h, is defined as the quantity of heat (Q) transferred during a unit of time (t). The SI units of heat flux are J/s.

Heat Flux Density

Heat flux density, H, is defined as the quantity of heat (Q) transferred during a unit of time (t) across a unit surface area (A) perpendicular to the direction of heat flow. The SI units of heat flux density are J/m^2·s or W/m^2.

Thermal Conductivity

Thermal conductivity, k, is the ability of a substance to transfer heat from molecule to molecule. It is defined as the quantity of heat (J) transmitted a distance of 1 m through a substance per unit cross section, per unit of temperature gradient per second. The units of k are

$$J/m·s·K = W/m·K$$

Actual heat conductance depends on thermal conductivity and on the thermal gradient. Thermal conductivity, k, is an empirically determined proportionality factor rather than a physical constant.

Thermal Diffusivity

Thermal diffusivity, α, is an expression of the rate at which a substance heats up as the result of a thermal gradient. It is the rate of change of temperature with time and is proportional to the thermal conductivity and inversely proportional to heat capacity on a volume basis. The units of α are m^2/s. Since α is the quotient of the thermal conductivity, k, and the volumetric heat capacity, C_v, it determines the temperature wave penetration into the soil; that is, it is a measure of the rate at which an imposed temperature gradient in the soil is dissipated by conduction.

Factors Affecting Soil Temperature

The temperature of soil in the field is determined by the interaction of numerous factors. Ultimately, all soil heat comes from two sources:

- radiation from the sun and sky
- conduction from the interior of the earth

Both external (or environmental) and internal (or soil) factors contribute in bringing about changes in soil temperature.

Environmental Factors

Factors that are external to the soil determine the amount of energy received at the soil surface.

Solar Radiation

The daily radiation flux density reaching the top of the atmosphere ranges between 1350 and 1400 W/m^2 (solar constant). Between 30 and 40% of the radiation reaching the earth is reflected back to the atmosphere or is lost by thermal radiation. The radiation flux that reaches the soil surface depends on (1) the angle at which the soil faces the sun and (2) insulation.

The *angle* at which the soil faces the sun varies with latitude, season, time of day, steepness and direction of slope (aspect), and altitude of the location. In general, the nearer the angle of incidence of the sun's rays to the perpendicular, the greater the absorption. Thus, the angle affects how much heat reaches the soil surface. In the Northern Hemisphere, southeastern slopes are warmest during the early summer, southern slopes are warmest at midsummer, and southwestern slopes are warmest during the fall. Correspondingly, a west-facing slope is normally somewhat warmer than an east-facing slope.

The earth is *insulated* by air, water vapor, clouds, dust, smog, snow, plants, and mulch. In the Temperate Zone, between 4.18 and 31.4 MJ/m^2 are received at the earth's surface per day. Approximately 24.3 MJ/m^2 are required to evaporate a layer of water 1 cm thick.

The following conversion is useful to calculate the amount of energy required to evaporate a 1 cm layer of water in an area of 1 m^2. Using the latent heat of vaporization, we calculate the following:

$$dALv\rho_wCF = (0.01 \text{ m})(1 \text{ m}^2)(580 \text{ cal/g})(4.184 \text{ J/cal})(1000 \text{ g/L})(1000 \text{ L/m}^3)$$
$$= 24.3 \text{ MJ}$$

where d is the depth of water, A is the area, Lv is the latent heat of vaporization, ρ_w is the density of water, and CF is the correction factor. Note the following: 1 calorie = 4.184 J; 1 cal/cm^2·min = 697.93 W/m^2.

Radiation from the Sky

Radiation from the sky contributes a relatively large amount of heat to the soil in areas where the sun's rays penetrate the earth's atmosphere at oblique angles. Under such conditions, much of the sun's energy is absorbed by the atmosphere and is radiated in all directions.

Conduction of Heat from the Atmosphere

The conduction of heat through air is small; therefore, it can have a substantial effect on soil temperature only by contact. This means that air convection (wind) is necessary to heat or cool the soil by conduction from the atmosphere.

Condensation

Condensation is an exothermic process. Whenever water vapor condenses from the atmosphere, it heats the soil because heat is transferred from the water to the soil. Under such conditions, increases of 5°C and more in soil temperature have been noted. In a similar way, the freezing of water releases heat from the water to the soil.

Evaporation

Evaporation is an endothermic process. The greater the rate of evaporation, the more the soil is cooled. Moist soils are seldom very hot. The energy required to change from liquid to vapor phase depends on the temperature and is approximately 2.45 MJ/kg at 20°C. This energy is known as the latent heat of vaporization. Thawing of ice also absorbs heat.

Rainfall and Irrigation

Depending on their temperature, rainfall and irrigation water can cool or warm the soil.

Insulation

The soil can be insulated from the environmental temperature factors by plant cover, mulches, snow, clouds, and fog. Insulation serves to maintain a more uniform soil temperature. During the summer, insulated soil is cooler than soil that is directly exposed to "the elements;" whereas, in the winter, the temperature situation is reversed.

Vegetation

Transpiration of water, reflection of incident radiation (back radiation), and energy used for photosynthesis by plants tend to decrease the temperature of the microclimate and indirectly, of the soil. A plant cover serves as insulation and consequently tends to reduce soil temperature fluctuations. Most of the energy that reaches the earth in the daytime is used as energy for evapotranspiration or is radiated back to the atmosphere. Only a small portion, perhaps as much as 10%, is used to heat the soil. At night, the soil loses heat, and some evaporation and thermal radiation take place.

Soil Factors

Six soil factors have significant influence on soil temperature. These factors—mineralogical composition, water content, color, heat capacity, thermal conductivity, and thermal diffusivity—are discussed below.

Mineralogical Composition

The mineralogical composition affects the heat capacity, thermal conductivity, and, therefore, the thermal diffusivity. These relationships are given in Tables 6.1–6.3.

Soil Water Content

An increase in soil water content in the dry range, from oven-dry to air-dry, increases the thermal conductivity of a soil, but only very little. As soon as the water starts to form bridges from particle to particle, the thermal conductivity increases markedly. This happens when the soil begins to look moist (about –3.0 MPa). For a DeWitt silt loam, these relationships are shown in Figures 6.1 and 6.2.

At a soil water pressure of around –1 MPa, the thermal conductivity of soil is similar to that of water. When the soil becomes wetter, its thermal conductivity rises considerably beyond this level due to water replacing air in the voids between soil particles.

It is fortunate that as thermal conductivity increases with water content, heat capacity

Table 6.1. Heat Capacities of Substances Related to Soil Composition

Material	Heat Capacity	
	Gravimetric, C_g (kJ/kg·K)	Volumetric, C_v (MJ/m^3·K)
Humus	1.67	2.34
Water	4.18	4.18
Ice	2.09	1.92
Air	1.05	0.0013
Clay	0.92	2.09
Quartz	0.80	2.09
Mica	0.88	2.47
Granite	0.80	2.09
CaCO$_3$	0.84	2.26
Fe$_2$O$_3$	0.63	3.14
Chalk	0.88	1.93
Wood	1.76	1.59

Sources: Adapted from Kohnke 1968, p. 177; Hillel 1982, p. 161.

Notes: The higher the heat capacity, the higher the amount of energy required to raise (or lower) the temperature of a substance by 1 K.

The C_g of humus is twice as high as that of quartz on a weight basis. However, since the bulk density of humus is lower, they are somewhat similar on a volume basis. Increasing water content increases the heat capacity of soil, whereas increasing air decreases the heat capacity of soil.

The following conversions are useful: to convert cal/g·K to J/kg·K, multiply by 4184; to convert cal/cm^3·K to J/m^3·K, multiply by 4.18 × 10^6; to convert cal/cm·s·K to J/m·s·K, multiply by 418.4.

Table 6.2. Thermal Conductivities of Some Substances

Material	Thermal Conductivity (W/m·K)
Air	0.021
Dry soil	0.125–0.209 (depends on compaction)
Snow	0.163
Water	0.544–0.586 (depends on temperature)
Building brick	0.586
Glass	0.418–1.05
Moist soil	0.836–1.674
Ice	0.836–2.09
Soil-forming minerals	1.255–12.5
Iron	66.9
Aluminum	209.2
Silver	418.4

Sources: Adapted from Kohnke 1968, p. 179; Hillel 1982, p. 162.

Table 6.3. Thermal Properties of Soil Constituents at 20° C and 1 atm

Material	Density, ρ (Mg/m^3)	Gravimetric Heat Capacity, C_g (kJ/kg·K)	Volumetric Heat Capacity, C_v (MJ/m^3·K)	Thermal Conductivity, k (J/m·s·K)	Thermal Diffusivity, α (m^2/s)
Quartz	2.65	0.732	1.92	8.368	43×10^{-4}
Many soil minerals	2.65	0.732	1.92	2.930	15×10^{-4}
Organic matter	1.3	1.925	2.51	0.251	1×10^{-4}
Water	1.0	4.184	4.18	0.594	1.4×10^{-4}
Air	0.0012	1.004	0.00121	0.026	2.1×10^{-5}

Sources: Adapted from van Wijk and DeVries 1963, p. 105.
Note: Both k and α are low for organic matter and air compared to soil minerals.

also increases. This means that the thermal diffusivity changes much less with soil water content changes than does the thermal conductivity. If it were the other way, soils would suffer from much larger temperature extremes.

Soil Color

Light-colored soils reflect more heat than darker soils. Although dark-colored soils absorb heat readily, these soils in the field often have high organic-matter contents and soil water contents, which may mean that if poorly drained, the soil will not warm up as rapidly in the spring as will well-drained, light-colored soils. Thus, several soil factors in addition to color have impact on the absorption and release of heat.

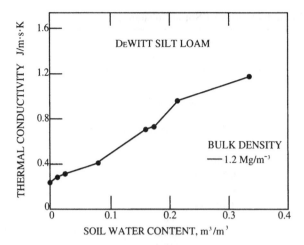

Figure 6.1. Relationship between the thermal transport coefficients and water content of a DeWitt silt loam.

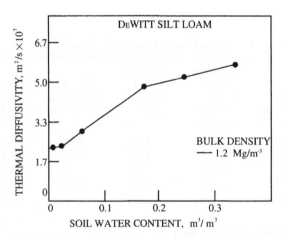

Figure 6.2. Relationship between the thermal diffusivity and water content of a DeWitt silt loam.

Heat Capacity

Heat capacity is defined as the quantity of thermal energy required to produce a 1 degree increase or decrease in temperature in a unit volume or mass. It quantifies the amount of heat a volume (or mass) of soil stores or releases with a unit change in temperature. Therefore, soil heat capacity governs how rapidly the change in soil temperature will occur in response to an absorption or dissipation of heat.

The heat capacities of the three soil phases are additive. For example, the volumetric heat capacity can be written as

$$C_v = SC_s + \theta_v C_w + f_a C_a \qquad [6]$$

where C_v is the volumetric heat capacity of the soil (J/m$^3 \cdot$K); S is the volume fraction of the soil solids (m^3/m^3), that is, ρ_b/ρ_p; θ_v is the volume fraction of the soil water (m^3/m^3), also called the volumetric soil water content; f_a is the volume fraction of the soil air (m^3/m^3), also called the aeration porosity; C_s is the volumetric heat capacity of the soil solids (J/m$^3 \cdot$K); C_w is the volumetric heat capacity of the soil water (J/m$^3 \cdot$K); and C_a is the volumetric heat capacity of the soil air (J/m$^3 \cdot$K).

We note that the volumetric heat capacity of the solid fraction (J/m$^3 \cdot$K) can be written as

$$SC_s = X_m C_m + X_o C_o \tag{7}$$

where S is the volume fraction of the solid component, X_m is the volume fraction of the mineral content, X_o is the volume fraction of the organic-matter component, C_m is the volumetric heat capacity of the mineral component, and C_o is the volumetric heat capacity of the organic component.

For most mineral soils, the air and organic contributions to the heat capacity are assumed negligible because of their low magnitudes. Using information in Table 6.1, we see that

$$C_a = 1.30 \text{ kJ/m}^3 \cdot \text{K} \quad \text{and} \quad C_o = 2.34 \text{ MJ/m}^3 \cdot \text{K},$$

but X_o usually varies from less than 0.01 to 0.03. Therefore, equation [6] becomes

$$C_v \approx X_m C_m + \theta_v C_w \tag{8}$$

Note that $X_m C_m$ can be written as

$$X_m = 1 - f = 1 - [1 - (\rho_b/\rho_p)] = \rho_b/\rho_p = S$$

where f is the total porosity and

$$C_m = C_{gm}\rho_p$$

where C_{gm} is the gravimetric heat capacity of the mineral phase. Thus, combining these two equations, we get for the solid phase, as presented in equation [8],

$$X_m C_m = SC_m = (\rho_b/\rho_p)(C_{gm}\,\rho_p) = \rho_b C_{gm}$$

For the aqueous phase, we can substitute

$$\theta_v C_w = (\rho_b/\rho_w)\theta_w C_w \rho_w = \rho_b \theta_w C_w = 4180\,\theta_w \rho_b$$

where C_w is the gravimetric heat capacity of water (4.18 $\times$ 10^3 J/kg$\cdot$K), and θ_w is the gravimetric soil water content (kg/kg).

After some rearranging, equation [8] can be written as

$$C_v = \rho_b C_{gm} + \theta_w C_w \rho_b = \rho_b[837 + (4180\theta_w)] \tag{9}$$

where 837 is the gravimetric heat capacity of the solid phase (J/kg$\cdot$K), and 4180 is the gravimetric heat capacity of water (J/kg$\cdot$K).

Since C_{gm} and C_w are generally recognized as constants (837 and 4180 J/kg$\cdot$K, respectively), C_v is directly dependent on the bulk density and the soil water content (equation [8]). At constant bulk density, the relation between C_v and θ_v is linear, as shown in Figure 6.3a. At constant water content, the relation between C_v and ρ_b is also linear (Figure 6.3b).

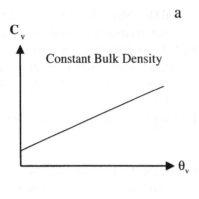

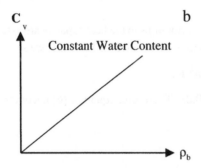

Figure 6.3. Relationship between the volumetric heat capacity and water content (a) and bulk density (b).

Example

If the gravimetric heat capacity of oven-dry mineral soil is near 837 J/kg·K, what is the volumetric heat capacity of 1 m^3 of oven-dry soil consisting of half solids and half pore space?

$$C_v = SC_s + \theta_v C_w + f_a C_a = S\rho_p C_{ms}$$

Since there is no water in the soil and the contribution from the soil air is negligible, we have

$$
\begin{aligned}
C_v = SC_s &= S\rho_p C_{ms} \\
&= (0.5 \text{ m}^3/\text{m}^3)(2650 \text{ kg/m}^3)(837 \text{ J/kg·K}) \\
&= 1.1 \times 10^6 \text{ J/m}^3\text{K} \\
&= 1.1 \text{ MJ/m}^3\text{K}
\end{aligned}
$$

The same soil filled with water would have a heat capacity of

$$
\begin{aligned}
C_v = SC_s + \theta_v C_w \\
&= 1.1 \times 10^6 + [(0.5 \text{ m}^3/\text{m}^3)(4.18 \times 10^6 \text{ J/m}^3\text{·K})] \\
&= 3.2 \times 10^6 \text{ J/m}^3\text{·K} \\
&= 3.2 \text{ MJ/m}^3\text{·K}
\end{aligned}
$$

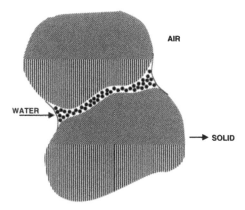

Figure 6.4. Soil particles and air spaces, showing the contact between the particles.

If only one-half of the pore space was filled with water, its heat capacity would be

$$C_v = SC_s + \theta_v C_w$$
$$C_v = 1.1 \times 10^6 + [(0.25 \ m^3/m^3)(4.18 \times 10^6 \ J/m^3{\cdot}K)]$$
$$\qquad + [(0.25 \ m^3/m^3)(1.21 \times 10^{-3} \ J/m^3{\cdot}K)]$$
$$= 2.15 \ MJ/m^3{\cdot}K$$

This example shows that C_v would be expected to range between 1 and 3 MJ/m³K (i.e., a threefold variation due to water content). Equation [8] and the results of these problems clearly show that soil water content and bulk density directly affect the magnitude of the heat capacity.

Thermal Conductivity

In general, the thermal conductivity, k, of a soil depends on its mineral and organic composition as well as its volumetric water content (Figures 6.1 and 6.4). At complete dryness, the heat flow passes mainly through the soil particles and has to bridge the air-filled gaps between the particles around their contact points. As a result, the magnitude of k at complete dryness is low. Note that the contact areas are small and the thermal conductivity of air is low compared to that of water or solids (Tables 6.2 and 6.3).

At low soil water content, the soil particles are covered by thin adsorbed water layers. The thickness of these layers increases with increasing water content. At a certain water content, liquid-water rings form around the contact points between soil particles; they have curved air-water interfaces.

An increase in soil water content causes the contact areas to expand, which results in an increase in k. With increasing water content, this expansion becomes smaller. According to Richter (1987), it is not the water content itself that determines the magnitude of k, but the soil water matric potential characterizing sites of contact between particles. The relationship between k and θ_v differs greatly among soils, but the relationship between k and matric potential is nearly identical among soils. Perhaps, the matric potential energy determines the size of the rings and wedges of water around the particles and, thus, the area of water bridging the soil particles.

Thermal conductivity is the amount of heat a material will conduct through a unit area in unit time under a unit temperature gradient. Fourier's law for steady-state heat flow can be rearranged to obtain

$$k = -H_s/(\partial T/\partial z) = -(Q/At)/(\partial T/\partial z) \tag{10}$$

where H_s is the soil heat flux density ($J/m^2 \cdot K$). As a result, k can be described as the steady-state thermal transport property characteristic for each soil medium within certain physical restrictions. We let

$$k = \alpha C_v \tag{11}$$

Values of k vary with

- soil phase: solid > liquid > air
- textural composition: sand > loam > clay > peat
- bulk density: directly related (see Figure 6.1)
- water status: directly related (see Figure 6.1)
- structure: similar to textural composition since aggregation of primary particles results in larger secondary particles
- surface contact between the particles

Several experimental methods have been developed for determining k in soil. Steady-state methods are used where heat is transferred through soil while the temperature at each position within the soil system remains constant with time. Transient-state methods are based on changing the temperature around a probe of a heating body for a short period of time, from which the temperature distributions around the probe and k are computed. The third method involves theoretical predictions.

Several mathematical models have been proposed for calculating thermal conductivity of soils differing in composition (Hillel 1982). Unlike heat capacity, which can be calculated as the sum of the heat capacities of the different soil components weighted by their respective volume fractions of the soil, the thermal conductivity of soils is also a function of the geometrical arrangement of the phases in the soil matrix. One useful approach, given by Bristow (1998) and Campbell (1985), and originating with McInnes (1981), gives

$$k = A + B\theta v - (A - D)\exp[-(C\theta_v)^E] \tag{12}$$

where A, B, C, D, and E are soil-dependent coefficients that have been related to soil properties by Campbell (1985). Another, somewhat more theoretical method is given by van Wijk and deVries (1963).

In general, k is not simply related with the thermal conductivities of the individual soil components, because the conduction of heat occurs through all kinds of sequences of the conducting materials, both in series and in parallel. The following equations assume only one situation in layered profiles, that is, in series.

Thermal Diffusivity

The thermal diffusivity, α, is defined as the measure of temperature change in a unit volume by the quantity of heat that flows through the volume, per unit time, when a unit temperature gradient is imposed across two opposite sides of the volume. The thermal diffusivity determines the heating and cooling rate accompanying a given temperature profile, that is, the soil's ability to warm or cool by conduction. The relation between k and α is

$$\alpha = k/C_v \qquad [13]$$

Example

Calculate the thermal diffusivity if the thermal conductivity is 1.2 J/m·s·K and the volumetric heat capacity is 2.26 MJ/m^{3}K.

Answer: $\alpha = 5.31 \times 10^{-7}$ m^2/s.

The thermal diffusivity is a transient-state transport coefficient for determining the temperature variation within a soil profile. Factors affecting the thermal diffusivity include the following:

1. Soil water status: A small amount of water reduces the insulating effect of the pore space filled with air. Further increases in water content markedly increase the heat capacity and thermal conductivity.
2. Bulk density: Compaction increases α by decreasing the total porosity.
3. Soil composition: See Table 6.3.
4. Organic matter: Lowers by increasing total porosity or decreasing mineral content.

Heat Conductance in Soil

Since the sun and thermal contributions from the interior of the earth are the primary sources of heat that affect soil temperature, it is obvious that transport of heat in soils is an important factor in determining soil temperature. Heat flow through porous materials such as soil is more complicated than flow through solids because of the varying combinations of phases and densities of soil components. Nevertheless, the same fundamental features of heat flow operate in both types of materials. The soil heat flux amounts to 10–30% of the net radiation that is received diurnally for a bare soil. On a daily basis, however, the net gain or loss is small because the heat entering the soil is nearly matched by the heat leaving the soil.

Heat Transport Mechanisms

Heat can be transferred in soils by three mechanisms: conduction, convective flow of water and of air, and a phase change. *Conduction* of heat occurs by transmission of thermal energy of motion from one particle to another. *Convection* of heat in soil can occur because of fluid flow. In unsaturated soil, convection of heat is quantified by the product of the flow of fluids such as water and air, the volumetric heat capacity of the fluid, and the difference in temperature. A *phase change* results in the absorption or release of heat, which is accompanied by a change in volume. For example, for liquid water to transition to ice, 334 J/g of heat must be liberated, and the volume is increased by about 9%. The thermal energy required for a phase change is referred to as latent heat. Of these transport mechanisms, transport of heat by conduction is the most important in soil.

Fourier's Law

The conduction of heat in solids was analyzed by Fourier (1822). The equation he developed is mathematically analogous to the diffusion equation (later developed by Fick), conduction of fluids in porous media (later developed by Darcy), and the conduction of electricity (developed by Ohm). Originally, the law was developed for solids in which, on a macroscopic level, all molecules were in contact with each other.

The first law of heat conduction is known as *Fourier's law*. In one dimension, this law is written as

$$H = -k\partial T/\partial z \tag{14}$$

where H is the heat flux density ($J/m^2 \cdot s$ or W/m^2). The parameter k is the thermal conductivity ($J/m \cdot s \cdot K$ or $W/m \cdot K$), and $\partial T/\partial z$ is the temperature gradient (K/m). Equation [14] serves as the defining equation for the thermal conductivity.

The negative sign in equation [14] indicates that flow of heat occurs in the direction opposite to the temperature increase; that is, heat flows from locations of higher to locations of lower soil temperature.

Fourier's law is sufficient to describe heat conduction under steady-state conditions, that is, where the temperature at each point in the soil and the flux density remain constant in time. Additional assumptions for the use of equation [14] include the following:

1. Soil heat flow is unidirectional, that is, only in the vertical direction.
2. No heat is generated or converted into other forms.
3. The medium is rigid, with no changes in space as a whole or in its constituents.
4. The thermal properties of the soil are the same with respect to the various axes; that is, the medium is isotropic.

Steady-State Heat Conductance in Composite Soils

Fourier's law as given by equation [14] can be used to predict the heat flux density under steady-state conditions in soil profiles having uniform properties. Unfortunately, these conditions seldom occur in agricultural soils, especially where Ultisols and Alfisols dominate the landscape. Therefore, there is a need to develop heat transport equations that can predict the flow of heat in soils that have differing physical properties in the profile. These soils are known as composite porous media.

Fourier's law for composite media can be written in incremental form as

$$H_s = -k\Delta T/\Delta z = -\Delta T/(\Delta z/k_{eq}) \tag{15}$$

where the quantity $\Delta z/k_{eq}$ is known as the resistance (R_h) to the flow of heat in the profile, and k_{eq} is the *equivalent* thermal conductivity for the profile. For a soil profile with two horizons of differing physical properties, the resistance in series can be written as

$$\Delta z/k_{eq} = L_1/k_1 + L_2/k_2 = R_{h1} + R_{h2} \tag{16}$$

where L and k are the thicknesses and thermal conductivities, and R_{h1} and R_{h2} are the resistances to heat flow of horizons 1 and 2, respectively. This equation can be used to calculate the *equivalent* thermal conductivity of the soil profile. Once the magnitude of this parameter and the soil temperatures of the boundaries are known, the heat flux density can be calculated for the profile. The temperature at any location within the soil profile can be estimated by substituting the heat flux density, equivalent thermal conductivity, and depth into equation [15].

The ratio of the thickness to the thermal conductivity for each horizon is known as the resistance to heat flow within that horizon. For a soil profile with numerous horizons, Fourier's law for series resistances can be rewritten as

$$H_s = -\Delta T / \Sigma R \qquad [17]$$

where ΣR is the sum of the resistances of heat in the profile. It is known as the *thermal resistance of the soil.*

Example

A uniformly textured silt loam soil has a compacted plow pan at the 5 cm depth that is 8 cm thick. If the soil temperature at 0600 is 18°C at the soil surface and 24°C at the bottom of the plow pan and the thermal conductivity is 1.2 and 1.5 W/m·K for the two layers, respectively, determine the following: (1) the equivalent thermal conductivity, (2) the heat flux density and direction of heat flow, and (3) the temperature at the boundary between the two layers.

The equivalent thermal conductivity is calculated from

$$(L_1 + L_2)/k_{eq} = L_1/k_1 + L_2/k_2 = 0.05/1.2 + 0.08/1.5 = 0.095 \text{ m}^2 \cdot \text{K/W}$$

Solving for k_{eq} gives 1.368 W/m·K. The heat flux density is

$$H_s = -k_{eq}\Delta T/\Delta z = -(1.368)[(24 - 18)/(-0.13 - 0)]$$
$$= -1.368(-46.15) = 63.16 \text{ W/m}^2$$

Since H_s is positive, the flow of heat is in the positive z direction (i.e., upward).

The temperature at the boundary with the plow pan (i.e., the 5 cm depth) is calculated from

$$\Delta T = T_2 - T_1 = 24 - T_1 = -H_s(\Delta z/k_2) = -63.16(-0.08/1.5)$$
$$T_1 = 24 - 3.37 = 20.63°C$$

The student should verify this temperature at the 5 cm depth using the thermal properties of the first layer.

Methods of Calculation of Soil Heat Flux Density

There are two methods of calculating the flux density of heat in soil. These are known as the calorimetric and temperature gradient methods.

Calorimetry

Calorimetry calculates heat flux density from the change in heat storage in the soil profile over a given time interval. Calorimetry utilizes the law of conservation of energy to give

$$H_{si} - H_{so} = \Delta H_s \qquad [18]$$

where H_{si} is the soil heat flux density going into the layer (J/m²·s), H_{so} is the soil heat flux density going out of the layer (J/m²·s), and ΔH_s is the rate of change in heat content per unit area (J/m²·s). Values of ΔH_s can be determined as follows. The amount of heat needed to change the temperature of a given soil volume is Q_q with SI units of J. Therefore, from equation [5], we write

$$Q_q = C_v V \Delta T \qquad [19]$$

Dividing by the cross-sectional area and time, we obtain

$$Q_q/A\Delta t = \Delta H_s = C_v V\Delta T/(A\Delta t) = C_v(\Delta T/\Delta t)\Delta z$$

or

$$\Delta H_s = C_v(\Delta T/\Delta t)\Delta z \qquad [20]$$

where C_v is the volumetric heat capacity of the layer (J/m³·K), Δz is the soil profile thickness (m), and $\Delta T/\Delta t$ is the change in temperature with change in time (K/s).

Example

A soil was found to have a net heat flux density, ΔH_s, of –2.13 MJ/m²·d. If the average volumetric water content was 0.25 m³/m³ and the bulk density was 1300 kg/m³, calculate the daily average temperature increase that occurs in 1 d in the top meter of soil assuming that the heat is evenly distributed in the profile.

$$\Delta H_s = Q_q/A\Delta t = C_v V\Delta T/(A\Delta t) = (C_v)(1.0)(\Delta T)$$
$$\Delta T = \Delta H_s/C_v = 2.13/[(1300 \times 837) + (0.25 \times 4.18 \times 10^6)]$$
$$= 2.13/2.13 = 1.0°C$$

The sign of ΔH_s is negative, indicating that heat is entering the soil. However, from a soil energy balance perspective, the heat flux density H_s is positive because it is entering the soil volume. Thus, it is an influx. This is a different sign convention than that used with the soil heat flow. In general, the uniform redistribution of heat throughout the profile does not occur in field soils.

Temperature Gradient Method

Fourier's law for steady-state heat flow can be rewritten as

$$H_s = -k\partial T/\partial z \approx -k(\Delta T/\Delta z) = -k[(T_2 - T_1)/(z_2 - z_1)] \qquad [21]$$

where H_s is the heat flux density and is equal to Q_q/At. The steady-state assumptions for heat conduction are as follows:

1. Heat flows in one direction only.
2. The flow of heat is invariant with time.
3. No heat coupling transport mechanism occurs.
4. The soil is considered to be a flat, solid slab.

In soil, the temperature gradient with the largest value is usually considered to be in the vertical direction. For this reason, we will use the coordinate z to indicate the vertical direction.

Example

The temperature of a DeWitt silt loam was measured at two depths twice in one day. The data are presented in the table below. If the thermal conductivity is 1.76 J/m·s·K, calculate the heat flux density and direction of heat flow at the two measurement times. Remember, we assume that z is positive in the upward direction and that the reference level is at the soil surface.

Sampling Time	Temperature (K)	
	Surface	0.10 m depth
0630	294	303
1630	327	309

At 0630 hours,

$$H_s = -1.76[(303 - 294)/(-0.1 - 0)] = -1.76(9/-0.1) = 158.4 \text{ J/m}^2\text{·s}$$

Since upward flow is considered to be positive, a positive H_s indicates upward flow of heat, and a negative H_s indicates downward flow of heat. At this time of the day, soil heat flow is toward the soil surface.

At 1630 hours,

$$H_s = -1.76[(309 - 327)/(0.1 - 0)] = -316.8 \text{ J/m}^2\text{s}$$

Thus, at this late-afternoon time, the flow of heat is vertically downward into the profile.

Equation of Continuity for One-Dimensional Conductance of Heat

In the previous chapter, we presented the energy balance or conservation law. The law of conservation of energy states that the rate at which energy enters a unit volume of soil minus the rate at which this energy leaves the volume of soil must equal the rate at which this energy is stored in the volume of soil. If the inflow of energy exceeds the outflow, there will be an increase in the amount of energy stored (or accumulated) in the volume of soil; if the converse is true, there will be a decrease in energy stored. If the inflow equals the outflow, a steady-state condition must prevail in which there will be no change in the amount of energy stored in the volume of soil.

The unit volume of soil for heat flow is shown in Figure 6.5. Here, $H_s = Q/At$ and $Q = H_sAt$. The equation of conservation of energy requires that the heat flowing into the unit volume of soil minus the heat flowing out of the same volume of soil is equal to the net change in heat storage. The amount of heat, with units of J, entering ABCD in a given time can be written as $H_s\Delta x\Delta y\Delta t$ where H_s is the heat flux density. Therefore, we write the inflow as $H_s\Delta x\Delta y\Delta t$. The amount of heat leaving EFGH in this same time interval is $H_s\Delta x\Delta y\Delta t + \Delta H_s\Delta x\Delta y\Delta t$. Therefore,

$$H_s\Delta x\Delta y\Delta t - (H_s\Delta x\Delta y\Delta t + \Delta H_s\Delta x\Delta y\Delta t) = -\Delta H_s\Delta x\Delta y\Delta t$$

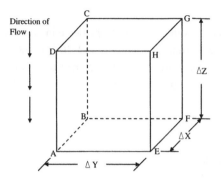

Figure 6.5. Schematic diagram of the unit volume of soil.

The change in heat content accumulated in the volume element is

$$C_v V \Delta T = C_v \Delta x \Delta y \Delta z \Delta T$$

Since the heat contents are the same, they must be equal; therefore,

$$C_v V \Delta T = -\Delta H_s \Delta x \Delta y \Delta t$$

After rearranging, the equation of conservation of heat in soil is

$$C_v (\Delta T / \Delta t) = -\Delta H_s / \Delta z \qquad [22]$$

The energy conservation equation in soil states that in the absence of any sources or sinks of heat, the time rate of change in heat content of a volume element of the conducting medium (soil) must equal the change of heat flux density with distance.

In differential form, the equation of conservation of energy, without consumption and generation, is written

$$C_v \partial T / \partial t = -\partial H_s / \partial z \qquad [23]$$

In the field, soil temperature is constantly changing, so the steady-state form of Fourier's law cannot accurately be used, particularly near the soil surface. As a result, the transient-state form of the heat flow equation was developed. After inserting Fourier's law for steady-state transport (equation [21]) into equation [23], the *transient-state* heat flow equation for conductance is

$$C_v \partial T / \partial t = \partial / \partial z (k \partial T / \partial z) \qquad [24]$$

If the soil and water content are homogeneous and isotropic, we may assume that $\rho_b C_w$ (or C_v) and k are independent of time and depth, z. Since $\alpha = k / C_v$, the transient-state heat flow equation in soil can be written as

$$\partial T / \partial t = \alpha (\partial^2 T / \partial z^2) \qquad [25]$$

This differential equation is known as the transient-state *heat flow equation* and can be solved for certain simple initial and boundary conditions. A demonstration of one of the simplest analytical solutions follows.

Under steady-state heat conduction, the temperature does not change with time. Therefore, $\partial T/\partial t = 0$. Since α, C_v, and k are constant and nonzero, equation [25] reduces to

$$\partial^2 T/\partial z^2 = 0 \qquad [26]$$

Integrating twice results in the general solution

$$T(z) = C_2 + C_1 z \qquad [27]$$

where C_1 and C_2 are constants that can be determined if the temperature is known for two values of z. The constants of integration are obtained from the boundary conditions

$$T(z = 0) = T(0) = T_0 \qquad [28]$$

and

$$T(z = L) = T(L) = T_1 \qquad [29]$$

where L is some depth in the profile. Substituting these two boundary equations into equation [27] gives

$$T(z) = T_0 + [(T_1 - T_0)/L]z \qquad [30]$$

Thus, under conditions where k is constant and there is no generation or consumption of heat, the temperature is linearly related to soil depth. Now that we have the temperature distribution in the soil profile, we use Fourier's law to determine the heat transport rate by conduction.

Heat Transport Equation with Conduction, Convection, and Latent Heat

A more general equation for the transport of heat in a rigid soil volume can be derived assuming transport by conduction and convection and with latent heat. Guymon (1994) derived this equation using the differential equation form with no radiative heat transfer and parallel conduction in the solid and aqueous phases. The macroscopic energy balance equation for a unit volume of soil is

$$\Delta h_s + \Delta e_a + Le = \Delta q_q \qquad [31]$$

where Δh_s is the net conducted heat transfer across the volume boundary, Δe_a is the net convected heat across the volume boundary, Le is the heat absorbed into the volume (i.e., the latent heat effects), and Δq_q is the change in stored energy in the soil volume. The units of each parameter are in differential units such as J/s. We will consider one-dimensional flow in a soil volume of constant cross-sectional area.

Conductive Heat Transport

The heat conducted in the soil volume is governed by Fourier's law. We denote the heat flux density in the soil matrix by H_m and in the water phase by H_w. It is assumed that the conducted heat in the soil air phase is negligible. The macroscopic cross-sectional areas for the matrix phase and the water phase are denoted by A_m and A_w, respectively.

The net total conducted heat is obtained by subtracting the heat output from the heat input to give

$$\Delta h_s = -[\partial(A_m H_m)/\partial z]\Delta z - [\partial(A_w H_w)/\partial z]\Delta z$$

Although A_m and A_w are assumed constant, they are kept inside the partial operator for further manipulation. Next, the constant Δz is moved inside the partial operator and both sides are divided by the unit volume V of soil to obtain

$$\Delta h_s/V = [\partial(SH_m)/\partial z] - [\partial(\theta_v H_w)/\partial z]$$

where S is the volume fraction of soil solids ($A_m\Delta z/V$) and θ_v is the volumetric soil water content ($A_w\Delta z/V$). Substitution of Fourier's law into the above equation gives

$$\Delta h_s/V = \partial/\partial z(Sk_m\partial T/\partial z) + \partial/\partial z(\theta_v k_w\partial T/\partial z)$$

where k_m is the thermal conductivity of the solid phase, and k_w is the thermal conductivity of the soil solution phase. If $\partial S/\partial z$ and $\partial\theta_v/\partial z$ are negligible and the solid and solution phases are in equilibrium, we have

$$\Delta h_s/V = \partial/\partial z(k\partial T/\partial z) \tag{32}$$

where k is the thermal conductivity of the soil water mixture, which has been defined as a linear additive combination of the thermal conductivities in the matrix and solution phases of the soil; that is,

$$k = Sk_m + \theta_v k_w \tag{33}$$

Convective Heat Transport

The net convective heat out of the volume element is

$$\Delta e_s = -\partial/\partial z(vA_w C_w T)\Delta z$$

where v is the soil water velocity in the soil pores (m^2/s), and C_w is the volumetric heat capacity of water. Moving the Δz constant inside the operator and dividing both sides by the constant total volume of soil gives

$$\Delta e_s/V = -\partial/\partial z(v\theta_v C_w T)$$

Since the water velocity in the soil pores, v, multiplied by the water content by volume, θ_v, is the soil water flux density in the soil, q, then

$$\Delta e_s/V = -\partial/\partial z(qC_w T) \tag{34}$$

Heat Storage

Assuming that the volumetric heat capacities of the solid and aqueous phases are not functions of temperature, the rate of change in heat stored in the soil volume is

$$\Delta q_q/V = SC_s\partial T/\partial t + C_w\partial(\theta_v T)/\partial t \tag{35}$$

where C_s is the volumetric heat capacity of the solid phase.

Derivation of the Soil Heat Transport Equation

The soil heat transport equation is derived by summing the net contributions from the conductive and convective transfers and heat storage. First, we divide each term by the volume, V, and substitute the conductive and convective transfer contributions to obtain

$$\partial/\partial z(k\partial T/\partial z) - \partial/\partial z(qC_wT) = SC_s\partial T/\partial t + C_w\,\partial(\theta_vT)/\partial t$$

The convective term and the last term of the heat storage change are expanded, C_w is constant, and assuming continuity of an incompressible fluid, two of the terms cancel (i.e., $\partial q/\partial z + \partial\theta_v/\partial t = 0$) to give

$$\partial/\partial z(k\partial T/\partial z) - qC_w\partial T/\partial z = SC_s\partial T/\partial t + \theta_vC_w\partial T/\partial t$$

Combining the last two terms gives

$$\partial/\partial z(k\partial T/\partial z) - qC_w\partial T/\partial z = C_v\partial T/\partial t \qquad [36]$$

where C_v is the volumetric heat capacity of the soil water mixture, and the thermal conductivity is estimated by equation [33].

Equation [36] can be extended to situations where the latent heat effects are significant, that is, where freezing and thawing terms are important. The resulting equation is

$$\partial/\partial z(k\partial T/\partial z) - qC_w\partial T/\partial z + Le(\rho_i/\rho_w)\partial\theta_{vi}/\partial t = C_v\partial T/\partial t \qquad [37]$$

where Le is the latent heat of fusion of water, ρ_i and ρ_w are the densities of ice and water, respectively, and θ_{vi} is the volumetric ice content. The soil heat transport equation is a linear, nonsymmetrical, partial differential equation with nonconstant coefficients k and C_v because they are dependent on the soil water content and soil depth.

Predictions of Soil Thermal Regimes

The temperature of the soil surface responds closely to the radiant energy budget. Daytime heating is by shortwave radiation (mostly shorter than 2 μm) from the sun and sky that has a spectrum at the earth's surface that would be appropriate to a radiator at about 6000 K. Nighttime cooling results from the loss of energy emitted from the surface of the soil as long-wave radiation (mostly longer than 4 μm) with a spectrum appropriate to a radiator at about 280 K. The balance between incoming and outgoing energy at the soil surface also includes other heat transport mechanisms as given by the energy balance equation.

Soil Surface

The temperature regime of the soil surface has two cyclical periods: diurnal and annual. Daytime heating and nighttime cooling are responsible for the diurnal period, and the annual period results from the variation in shortwave radiation throughout the year. This radiation varies little at the equator but varies greatly at latitudes such as 36° N in Fayetteville, Arkansas. Following the winter solstice on December 22 in the Northern Hemisphere, shortwave radiation increases and soon begins to provide an energy surplus sufficient for warming of the surface to commence. Warming then continues for the next

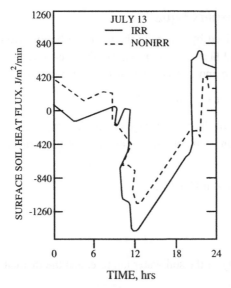

Figure 6.6. Soil surface heat flux density on July 13, 1978, as observed in irrigated and nonirrigated DeWitt soil (Erstein 1979).

6 months until incoming shortwave radiation begins to decrease. The reverse then holds for the half year between the summer and winter solstices.

The diurnal distribution of the heat flux density at the soil surface of a DeWitt silt loam on July 13 was calculated by Erstein (1979) and is shown in Figure 6.6. The heat flux density was negative during the daylight hours, indicating that heat flow was entering the soil. At night, the heat flux density was positive, indicating that heat was leaving the soil profile.

In order to examine the diurnal and seasonal penetration of heat into the soil profile, we assume that the soil profile oscillates around an average temperature as a function of time. We choose a starting time such that at t = 0, the soil surface is at the average temperature. The temperature at the soil surface can be approximated by

$$T(0, t) = T_a + A_0 \sin \omega t \qquad [38]$$

where T(0, t) is the temperature of the soil surface as a function of time, T_a is the average surface temperature, and A_0 is the amplitude of the surface temperature fluctuation. The amplitude is the temperature difference between the maximum (or minimum) and the average soil temperature.

The maximum daytime temperature is $T_a + A_0$, and the minimum nighttime temperature is $T_a - A_0$. The parameter ω is known as the angular or radial frequency of the oscillation and is given by

$$\omega = 2\pi/t_p \qquad [39]$$

The parameter ω is measured in radians and not degrees, where 1 radian is equal to 57.3 degrees. A radian is that angle subtended at the center of a circle by an arc equal to the radius. The parameter t_p is the period of cycle or oscillation (e.g., 24 h for a daily cycle and either 12 mo or 365 d for an annual cycle). The parameter t is the time, which is taken as zero when $T = T_a$ and is increasing if the temperature at the soil surface is approximated by equation [38]. Equation [38] is often used as a boundary condition for describing temperature variations of the soil surface.

Example

Draw a graph of the temperature at the soil surface, $T(0, t)$, as a function of time if T_a is 20°C, A_0 is 12°C, and ω is $2\pi/24$ radians per hour.

The working equation is $T(0, t) = T_a + A_0 \sin(\omega t)$, where $\omega = 2\pi/24 = \pi/12$. A table of the values of $T(0, t)$ is given below.

t (h)	ωt	$\sin(\omega t)$	$T(0, t)$
0	0	0	20
2	$\pi/6$	0.5	26
4	$\pi/3$	0.866	30.4
6	$\pi/2$	1.00	32
8	$2\pi/3$	0.866	30.4
10	$5\pi/6$	0.50	26
12	π	0	20
16	$4\pi/3$	−0.866	9.6
18	$3\pi/2$	−1.0	8
20	$5\pi/3$	−0.866	9.6
24	2π	0	20

Under field conditions, the average soil temperature does not occur at zero time. Therefore, some modification of equation [38] is necessary to predict more accurately the temperature variations during the day. This introduces the phase shift constant equal to $-\omega t_0$, where t_0 is the time when the average temperature occurs. Therefore, the working equation to describe the soil temperature at the soil surface now becomes

$$T(0, t) = T_a + A_0 \sin(\omega t - \omega t_0) \qquad [40]$$

Example

Draw a graph of the temperature at the soil surface using the same maximum and minimum temperatures as above but with the daily average soil temperature at 1000 h. What are the differences in the two curves?

For this problem, calculations using equation [40] result in the following:

t	ωt	ωt_0	$\sin(\omega t - \omega t_0)$	$T(0, t)$
8	$2\pi/3$	$5\pi/6$	$\sin(-\pi/6) = -0.5$	14
10	$5\pi/6$	$5\pi/6$	$\sin(0) = 0$	20
16	$4\pi/3$	$5\pi/6$	$\sin(\pi/2) = 1.0$	32
18	$3\pi/2$	$5\pi/6$	$\sin(2\pi/3) = 0.866$	30.4
20	$5\pi/3$	$5\pi/6$	$\sin(5\pi/6) = 0.5$	26
24	2π	$5\pi/6$	$\sin(7\pi/6) = -0.5$	14

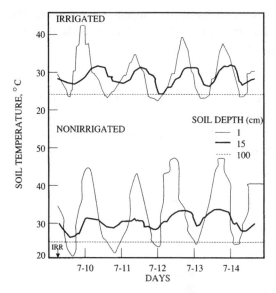

Figure 6.7. Diurnal variation in the irrigated- and the nonirrigated-soil temperatures at the 1, 15, and 100 cm depths on July 10–14 (Erstein 1979).

A comparison of the calculations from this problem with those from the previous problem indicates that the computed temperatures are shifted to the right.

Soil Profile

If we assume that at some infinite depth the soil temperature is constant and equal to T_a, substitution of these initial and boundary conditions into equation [25] and with some manipulation gives the following working equation for the diurnal temperature at any depth:

$$T(z, t) = T_a + A_0 e^{-z/d} \sin[\omega(t - t_0) - z/d] \qquad\qquad [41]$$

where d = damping depth (m), that is, the depth at which the temperature amplitude decreases to the fraction 1/e or 0.37 of the amplitude at the soil surface, A_0. Equation [41] predicts that at a given depth z, the amplitude of the temperature fluctuation is smaller than A_0 by the factor $e^{-z/d}$, and that there is a phase shift (i.e., a time delay of the temperature peak) equal to $-z/d$. This indicates that during the propagation of a periodic temperature wave in the soil the amplitude of temperature decreases and the phase lag increases with depth in the profile. Equation [41] also predicts that the average temperature is constant with depth in the profile.

The diurnal temperature distribution of three depths of a bare DeWitt silt loam is shown in Figure 6.7. These diurnal curves show that the nonirrigated soil had a larger temperature amplitude near the soil surface than did the irrigated soil. At the 100 cm depth there were no differences in soil temperature due to water management. These curves also show the damping of the soil temperature with depth.

The damping depth can be calculated from the ratio of the thermal diffusivity and radial frequency as

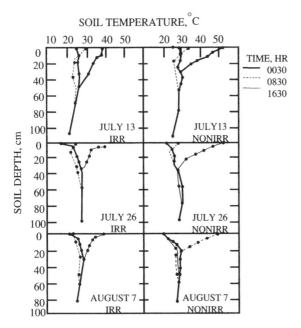

Figure 6.8. Diurnal variation in irrigated- and nonirrigated-soil temperatures as a function of depth and time at three selected dates (Erstein 1979).

$$d = (2\alpha/\omega)^{1/2} = (2k/C_v\omega)^{1/2} \qquad\qquad [42]$$

At $z = 2d$, the quantity $A(z) = A(0)e^{-2}$ or $A(z)/A(0) = 0.14$ or only 14% of the surface temperature variation.

As equation [42] suggests, the square of the damping depth is directly related to the thermal diffusivity and inversely related to the radial frequency. In particular, the magnitude of d depends directly on those soil properties that impact C_v and k, as well as the period of the temperature fluctuation chosen. For example, the damping depth for the annual temperature is $(365)^{1/2} = 19$ times larger than the diurnal variation in the same soil.

Soil profile temperature distributions for three times during three days are presented in Figure 6.8. As expected, the greatest change in temperature during a day occurred at the soil surface; practically no change in temperature occurred at the 100 cm depth. Changes in the slope of the curves are indicative of the direction of flow of heat. Since heat flows from locations of higher temperature to locations of cooler temperature, profile temperature curves that slope to the right indicate downward heat flow. Thus, as expected, heat flowed into the soil during the day and out of the soil at night.

Erstein (1979) found the damping depth in a DeWitt silt loam to be dependent on the soil water status. Under irrigated conditions, d averaged 10.3 cm, whereas under nonirrigated conditions, a lower d was found and averaged 7.8 cm. The higher d in the wet soil was attributed to the uniformity effects of water on heat transport in the profile and to the increase in the thermal transport parameters k and α with increased water content.

The temperature amplitude is reduced in the profile because of the absorption and release of heat as the soil temperature changes in the layers or horizons.

Example

The daily maximum soil surface temperature is 32°C, and the minimum is 18°C. Assume that the diurnal temperature wave is symmetric, that the mean temperature in the profile is equal to the average of the temperatures at 0800 and 2000 hours, and that the damping depth is 10 cm. Calculate the soil temperatures at 1400 and 0200 h for soil depths of 0 and 50 cm.

We begin by writing equation [41] as

$$T(z, t) = T_a + A_0 e^{-z/d} \sin(\omega t - \omega t_0 - z/d)$$

where $T_a = (32 + 18)/2 = 25°C$, $A_0 = 32 - 25 = 7°C$, $d = 10$ cm, $\omega = 2\pi/24$ h, and $t_0 = 8$ h. Note that we can write $\omega t - \omega t_0$ as $\omega(t - t_0)$. Therefore, 1400 h – 0800 h = 0600 h, and 0200 h – 0800 h = –600 h.

At $z = 0$ and time = 1400, the soil temperature is

$$T(0, 6) = 25 + (7)(e^0)\sin[(2\pi/24)(6 - 0)] = 32°C$$

At $z = 0$ and time = 0200, the soil temperature is

$$T(0, 18) = 25 + (7)(e^0)\sin[(2\pi/24)(-6 - 0)] = 18°C$$

At $z = 50$ and time = 1400, the soil temperature is

$$T(50, 1400) = 25 + (7)(e^{-5})\sin[(2\pi/24)(6 - 5)] = 25.0°C$$

At $z = 50$ and time = 200, the soil temperature is

$$T(50, 2000) = 25 + (7)(e^{-5})\sin[(2\pi/24)(-6 - 5)] = 25°C$$

These calculations show that soil temperatures vary with depth in the profile and are damped with depth. At the lower depths the temperatures are not as variable on a daily basis.

The annual fluctuations of soil temperature penetrate much deeper than the daily fluctuations. The mean temperature at about 2 m depth is approximately the same as the annual mean of the temperature of the air above the soil surface.

Experimental Estimation of Soil Temperature

The temperature of a soil cannot be directly measured. However, there are several types of instruments used to estimate the influence of temperature on some property of a body that responds to changes in its heat content. These instruments include the liquid-in-glass thermometer, bimetallic thermometer, resistance thermometer, thermocouple, and remote-sensing thermometer.

Liquid-in-Glass Thermometer

Probably the most common instrument for measuring soil temperature is the liquid-in-glass thermometer. This type of thermometer actually responds to the volume changes resulting from variations in the temperature of the liquid reservoir. A good contact is required between the reservoir and the soil of which the temperature is to be measured. Therefore, the resistance for the transfer of heat between the reservoir and the soil must be low. At the same time, the exchange of heat between the reservoir and other objects must be minimized.

When a liquid-in-glass thermometer is placed in close contact with soil or immersed in a liquid, the conduction of heat between the thermometer bulk and its surroundings is usually sufficient for ensuring a reliable reading of the temperature of the medium near the reservoir. However, small errors can arise from heat conduction along the stem of the thermometer and from the fact that part of the liquid may be at a temperature different from that of the bulb. Soil thermometers should not be placed in a metal casing. However, radiation errors are usually negligible when a thermometer is placed in the soil.

When liquid-in-glass thermometers are placed in air, ventilation is generally needed to ensure sufficient thermal contact between the thermometer and its surroundings. In addition, the thermometer should be shielded against solar and terrestrial radiation, because air is largely transparent for both. A simple and cheap way of ventilating is by swirling the thermometer.

Alcohol and toluene, besides the familiar mercury, are used as thermometric liquids, particularly in thermometers that measure only the maximum and minimum temperatures. The degree of accuracy attained with liquid-in-glass thermometers is generally a few tenths of a degree Celsius.

Bimetallic Thermometer

The sensitive element in bimetallic thermometers consists of two strips of metals that have different thermal expansion coefficients and are firmly welded together. When the temperature changes, the strips are deformed and this deformation is used to move the pointer over a temperature scale. Linear or slightly curved strips and coils are commonly used. The degree of accuracy attainable is also a few tenths of a degree Celsius. Bimetallic thermometers are easily made for recording and are frequently used in thermographs.

Resistance Thermometer

Electrical methods are based on the change of resistance with temperature or on the thermoelectric effect. The advantages for specialized work are that electrical methods can be used at a distance for reading and recording, and they can measure rapid changes in temperature. The disadvantage is that resistance thermometers are not easily standardized and require frequent recalibration.

Some resistance thermometers consist of a thin wire that is mostly platinum or nickel and is often spiraled on a cylinder. A bridge circuit is generally used for measuring the resistance. The resistance of platinum and of other pure metals increases by 0.4–0.5% of its value for each degree Celsius rise in temperature at ordinary temperatures.

A special type of resistance thermometer is the thermistor. This is a semiconductor that changes its resistance, R, with temperature. The simplest equation is

$$R = B\exp(a/T) = \exp[a(1/T - 1/T_0)] \qquad [43]$$

where a and B are positive constants, T is the absolute temperature, and T_0 is the standard reference temperature. This shows that the resistance of a thermistor decreases exponentially with increasing temperature. The temperature coefficient can be written as

$$(1/R)(dR/dT) = a/T^2 \tag{44}$$

Thermistors can be made in various shapes, such as rods, plates, and small beads. Because they are semiconductors, their resistance is comparatively high. The resistance of a thermistor at a given T is more liable to change with time than the resistance of a pure metal. Therefore, more frequent calibration is required.

Thermocouple

The thermocouple is one of the most widely used instruments for measuring soil temperature. When two wires of different metals are welded together at two points and the welds are kept at different temperatures, an electrical potential difference between the welds is established, and current flows through the circuit formed by the two wires. This effect is called the thermoelectric, or Seebeck, effect. When this effect is used for temperature measurements, one of the welds is usually kept at a reference temperature while the other is placed in contact with the object of study.

The thermoelectric potential difference, which is roughly proportional to the temperature difference between the two welds, is measured by a compensation method, or the thermoelectric current is measured with a galvanometer placed in the circuit. Combinations of metals frequently used at ordinary temperatures include (1) copper-constantan, which produces approximately 40 μV/°C difference, (2) iron-constantan, which produces approximately 50 μV/°C difference, and (3) chromel-constantan, which produces approximately 60 μV/°C difference. Constantan refers to a family of copper-nickel alloys containing anywhere from 45 to 60% copper. These alloys also typically contain small percentages of manganese and iron along with other trace impurities such as carbon, magnesium, and silicon. Chromel is a nickel-chromium alloy usually containing 90% nickel.

In many applications, the ends of the two wires constituting a thermocouple are connected directly to a potentiometer. The temperature of the potentiometer serves as the reference temperature. Most modern instruments contain built-in electrical compensation for changes in the temperature of the instrument.

Remote-Sensing Thermometer

In recent years one of the most popular methods of estimating the temperature of the surfaces of soil, plant leaves, and crop canopies is infrared thermometry. These sensors measure the infrared radiation emitted by the surface according to the Stefan-Boltzmann equation

$$R_l = e\sigma T^4 \tag{45}$$

where R_l is the longwave radiation in the band of interest, e is the emissivity (or the reflectance), σ is the Stefan-Boltzmann constant (5.675×10^{-8} W/m^2·K^4), and T is the absolute temperature. For most soil and plant surfaces, the emissivity is close to 1.0 so that no correction is made. For highly reflective surfaces, however, corrections of the emissivity should be made.

The sensor is sensitive to a given band of infrared radiation, usually between 8 and 10 μm. The interference from air containing a low percentage of water vapor in a path of several hundred meters is minimal; however, in air with a high water vapor content, interference can become a problem at distances greater than 10 m.

Cited References

Bristow, K. L. 1998. Measurement of thermal properties and water content of unsaturated sandy soil using dual-probe heat-pulse probes. *Agricultural and Forest Meteorology* 89:75–84.

Campbell, G. S. 1985. *Soil Physics with Basic-Transport Models for Soil-Plant Systems.* Elsevier. New York.

Erstein, P. D. 1979. Heat transfer in Crowley silt loam. M.S. thesis. Department of Agronomy, University of Arkansas, Fayetteville.

Guymon, B. L. 1994. *Unsaturated Zone Hydrology.* Englewood Cliffs, NJ: Prentice-Hall.

Hillel, D. 1982. *Introduction to Soil Physics.* Academic Press. New York.

Kohnke, H. 1968. *Soil Physics.* McGraw-Hill. New York.

McInnes, K. J. 1981. Thermal conductivities of soils from dryland wheat regions of eastern Washington. M.S. thesis. Washington State University, Pullman.

Richter, J. 1987. *The Soil as a Reactor: Modelling Processes in the Soil.* Catena Verlag. Cremlingen, Germany.

Soil Science Society of America. 1987. *Glossary of Soil Science Terms.* Madison.

van Wijk, W. R., and D. A. DeVries. 1963. Periodic temperature variations in a homogeneous soil. In W. R. van Wijk and D. A. DeVries (eds.), *Physics of the Plant Environment.* North-Holland. Amsterdam.

Additional References

Hanks, R. J. 1992. *Applied Soil Physics: Soil Water and Temperature Applications.* 2d ed. Springer-Verlag. New York.

Hanks, R. J., and G. L. Ashcroft. 1980. *Applied Soil Physics.* Springer-Verlag. New York.

Horton, R., and S. Chung. 1991. Soil heat flow. In J. Hanks and J. T. Ritchie (eds.), *Modeling Plant and Soil Systems,* pp. 397–438. Agronomy Monograph 31. American Society of Agronomy. Madison.

Kirkham, D., and W. L. Powers. 1972. *Advanced Soil Physics.* Wiley-Interscience. New York.

Koorevaar, P., G. Menelik, and C. Dirksen. 1983. *Elements of Soil Physics.* Elsevier. New York.

Marshall, T. J., and J. W. Holmes. 1979. *Soil Physics.* Cambridge University Press. New York.

Parikh, R. J., J. A. Havens, and H. D. Scott. 1979. Thermal diffusivity and conductivity of moist porous media. *Soil Science Society of America Journal* 43:1050–1052.

Rosenberg, N. J., B. L. Blad, and S. B. Verma. 1983. *Microclimate: The Biological Environment.* John Wiley and Sons. New York.

Problems

1. A soil at 25°C has a volumetric water content of 0.3 m³/m³ and a bulk density of 1.42 Mg/m³. Find the amount of heat released by a unit area of soil when the temperature is lowered to 20°C. Assume that the depth of the soil profile is 0.75 m and that the heat is evenly distributed in the profile.

2. The heat flux density, H_s, across the soil surface into a moist soil having a water content of 0.25 kg/kg and a bulk density of 1.42 Mg/m³ was equal to 2.09 MJ/m²/d. Find the average temperature increase that occurs in 1 d in the top 1.2 m of the soil profile assuming that the heat is evenly distributed in the profile.

3. A soil profile has an average temperature of 20°C, an average water content of 0.25 m³/m³, and an average bulk density of 1.4 Mg/m³. Find the amount of heat required to raise the temperature of a unit area of soil to 25°C. Assume that the depth of the profile is 2.0 m.

4. Calculate the amount of energy required to evaporate 1 cm of water in 1 ha if the soil temperature is 50°C. Make the same calculations assuming the soil temperature is 25°C.

5. A soil profile was monitored for its temperature variation during the day. The results are presented in the table below. Find the quantity of heat stored in the soil profile per unit surface area for the period from 0600 to 1500.

Soil Depth (cm)	Soil Temperature (°C) 0600	Soil Temperature (°C) 1500	Bulk Density (Mg/m³)	Water Content (m³/m³)
0–5	18	35	1.21	0.09
5–20	21	31	1.45	0.14
20–45	23	29	1.56	0.16
45–60	24	26	1.50	0.18
60–100	25	25	1.45	0.25

6. Draw a graph of the temperature of the soil surface, $T(0, t)$, as a function of time if T_a is 20°C, A_0 is 10°C, and $\omega = 2\pi/24$ radians per hour ($t_p = 24$ h). Assume that the average temperature occurs at 9:00 A.M.

7. A soil has a thermal conductivity of 0.62 J/m·s·K and a thermal heat capacity of 1.67 MJ/m³·K. Calculate the damping depth for a diurnal and for an annual temperature fluctuation.

8. Using the information in problem 7, calculate the amplitude of the daily temperature fluctuation at 0.3 m depth if the amplitude at the soil surface is 12°C.

9. What is the phase difference between the temperature fluctuation at the soil surface and the damping depth in problem 8?

10. Explain why heat flow is slower in soil than in the atmosphere.

11. Contrast the effects on soil temperature of
    a. clear plastic surface cover
    b. black polyethylene mulch
    c. south-facing ridges
    d. snow

12. A soil column contains 0.4 m of fine sand over 0.3 m of sandy loam. If the thermal conductivity is 0.2 and 0.12 J/m·s·K in the sand and loam soils, respectively, calculate the steady-state heat flux density through the column assuming that the temperature is 35°C at the soil surface and 15°C at the bottom of the column. What is the temperature at the boundary between the two layers of soil? What would the heat flux density and temperature at the boundary be if the thermal conductivity of the first layer was twice as great?

CHAPTER SEVEN

Soil Aeration

Introduction

Soil aeration is the process by which gases consumed or produced in the soil are exchanged for gases in the atmosphere. From an agricultural point of view, it is one of the most important determinants of soil productivity. Soil aeration is a vital process because it controls the soil status of two life-sustaining processes: respiration and photosynthesis. Plant roots absorb O_2 and release CO_2 during respiration. Similarly, soil microorganisms also respire and under certain conditions might compete for O_2 with roots of higher plants.

Respiration involves the oxidation of organic compounds and can be represented by the general reaction in which glucose is oxidized:

$$C_6H_{12}O_6 + 6O_2 \rightarrow 6CO_2 + 6H_2O \qquad [1]$$

This reaction shows that carbon dioxide is formed and released to the soil atmosphere by respiring organisms. For respiration to continue in soil, however, transport is involved in supplying O_2 and removing CO_2 from the system. In photosynthesis, the reaction is reversed.

Atmospheric oxygen is essential to all forms of life. The atmosphere is a source of oxygen and constantly supplies O_2 to the soil, which acts as a sink. Oxygen in soil supports the various organisms and roots living in the soil system because of the low rates of diffusion of O_2 through water films. In the cases of CO_2 and water vapor, the soil is the source and the atmosphere is the sink.

In the field, restricted soil aeration results from excess water, which may be caused by poor drainage (high soil water content resulting from shallow water tables, perched water tables, and/or ponded water on the soil surface); by waterlogging due to small pores, mechanical compaction of the soil, or a high content of swelling clays; or by organic-matter decomposition by soil microorganisms that use oxygen in soils with low rates of replenishment. Waterlogged conditions typically occur in wetlands and low, depressional areas or areas where water is ponded for some time. Plants differ in their sensitivity to poor aeration.

In this chapter we examine the principles associated with the presence and transport of gases in soil. We will emphasize the transport mechanisms, the concentration distributions of gases in the soil profile, and methods of characterization of soil aeration.

Gaseous Composition of the Cosmos and Soil Atmosphere

The chemical composition and mobility of elements in the cosmos, atmosphere, and soil vary. We begin by examining the elemental compositions of various media.

Gaseous Composition of the Cosmos

Oxygen accounts for about 0.1% of the cosmic atoms and ranks third among the elements in abundance. The predominant element in the cosmos is hydrogen; together with helium, it composes 99.86% of the total. The elements H, He, O, C, and N make up 99.99% of all atoms (Stolzy 1974).

Table 7.1. Volume Percentage of CO_2 and O_2 in the Soil Gaseous Phase at Two Sampling Times and Five Depths

Sample at	Winter		Summer	
Soil Depth (cm)	CO_2	O_2	CO_2	O_2
30	1.2	19.4	2.0	19.8
61	2.4	11.6	3.1	19.1
91	6.6	3.5	5.2	17.5
122	9.6	0.7	9.1	14.5
152	10.4	2.4	11.7	12.4

Source: Adapted from Wild 1993.

Oxygen is the most abundant atom in the earth's crust (53.77%), primarily because it combines with other elements, such as silicon. Only about 0.01% of all oxygen is molecular oxygen in the atmosphere. A large exchange of O_2 between the hydrosphere and the atmosphere occurs due to the exchange of gaseous oxygen in the atmosphere with dissolved oxygen in the ocean. Photosynthesis in the biosphere results in an atmospheric turnover of O_2 about every 5400 years (Stolzy 1974).

The most important source of O_2 for land organisms is the atmosphere, where the concentration is buffered at about 21% because of the great volume and mobility of the air medium. Therefore, most of the land environment is provided with a uniform and adequate supply of oxygen. At high altitudes, the percentage of oxygen relative to other gases remains about 21%, but as atmospheric pressure decreases, the volume concentration of oxygen in air decreases, and the concentration gradient across membranes is reduced. Therefore, oxygen supply to respiring cells is limiting at higher altitudes.

Gaseous Composition of the Soil Atmosphere

The chemical composition of the soil atmosphere depends on conditions such as water content, temperature, porosity, microbial activity, and structure. Generalizations are difficult because both plant roots and soil microorganisms consume oxygen and produce carbon dioxide. A comparison of the compositions of the soil atmosphere in a poorly drained silt loam is given in Table 7.1.

These data show that soil O_2 and CO_2 concentrations are inversely related. Concentrations of CO_2 increase with depth while O_2 concentrations decrease with depth. The low concentrations of O_2 at the deeper depths during the winter can be attributed to higher soil water content, which restricts the transport by diffusion from the atmosphere.

In general, an increase in CO_2 concentration in soil is associated with a decrease in O_2 concentration. The higher concentration of CO_2 in the soil results from the respiration of microorganisms and plant roots, during which O_2 is consumed and CO_2 is released. Respiration provides energy for various metabolic processes. The actual composition of soil gases depends on temporal and spatial conditions such as season of the year, soil, crop and tillage, especially as related to the activity of soil microorganisms and plant roots, and depth in the profile. Therefore, there is considerable variation in gas composition within pores of different soils.

Trace gases, such as methane, nitrous oxide, hydrogen sulfide, and ammonia, are also present in soil. These compounds are chemically and metabolically formed and can be transformed into organics that can be absorbed by roots. Gases commonly found in swamps, soils cropped to rice, and waterlogged paddy soils are CH_4 (methane) and small amounts of CO_2, H_2, and N_2. Methane is a typical gas produced by the anaerobic fermentation of organic matter under neutral or slightly alkaline conditions.

Soil structural units largely determine how air moves into the soil profile. The pore space in soil can be filled with gases when the soil is dry, or with water when the soil is saturated or waterlogged. Soils that are poorly drained and slowly permeable and have shallow water tables can be expected to have poor aeration at some time during the year. The spatial extent of soils with these characteristics is extensive in many fine-textural fluvial flooded plains and deltas.

Soil gases can exist in three states: (1) a free state, filling the empty soil pores or voids, (2) dissolved in the water phase, and (3) adsorbed on the solid phase. The quantity of a gas dissolved in the aqueous phase depends on the type of gas, temperature, salt concentration in the aqueous phase, and the concentration of that gas in the gaseous phase. Gases such as CO_2, NH_3, and H_2S become ionized in water and are more soluble in water than the more abundant gases of O_2 and N_2. Equilibrium conditions among water, water vapor, oxygen, and carbon dioxide in pore spaces are greatly affected by diurnal changes in temperature.

Volume Fraction of Soil Air

From a physical point of view, soil can be thought of as a three-phase system: solid, liquid, and gas. Parameters such as bulk density, water content, and gas content are expressed in terms of mass and volume ratios of the three phases.

Since the solid phase is essentially constant over time, the fractions of water and air are dynamic, and their ratios on a volume basis are inversely related. Two equations used to calculate the volume fraction of soil air are

$$f_a = f - \theta_v \qquad [2]$$

and

$$f_a = 1 - \theta_v - S \qquad [3]$$

where f_a is the volume fraction of air (aeration porosity), f is the total porosity, θ_v is the volumetric water content, and S is the volume fraction of the solid phase (ρ_b/ρ_p).

Computations of Gas Concentration

The composition of soil air is generally reported in terms of the percentage of the total volume that would be occupied by each gas in its pure state, called the *partial pressure* of that gas. The *ideal gas law* can be used to convert gas partial pressure to gas concentration. On a mass basis,

$$C = (M/RT)p \qquad [4]$$

where p is the partial pressure of the gas (Pa), M is the molar mass (kg/mol), R is the gas constant (8.314 J/molK), T is the absolute temperature (K), and C is the gas concentration (kg/m^3 of soil air).

The partial pressure of a gas is the volume fraction (0.21 for O_2; 0.79 for N_2) multiplied by atmospheric pressure (1.013 × 10^5 Pa at sea level). It is the pressure that a gas would exert if it alone were present in the volume occupied by the mixture. Two useful conversions to remember are J = N·m and Pa = N/m^2. Thus, Pa = J/m^3.

Example

Determine the gaseous concentrations of O_2, N_2, and CO_2 at standard temperature (25°C) and pressure (101.3 kPa).

$$C(O_2) = [0.032/(8.314 \times 298)](0.21 \times 1.013 \times 10^5)$$
$$= 0.275 \text{ kg/m}^3$$
$$C(N_2) = [0.028/(8.314 \times 298)](0.79 \times 1.013 \times 10^5)$$
$$= 0.904 \text{ kg/m}^3$$
$$C(CO_2) = [0.044/(8.314 \times 298)](0.0003 \times 1.013 \times 10^5)$$
$$= 0.00054 \text{ kg/m}^3$$

On a molar basis, equation [4] becomes

$$C = (1/RT)p \tag{5}$$

where the units of C are in moles per cubic meter.

If we assume that the amount of gas adsorbed by the solid phase is negligible, then the gas concentration in the soil, C, is related to the concentrations of the gas in the soil air, C_a, and in the soil solution, C_l. Mathematically, this can be written as

$$C = C_a(f_a + a_b\theta_v) \tag{6}$$

where f_a is the aeration porosity (m^3/m^3), a_b is known as Bunsen's coefficient of solubility (m^3/m^3), and θ_v is the volumetric soil water content (m^3/m^3). *Bunsen's solubility coefficient* is the solubility of the gas per unit volume of water at the given temperature and normal pressure (101.3 Pa). In soil, it is the ratio of the solubility of the gas in water to the solubility coefficient in air:

$$a_b = a_w/a_a \tag{7}$$

where a_w is the solubility coefficient in water (mol/m^3·Pa) and a_a is the solubility coefficient in air (mol/m^3·Pa). Values of a_b for O_2 and CO_2 decrease with increases in temperature and are given in Table 7.2.

The solubility coefficient of the gas in water is determined from *Henry's law* as

$$a_w = C_w/p \tag{8}$$

where C_w is the concentration of the gas in water and p is the partial pressure. For air, the relationship is similar and is given as

$$a_a = C_a/p \tag{9}$$

Table 7.2. Solubility Coefficients and Bunsen's Coefficients of O_2 and CO_2 in Air and Water at Different Temperatures

| Temp. (°C) | Solubility Coefficient | | | | Bunsen's Coefficient, a_b | |
| | Air ($\times\ 10^{-2}$ g/m³·PA) | | Water ($\times\ 10^{-4}$ g/m³·PA) | | | |
	O_2	CO_2	O_2	CO_2	O_2	CO_2
0	1.410	1.95	6.9	334	0.0489	1.713
5	1.385	1.92	6.0	278	0.0436	1.450
10	1.360	1.88	5.4	233	0.0394	1.238
15	1.337	1.85	4.8	199	0.0360	1.075
20	1.314	1.82	4.4	171	0.0333	0.942
25	1.292	1.79	4.0	148	0.0309	0.829
30	1.270	1.76	3.7	130	0.0290	0.738

Source: Adapted from Glinski and Stepniewski 1985, pp. 46–47.

Application of equation [6] to soils shows that the contribution of the aqueous phase to the concentration of oxygen in soil is low. Therefore, changes in the term f_aC_a are closely related to the soil gas concentration.

Oxygen Concentrations in Soil Aggregates

Soil oxygen concentration varies with soil physical factors such as pore size, depth in the profile, and size of the aggregate, as well as with biological activity. After wetting by rain or irrigation, the interiors of soil aggregates may become anaerobic. The spatial concentration distributions of oxygen in spherical aggregates of a silty clay loam and a silt loam from Iowa were measured by Sextone et al. (1985). Steep oxygen gradients were found in the aggregates, with the highest concentration at the exterior of the aggregate and lowest in the interior of the aggregate (Figure 7.1). Different oxygen gradients were found within and among the aggregates; aggregates larger than 20 mm diameter often had anaerobic interiors.

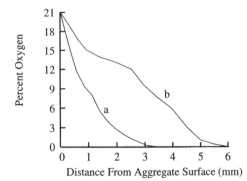

Figure 7.1. Oxygen concentrations as a function of distance from two aggregate surfaces. Aggregates a and b were from silt loam soils from a cultivated field and had radii of 7 and 8 mm, respectively. (Redrawn from Sextone et al. 1985.)

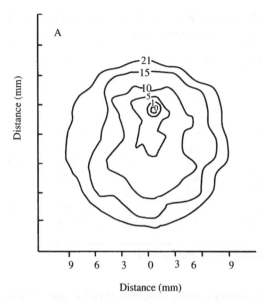

Figure 7.2. Contour map of the oxygen concentrations within an aggregate from a cultivated field. (Redrawn from Sextone et al. 1985.)

Contour plots showed that oxygen distributions within the aggregates were asymmetric and zero oxygen concentration did not always occur at the center of the aggregate, which suggested a nonuniform distribution of sites of oxygen consumption (Figure 7.2). Sextone et al. (1985) attributed this to a nonuniform distribution of organic C within the aggregate. They concluded that in general, individual aggregates had to be wet and greater than 9 mm in radius to contain significant anaerobic centers. However, one aggregate of only 8 mm in diameter also had an anaerobic zone. Every soil aggregate involved in denitrification also had an anaerobic zone, that is, a zone where the oxygen concentration was negligible. Of the 57 aggregates examined whose radii ranged from 3 to 12 mm, 12 aggregates exhibited at least one profile showing complete anaerobiosis. Sextone et al.'s results are particularly applicable in poorly structured clayey soils, especially subsoils, that have large peds (50–100 mm) and in structured soils that have poor aeration porosity and are temporarily anaerobic after wetting.

Mechanisms of Soil Gas Exchange

Soil aeration involves the exchange of gases between the soil and the atmosphere. The upper surface of the soil is in contact with the atmosphere, and the soil's lower boundary may be in contact with a water table or some impervious material. Therefore, the exchange of gas involves the two mass transport mechanisms of mass flow due to total pressure differences and diffusion due to partial pressure differences.

Convective Flow

There are four conditions to be considered in evaluating convective flow of gases in soil:

1. Changes in soil temperature: Early ideas were that differences in temperature of soil horizons and also temperature differences between the soil and atmosphere can contract or expand the volume of soil air and, thus, create air convection currents. Calculations indicate that although diurnal soil temperature changes can affect the volume of soil air in the surface horizon, they have little significance for the whole profile.

2. Changes in the volume of air space due to changes in soil water content: During rainfall and irrigation, water infiltrates the soil surface and forces out an equivalent volume of air. During evaporation and drainage, the volume of water lost is replaced by atmospheric air. Thus, the movement of water causes air to move into or out of the profile, depending upon whether the soil is gaining or losing water.

3. Changes in barometric pressure in the atmosphere: As the barometric pressure increases, air is forced into the profile. As barometric pressure decreases, air is removed from the soil. However, diurnal changes in barometric pressure are usually small.

4. Wind gusts over the soil surface: When wind blows over irregular surfaces, eddies are formed. The general effect is to create an increased pressure on the windward side of an obstruction and reduced pressure on the leeward side.

In general, convective flow of soil air results in a mass of air moving with a zone of higher *total* gas pressure to a zone of lower *total* gas pressure.

Convective flow of air is similar to the convective flow of water in that the flow of both fluids is proportional to a pressure gradient. The process is dissimilar in that water is relatively incompressible in comparison to air, which is highly compressible. Thus, air density and viscosity are strongly dependent on pressure and temperature. Water also has greater affinity to the surfaces of mineral particles.

In one dimension (z), the volume convective flux density of air is

$$J_v = -(k/\eta)\partial P/\partial z = -K\partial P/\partial z \qquad [10]$$

where J_v is the volume convective flux density of air (m^3/m^2s or m/s), k is the intrinsic permeability of the air-filled pore space (m^2), η is the viscosity of soil air (Pa·s), P is the total air pressure (Pa), and K is the air conductivity (m^2/Pa·s). The quantity $\partial P/\partial z$ is the total air pressure gradient (Pa/m). Assuming the total pressure gradient is constant, equation [10] indicates that increases in K result in increases in the flux density of air.

The intrinsic permeability, k, depends only on the geometry of the soil pores and not on the gas. Geometry factors include volume fraction, size, and continuity of the pores. The viscosity of water is 50 times greater than air:

viscosity of air ~ 20 µPas
viscosity of water ~ 1.0 mPas

Air conductivity, K, has the same units as water conductivity expressed on a volume basis (m^2/Pa·s) and depends on the aeration porosity, size, continuity, and geometry of gas-filled soil pores. It is highly dependent on the volumetric water content of the soil.

If gas flux density is expressed in terms of mass/(area $\times$ time), the convective transport equation is

$$J_m = -(Ck/\eta)\partial P/\partial z \qquad [11]$$

where J_m is the mass convective flux (kg/m^2·s), and C is the concentration of soil air (kg/m^3). If we assume that soil air behaves as an ideal gas, then the ideal gas law is

$$PV = nRT \qquad\qquad [12]$$

where P is the total pressure, V is the volume, n is the number of moles of gas, R is the universal gas constant per mole (8.314 J/molK), and T is the absolute temperature. Since the gas concentration can be written as

$$C = m/V \qquad\qquad [13]$$

and m is the mass that is equal to nM, where M is the molar mass, from equation [4] we have the gaseous concentration as related to the partial pressure, p.

In three dimensions, the volume convective flux density of air can be written as

$$J_v = -(k/\eta)\nabla P \qquad\qquad [14]$$

where ∇P is the three-dimensional gradient of soil total air pressure (Pa) (Hillel 1982), and k is assumed to be isotropic.

These equations show that mass flow of gas results from a gradient of total gas pressure. Although all four factors causing convective flow of gas in soil have an effect, the result is that convective flow of gases in soils is not continuous or usually of sufficient magnitude to overshadow transport by diffusion, which is believed to be the principal transport mechanism for gaseous transfer between the soil and the atmosphere.

Diffusion

The rate of diffusion of gas, either from the soil to the atmosphere or from the atmosphere to the soil, depends on both soil and gas properties. Particularly important properties include (1) the difference in gas concentration between two regions, (2) the length of the diffusion path, (3) the diffusion coefficient of the gas, and (4) the shape and continuity of the air-filled pores.

Transport of gas by diffusion occurs as a result of gradients of *partial* pressure (or concentrations of the various constituents in soil air). These gradients are a result of the respiration processes of roots and microorganisms and are a direct consequence of the random thermal motion of molecules, which causes the gaseous molecules to migrate from a zone of higher to a zone of lower concentration. Since soil usually contains more CO_2 and less O_2 than the atmosphere, diffusion moves CO_2 out of the soil into the atmosphere and O_2 from the atmosphere into the soil. Diffusive transport of gas can occur even though the total gas pressure may remain the same.

Diffusion of gases such as O_2 and CO_2 in soil occurs partly in the gaseous phase and partly in the liquid phase. For steady-state conditions, the diffusive transport can be described by Fick's law and in one direction is governed by

$$J_d = -D\partial C_a/\partial z \qquad\qquad [15]$$

where J_d is the diffusive flux density (kg/m^2·s), D is the apparent gas diffusion coefficient in soil (m^2/s), C_a is the gas concentration (kg/m^3 of soil air), and z is the distance (m). If partial pressure, p, is used instead of concentration of the diffusing component, equation [15] becomes

$$J_d = -(D/b)\partial p/\partial z \tag{16}$$

where b is the ratio of the partial pressure to the concentration ($Pa \cdot m^3/kg$). The quantity D/b is the inverse of the diffusion impedance and may be designated as the diffusion constant K. Diffusion of gases in soil due to a partial pressure gradient of one component is always accompanied by a counterdiffusion of other gas components.

The Gas Balance Equation

The law of conservation of mass governs the fate and transport of gases in soil. If we consider the unit volume of soil as the system, the conservation equation can be derived in the same manner as with other substances. We will derive the governing equation using the incremental approach in one dimension and with O_2 as the molecule of interest.

The change in storage of O_2 (kg) in the volume element is the difference in the inflow and outflow at the boundaries:

$$J\Delta x\Delta y\Delta t - (J\Delta x\Delta y\Delta t + \Delta J\Delta x\Delta y\Delta t) = -\Delta J\Delta x\Delta y\Delta t$$

This change in storage is equivalent to the change in concentration of O_2 in the volume element, that is, ΔC. Therefore, we write

$$\Delta C\Delta x\Delta y\Delta z = -\Delta J\Delta x\Delta y\Delta t$$

which upon rearrangement and canceling terms becomes

$$\Delta C/\Delta t = -\Delta J/\Delta z$$

which in differential form becomes

$$\partial C/\partial t = -\partial J/\partial z \tag{17}$$

Equation [17], which is the continuity equation for O_2, was derived assuming that there was no consumption or generation of O_2 in the volume element. This assumption is not valid for most field conditions. In order to account for generation and consumption in the volume element, we add the following term

$$\partial C/\partial t = -\partial J/\partial z \pm S(z,t) \tag{18}$$

where $S(z,t)$ is positive for generation processes (such as production of CO_2) and negative for consumption processes (such as absorption of O_2). A positive sign for S represents a source such as the addition of CO_2 from respiration; a negative sign for S indicates a sink such as the extraction of O_2 by roots and microbes. The units of S are the same as those of $\partial C/\partial t$ and $\partial J/\partial z$ ($kg/m^3 \cdot s$). It is well known that C, D, and S vary with space and time even in homogeneous soil.

Under field conditions, mass flow of soil air is laminar and transient; therefore, substituting equation [10] into equation [17] gives the transient-state form for laminar convective flow of a gas in soil:

$$(M/RT)\partial P/\partial t = (\partial/\partial z)(Ck/\eta)(\partial P/\partial z) \tag{19}$$

Although unlikely, if the ratio Ck/η is constant with depth in the soil profile, equation [19] becomes

$$\partial P/\partial t = a\partial^2 P/\partial z^2 \qquad [20]$$

where the parameter a is $(RTkC/M\eta)$. This equation has the form of Fick's transient-state diffusion equation.

To account for transient diffusion conditions in the field, the equation of continuity is introduced. With generation or consumption, we have

$$\partial C/\partial t = -\partial J/\partial z \pm S(z, t) \qquad [21]$$

where $S(z, t)$ is the source or sink term describing the rate of production or absorption of the gas in the soil $(mg/m^3 \cdot s)$. Substituting equation 15 into equation

$$\partial C/\partial t = (\partial/\partial z)(D\partial C/\partial z) \pm S(z, t) \qquad [22]$$

If D is known to be constant with respect to depth, equation [22] simplifies to

$$\partial C/\partial t = D\partial^2 C/\partial z^2 \pm S(z, t) \qquad [23]$$

which can be solved analytically for several initial and boundary conditions to give the gas concentration as a function of depth in the profile and time. This will be illustrated later.

Example
Assume that a soil is consuming 7 $g/m^2 \cdot d$ of oxygen and that this extraction occurs only in the surface 25 cm of soil. The soil volume is composed of 20% air, and this air contains 20% oxygen. How much oxygen (g) is contained in the top 25 cm of soil? Assuming no replacement, how many days will this stored amount of oxygen last if the consumption rate is constant?

rate of oxygen extraction = S = 7 g oxygen/$m^2 \cdot d \times (1/0.25 \text{ m})$
      = 28 $g/m^3 \cdot d$
volume of soil = 0.25 m $\times$ 1 m^2 = 0.25 m^3
volume of soil O_2 = 0.25 $m^3 \times 0.20$ $m^3/m^3 \times 0.2 = 0.01$ m^3
mass of soil O_2 = 0.01 $m^3 \times 1000$ L/$m^3 \times 32$ g/22.4 L
      = 14.3 g of O_2 in layer

Note: There are 32 g/mole of O_2 and 1 mole of O_2 occupies 22.4 L.

density of soil O_2 = 14.3 g/0.25 m^3 = 57.1 g/m^3
available time = $(57.1 \text{ g}/m^3)/(28 \text{ g}/m^3 \cdot d)$ = 2 d

This calculation shows that there is only a 2-day supply of O_2 in the profile. This is why continuous flooding for 2 days is the usual rule of thumb (i.e., the point of no return) for crop flood damage. Therefore, the importance of the exchange of O_2 between the atmosphere and the soil cannot be overemphasized. Also, the calculation suggests that the more rapid the extraction of O_2, the greater the need for the exchange between the atmosphere and the soil.

Table 7.3. Molecular Diffusion Coefficients of O_2 and CO_2
as a Function of Temperature

| Absolute Temperature (K) | Diffusion Coefficient (m²/s) | | | |
| | Air (D × 10⁻⁵) | | Water (D × 10⁻⁹) | |
	O_2	CO_2	O_2	CO_2
273	1.78	1.38	0.99	1.15
278	1.84	1.43	1.27	1.30
283	1.89	1.48	1.54	1.46
288	1.95	1.53	1.82	1.63
293	2.01	1.59	2.10	1.77
298	2.07	1.64	2.38	1.92
303	2.14	1.70	2.67	2.08

Source: Adapted from Glinski and Stepniewski 1985, pp. 46–47.

Diffusion of Gases in Soil

The diffusion of gases in soil depends on the magnitude of the *molecular diffusion coefficient* and concentration (or partial pressure) gradient. It is of interest to examine those soil factors that affect the molecular diffusion coefficient (i.e., the gas transport coefficient).

Diffusion Coefficient

The magnitude of D (m²/s) depends upon properties of both the gas and the medium and is affected by temperature and pressure (Table 7.3). The magnitude of D increases with temperature and is roughly 10,000 times greater in soil air than in soil water.

The effects of both temperature and pressure on the molecular diffusion coefficient have been expressed by

$$D(T, P) = D_{o(NTP)}(T/T^o)^N(P^o/P) \tag{24}$$

where $D_{o(NTP)}$ is the diffusion coefficient in the pure phase at standard temperature and pressure, T is the absolute temperature (K), P is the pressure (kPa), and the superscript o indicates temperature or pressure at standard temperature and pressure (STP; 273.16 K and 101.3 kPa). The exponent N has a value of 2 for O_2 and 1.75 for CO_2. The diffusivity increases at a rate of about 0.7% per kelvin with temperature and decreases at a rate of about 1% per kilopascal with increasing atmospheric pressure.

The diffusion coefficient in soil, D, is related to the diffusion coefficient in air, D_a, by

$$D = f(D_a)$$

One equation often used to approximate the magnitude of D relates the ratio D/D_a to the aeration porosity:

$$D/D_a = -0.12 + 0.9f_a \tag{25}$$

The ratio D/D_a is known as the *relative diffusion coefficient*. The advantage of using the ratio is that the effects of state variables such as temperature, pressure, and type of gas are canceled and only soil factors are important. Equation [25] shows that D/D_a is linearly related to the aeration porosity and is zero at $f_a = 0.133$ m^3/m^3. This suggests that soils with low f_a values have restricted aeration.

Since O_2 and CO_2 can diffuse in both the aqueous and gaseous phases, a diffusion constant K can be defined that separates the contributions from the two phases of the soil. The diffusion constant in air is

$$K_a = a_a D_a \qquad [26]$$

and the diffusion constant in water, K_w, is

$$K_w = a_w D_w \qquad [27]$$

where D_a and D_w are the diffusion coefficients in air and water, respectively, and a_a and a_w are the solubility coefficients in air and water, respectively. The diffusion coefficient is used in Fick's first law as the transport coefficient. Taking the ratio K_w/K_a, we get

$$K_w = [(a_w D_w)/(a_a D_a)]K_a \qquad [28]$$

Equation [28] can be rewritten as

$$K_w = a_b(D_w/D_a)K_a \qquad [29]$$

where a_b is Bunsen's solubility coefficient. Equation [29] shows that at the same gradient of partial pressure, gas diffusion in water takes place $a_b(D_w/D_a)$ times slower than in air. Substituting values at 293 K into equation [29] shows that the diffusion of O_2 in water is about 300,000 times slower than in air. In the case of CO_2, diffusion in water is about 10,000 times slower than in air.

To illustrate this for O_2 and CO_2, we let the temperature be 293 K and use equations [27] and [28]. For O_2, the calculations are as follows:

$$K_w/K_a = 0.0333 \times (2.1 \times 10^{-9})/(2.01 \times 10^{-5}) = 3.48 \times 10^{-6}$$

Therefore,

$$K_a = 287,430 K_w$$

For CO_2, the calculations are as follows:

$$K_w/K_a = 0.942 \times (1.77 \times 10^{-9}/1.59 \times 10^{-5}) = 1.05 \times 10^{-4}$$

Therefore, solving for K_a gives

$$K_a = 9536 K_w$$

These calculations show that at the same gradient of partial pressure, the mobilities of O_2 and CO_2 in air are several orders of magnitude greater than in water.

Influence of the Coordinate System on the Transport Coefficient

Campbell (1985) showed that, under steady conditions, the flux of gas across any surface is constant; however, the flux density may change if the area available for flow changes. The area can vary depending on the geometry.

Steady-state diffusion of gases in soil follows Fick's law and is written as

$$J = -D\partial C/\partial z = Q/At = j/A \tag{30}$$

where J is the flux density (kg/m$^2\cdot$s), j is the flux (kg/s), A is the cross-sectional area (m^2), Q is the quantity of gas (kg), D is the diffusion coefficient (m^2/s), C is the gas concentration (kg/m^3), t is time, and z is the distance (m). This steady-state equation [30] can be rearranged and integrated to give

$$j = K(C_2 - C_1) \tag{31}$$

where K is the conductance (i.e., the reciprocal of the resistance) and is equal to $AD/\Delta z$ in the planar case, and the subscript indicates where the concentration is measured.

To see how the flux density changes with geometrical shapes, equation [31] may be rearranged and integrated to obtain

$$j\int[1/A(z)]dz = D\int dC = D(C_2 - C_1) \tag{32}$$

Comparing equations [31] and [32], the general form for the conductance becomes

$$K = D/\int[1/A(z)]dz \tag{33}$$

When diffusion is only one-dimensional (i.e., planar), the area available for flow remains constant with distance, so we write

$$\int[1/A(z)]dz = 1/A\int dz = (z/A)\Big|_{z_1}^{z_2} = (1/A)(z_2 - z_1) \tag{34}$$

and when $A = 1$ m^2, the conductance (m/s) for the planar case becomes

$$K = D/(z_2 - z_1) \tag{35}$$

When diffusion is spherical, as it may be to or from an aggregate or microorganism colony in soil, $A = 4\pi r^2$ where r is the radial distance from the center of the sphere. Graphically, this system is shown in Figure 7.3. Integration of equation [32] gives the conductance (m^3/s) for the spherical case:

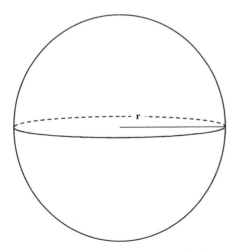

Figure 7.3. Spherical coordinates for diffusion. The symbol r represents the radius of the sphere.

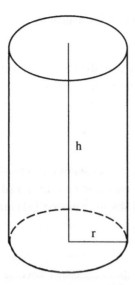

Figure 7.4. Cylindrical coordinates for diffusion. The symbols h and r represent the height and radius of the cylinder, respectively.

$$K = D/\!\int(1/4\pi r^2)dr = 4\pi D(1/r_2 - 1/r_1)$$

or

$$K = 4\pi D r_1 r_2/(r_2 - r_1) \tag{36}$$

Note that when r_2 is much, much larger than r_1, the conductance becomes equal to $4\pi D r_1$.

When diffusion occurs in cylindrical coordinates, $A = 2\pi r$, per meter of root length, where r is the distance from the center of the cylinder. Graphically, this coordinate system is shown in Figure 7.4.

Integration of equation [33] gives the conductance (m^2/s) for the cylindrical case:

$$K = D/\!\int(1/2\pi r)dr = 2\pi D/\!\int(1/r)dr$$

or

$$K = 2\pi D/\ln(r_2/r_1) \tag{37}$$

The units of conduction, K, vary depending on the coordinate system used. For example, we summarize as follows:

Coordinate	Units of K	Flux
Planar	m/s	per unit area
Spherical	m^3/s	per sphere
Cylindrical	m^2/s	per meter of length

Steady-State Concentration Profiles

We now examine the concentration distributions of gases in the soil profile given either extraction or generation. To simplify, we assume that D and S are constant with depth.

With Extraction

Under steady-state conditions, $\partial C/\partial t$ equals zero, and the diffusion equation with extraction can be expressed as

$$D\partial^2 C/\partial z^2 - S = 0 \qquad [38]$$

where D is the diffusion coefficient, z is the spatial coordinate (positive downward), and S is the rate of extraction, which for this situation is assumed to be constant with depth and time. This equation can be used for predicting the movement of O_2 in the soil profile under steady-state conditions. Equation [38] can be written as

$$\partial^2 C/\partial z^2 = S/D \qquad [39]$$

The boundary conditions are

$$C(0, t) = C_0 \qquad [40a]$$
$$D\partial C/\partial z(L, t) = 0 \qquad [40b]$$

where L is a specified depth at which there is no diffusive transfer across this boundary. Equation [40a] indicates that the concentration of O_2 at the soil surface remains equal to C_0 for all time. Equation [40b] indicates that the flux of O_2 across the boundary at $z = L$ is zero. This condition may occur at a water table or at some impermeable boundary.

The first integration of equation [39] with respect to z gives

$$\partial C/\partial z = Sz/D + I_1 \qquad [41]$$

where I_1 is a constant of integration. Using equation [40b], we find

$$I_1 = -SL/D \qquad [42]$$

Substituting this into equation [41], we obtain

$$\partial C/\partial z = Sz/D - SL/D \qquad [43]$$

We integrate again to find the working equation that describes the concentration of O_2 in the soil profile:

$$C(z) = Sz^2/2D - SLz/D + I_2 \qquad [44]$$

Applying the boundary condition [40a] to equation [44] again and solving for I_2, we get

$$C_0 = I_2 \qquad [45]$$

Inserting this into equation [44], we obtain the particular solution

$$C(z) = C_0 + Sz^2/2D - SLz/D \qquad [46]$$

which can be rearranged to obtain the working equation

$$C(z) = C_0 + (S/D)[(z^2/2) - Lz] \qquad [47]$$

This analytical solution to the diffusion equation shows that at $z = 0$, $C = C_o$. For given values of S/D, L, and C_o, the concentration of O_2 in the profile decreases curvilinearly to a minimum at L. Also, note that as D increases at constant C, the O_2 concentration at depths other than zero increases toward C_o.

Example

Calculate the concentrations of O_2 at 1 and 10 cm depths if $C_o = 0.2$ kg/m^3, L = 1.0 m, and D = 1×10^{-7} m^2/s. First, assume that S = 0, then assume that S = 0.5 mg/m^3·s. The working equation is

$$C(z) = C_o + (S/D)[(z^2/2) - Lz]$$

When S = 0, then the concentration of O_2 throughout the profile is 0.20 kg/m^3. Next, we assume that S = 0.5 mg/m^3·s. At z = 0.01 m, we have

$$C(0.01) = 0.20 + \{[(0.5 \text{ mg/m}^3\text{·s})(1 \text{ kg}/10^6 \text{ mg})]/(1 \times 10^{-7} \text{ m}^2\text{/s})\}$$
$$[(0.0001/2) - (1.0 \times 0.01)]$$
$$= 0.2 + 5(-0.0099)$$
$$= 0.1505 \text{ kg/m}^3$$

At z = 0.1 m, we have

$$C(0.10) = 0.20 + 5(0.005 - 0.1)$$
$$= -0.275 \text{ kg/m}^3$$

The negative O_2 concentration calculated at z = 0.1 m is physically impossible and suggests that S, D, or both are not constant with depth in the soil profile or that the boundary conditions have been violated. Therefore, one or more of the assumptions used in developing the working equation were violated.

With Generation

Frequently, CO_2 is generated in the soil profile due to the respiration of plant roots and microorganisms. Under steady-state conditions, equation [23] becomes

$$D\partial^2 C/\partial z^2 + S = 0 \tag{48}$$

Using the same boundary conditions as in equation [40] and integrating twice will give the CO_2 concentration distribution equation in the soil profile:

$$C(z) = C_o - (S/D)[(z^2/2) - Lz] \tag{49}$$

This equation shows that with generation, the concentration of a gas such as CO_2 increases with depth in the profile. This is one of the reasons root growth and development are limited to near the soil surface. The shapes of the curves predicted by equation [49] should be contrasted with the shapes of those curves predicted by equation [47].

The production rate of CO_2 in soil varies between 0.2 and 0.6 mg/m^3·s (Koorevaar et al. 1983) and is affected by all those factors that influence biological activity, including soil temperature and water status.

Soil Air Pressure under Field Conditions

Soil gases are characterized by the amount or volume of the gas phase, gas composition, and gas mobility. The amount or volume of gas in the soil is represented by the aeration porosity, f_a. Values of f_a generally decrease with depth in the soil profile due to decreasing total porosity as a result of compaction and to increasing soil water content. The composition of soil air also varies with depth and time. With depth, the concentration of CO_2 increases and that of O_2 decreases because of biological activity in the profile.

Under equilibrium and isothermal conditions, the air pressure can be expressed by

$$P = P_o \exp(-Mgz/RT) \tag{50}$$

where P is the absolute pressure, and P_o is the standard pressure at STP. Substitution of z = 1 m and T = 290 K into equation [50] gives $P = 0.99988P_o$. Thus, air pressure changes by only 0.012% over a distance of 1 m, a small percentage. This also can be applied to the partial gas pressures of the various gas components. These calculations show that soil air and the atmosphere at the soil surface are in equilibrium with each other *only* if the total gas pressure and the composition of the soil air are the same as in the atmosphere.

Characterization of Soil Aeration

The aeration status of a soil can be characterized by measuring aeration porosity, composition of soil atmosphere, gaseous diffusion rates (primarily the oxygen diffusion rate), oxidation-reduction potential (Eh), and perhaps oxygen uptake and respiratory quotient and reduced chemical forms in the soil solution.

Aeration Porosity

The air-filled porosity is expressed as a fraction or percentage of the total soil volume that is occupied by air. Since water and air compete for the same pore space, values of f_a are indirectly related to the volumetric soil water content. Mathematically, the relationships between these parameters were given by equations [2] and [3].

Values of f_a depend on soil texture, water content, and structure. For a given soil, f_a is usually determined by taking an "undisturbed" sample and determining the difference between the total porosity and the volumetric water content. Some soil physicists consider that soils with values of $f_a < 0.1$ are likely to be poorly aerated.

Another term of some applied interest is the field air capacity. It is defined as the fractional volume of air in a soil at the "field capacity" water content. Air capacity depends on texture (sands, >25%; loams, 15–20%; clayey soils, <10%) and structure; strongly aggregated soils generally have an air capacity between 20 and 30%. As the aggregates are broken down, the macroscopic pores tend to disappear, so that a strongly compacted soil may contain <5% by volume of large pores.

Other parameters that have been used to describe soil aeration include porosity, void ratio, and degree of saturation. All of these parameters describe the capacity of the soil to contain air but reveal nothing about the composition of the air or its rate of transport. As a result, they are considered to be static indicators of soil aeration.

Composition of Soil Atmosphere

The composition of the gas phase is also a static indicator of soil aeration. However, it reveals more information on problem conditions, particularly whenever the O_2 concentration of the soil air falls significantly below that of the atmosphere. Under these conditions, it indicates restricted gas exchange across the air-soil boundary.

Measurement of the composition of soil gases was the earliest method used to characterize soil aeration. Over the years, measurements of soil gases have been taken in situ and from gas samples removed from the soil profile. Both measurement techniques have limitations caused by the volume of the gas sample required, the volume of soil sampled, and the time required to take the sample.

The concentration of a mixture of gases can be expressed in several ways. Therefore, it is important to keep these terms straight when using gas concentrations to predict the status of the gas in soil at various times and depths. These terms apply to any gas and not just to oxygen.

The *mole fraction* (F) of a gas such as O_2 is the ratio of moles of oxygen to the total moles of gas. Mathematically, this is expressed as

$$F = N_O/(N_O + N_n) \tag{51}$$

where N_O is the moles of oxygen, and N_n is the moles of the other gas (such as N_2).

The *partial pressure* of O_2 (p_O) is the pressure of oxygen in a gas mixture. For a mixture that obeys the ideal gas law, p_O is given by

$$p_O = P_t F = P_t[N_O/(N_O + N_n)] \tag{52}$$

where P_t is the total pressure of the gas mixture. Since P_t essentially equals $p_O + p_n$,

$$F = p_O/P_t \tag{53}$$

which states that the mole fraction is equal to the ratio of the partial pressures. Therefore, the mole fraction F, the partial pressure ratio, and the partial pressure p_O are all measures of gas concentration. As a result, each parameter can be used in the diffusion equation in the place of concentration.

The concentration of gases generally increases with pressure and decreases with temperature. The mass concentration of a dissolved gas, C_m, is proportional to the partial pressure of the gas p_O and according to Henry's law can be written as

$$C_m = a_m p_O/P_t \tag{54}$$

where a_m is the solubility coefficient of the gas in water, and P_t is the total pressure of the atmosphere. The volume concentration of a dissolved gas can be similarly written as

$$C_v = a_v p_O/P_t \tag{55}$$

where a_v is the solubility expressed in terms of the volume ratios (i.e., a_v is the volume of dissolved gas relative to the volume of water). Solubilities of gases in water are a function of temperature (Table 7.2). The solubility of CO_2 is about 50 times greater than that of N_2 and about 25 times greater than that of O_2. This high solubility of CO_2 in water can be attributed to its hydrolysis with water and results in the formation of hydrogen and bicarbonate ions. The extent of this reaction is determined by the partial pressure of CO_2 in the ambient atmosphere, which can be considered as constant for most practical purposes. Other gases, such as NH_3 and H_2S, form acids and bases in aqueous solution. Therefore, they do not behave as ideal gases, because the partial pressures and concentrations are not linearly related. It is generally thought that soil O_2 concentrations less than 5% are likely to affect plant growth. Even soils with higher concentrations of O_2 may have zones that are deficient.

A soil oxygen sensor is frequently used to measure O_2 concentrations in the air and in the soil. This sensor is based upon the electrochemical reduction of oxygen at the exposed end of a small gold cathode. The cathode is maintained at approximately –0.8 volts relative to a silver anode. At this potential, the magnitude of the electrical current flowing through the KCl solution between the electrodes is governed by the rate at which oxygen diffuses from the soil, through the inner membrane, to the cathode for reduction. The rate of diffusion is proportional to the oxygen concentration in the soil. Consequently, the potential associated with the reduction current will also be proportional to the oxygen concentration.

The output of the sensor is an electrical potential that is directly proportional to the partial pressure of oxygen at the surface of the sensor. A voltmeter is used to measure the electrical potential.

When the calibration is done in air, corrections must be made for the ambient humidity and pressure. The partial pressure (p_w) of the water vapor in the air must be subtracted from the total barometric pressure (P) to calibrate in millibars of oxygen. The equation to be used is

$$V = (\%O_2/100)(P - p_w) \tag{56}$$

where V is in millivolts and the pressure is in millibars. One useful unit transformation is 1 bar = 750.1 mm Hg. In order to accurately use atmospheric air for calibration, one needs to know the current barometric pressure, relative humidity, and temperature of the air. From the air temperature, the saturated vapor pressure of water can be determined with a table supplied by Jensen Instruments.

Oxygen Diffusion Rate

The aeration status of a soil can be characterized by the oxygen diffusion rate (ODR) in the soil. The ODR method has been found to be satisfactory in the higher range of soil aeration and to be less sensitive in the lower range of soil aeration (poorly drained and flooded soils). This method is predicated upon the concept that the limiting factor in the O_2 supply to surfaces of plant roots and microorganisms is the rate of diffusion of O_2 through the moisture films around the organism.

In the ODR method, the oxygen reduction rate is measured at a cylindrical Pt electrode surface when a constant electrical potential is applied. Once the oxygen present at the electrode surface is reduced, further reduction depends on the diffusion of O_2 to the electrode surface. The resulting current measures the flux of O_2 to the water-covered electrode, which acts as a cylindrical sink. Oxygen flow to the cathode is similar to the maximum O_2 available to a respiring root or to respiring microorganisms. The current is proportional to the rate of O_2 flux at the electrode surface and is related by

$$ODR = 60MI/nFA \qquad [57]$$

where M is the molar mass of oxygen (32 g), I is the electrical current in microamperes, n is the number of electrons required to reduce one molecule of O_2 (equals 4), F is the Faraday constant (96,500 coulombs), and A is the electrode surface area (cm^2). The units of ODR are those of flux density (i.e., micrograms of oxygen per square centimeter per minute).

Many studies have shown that there is a fairly good relationship between ODR and plant response. Values less than 0.20 µg/cm^2·min often indicate inadequate aeration for most plant species. Values greater than 0.40 µg/cm^2·min are considered sufficient for optimum growth of most plants. Under prolonged conditions in the field, the Pt electrodes may be poisoned in aqueous and soil environments due to the formation of oxide and sulfide coatings, which cause inaccurate measurements.

Oxidation-Reduction Potential

Oxidation-reduction potential, or redox, is a measure of electron availability (or activity) potential in a chemical or biological system. Redox potentials of soil can help to identify changes in oxygen availability because they are closely linked to the presence or absence of oxygen, particularly at low oxygen contents. This method has been used to characterize the intensity of reduction and to identify different forms of redox couples (e.g., Fe^{2+} and Fe^{3+}) in flooded soils and sediments.

A relatively inert electrode, usually platinum, is used to measure the potential of all redox couples. The redox of the soil solution is measured by inserting the inert Pt electrode and a reference electrode (such as saturated calomel or Ag:AgCl) into the aqueous solution and measuring the difference in potential between the two electrodes. When inserted into soil, the Pt can either give up electrons to the soil solution or take up electrons from it. The potential of the Pt electrode is compared with the constant potential of the reference electrode. The measured redox potential can be represented by

$$E_h = E_0 - (RT/nF)\ln(\text{reductant/oxidant}) \qquad [58]$$

where E_h is the potential difference between the reference electrode and the Pt electrode (mV), E_0 is the reference electrode potential, R is the gas constant, n is the number of electrons transferred in the reaction, T is the absolute temperature, and F is the Faraday constant. The activity of equation [58] indicates that the potential of the system (E_h) is proportional to the natural logarithm of the ratio of the reduced to oxidized products.

Studies have shown that values of E_h vary with soil water status. Under well-drained conditions (oxidized systems), low concentrations of redox couples reduce the stability and reproducibility of E_h measurements, whereas in flooded soils and in soils with low oxygen levels, the higher concentrations of redox couples increase the sensitivity of E_h measurements, thus making this technique more applicable to anaerobic systems. Well-drained soils have characteristic E_h values of greater than 400 mV, and as the soil oxygen decreases, soil E_h decreases. In some soils with high carbon contents, E_h values as low as −350 mV can be observed. Oxygen disappears at E_h values of approximately 300 mV; NO_3 is removed between 200 and 300 mV, followed sequentially by the reduction of Mn^{4+}, Fe^{3+}, and SO_4^{2-}.

Respiratory Quotient and Respiration Rate

Aerobic respiration involves the decomposition of complex carbon molecules, the consumption of oxygen, and the release of CO_2, water, and energy for cellular growth. Under these conditions, the *respiratory quotient*, RQ, is defined as

$$RQ = \text{volume of } CO_2 \text{ released/volume of } O_2 \text{ consumed} \qquad [59]$$

and is equal to 1.0. However, when respiration is anaerobic, RQ increases since O_2 is no longer consumed but CO_2 continues to be evolved. The respiratory activity of the soil organisms, or respiration rate, is measured as the volume of O_2 consumed (or CO_2 released) per unit soil volume per unit time.

Soil respiration is a combined effect of respiration from microorganisms and from plant roots. It is greatly stimulated by the abundance of carbon materials in the soil immediately around the roots. Respiration rates are spatially and temporally dependent. Factors affecting the respiration rate include

1. soil conditions such as organic matter and water content,
2. cultivation and cropping practices,
3. atmospheric factors such as air temperature, and
4. the presence of biological organisms.

Consumption of O_2 and production of CO_2 in soil will tend to be highest when plant root and microbial activity is highest. This will depend upon soil, crop, and climate conditions. Consumption and production fluxes up to 20 g/m²·d are not uncommon during the summer if the soil water status is optimum for rapid plant growth and microbial activity.

The effect of temperature on the flux density of CO_2 at the soil surface can be expressed by the Q^{10} equation

$$S = S_0 Q^{T/10} \qquad [60]$$

where S and S_0 are the flux densities at temperatures T (°C) and 0°C, respectively, and Q is the magnitude of the increase in S for a 10°C increase in temperature, called the Q^{10} factor. Normally, values of Q for biological processes lie between 2 and 3. Temperature changes are one of the principal causes of the large seasonal fluctuations in soil respiration rate in temperate climates.

Example

Using equation [60], plot the relationship between the CO_2 flux density at the soil surface and the mean soil temperature assuming S_0 is 1.2 g/m²·d and Q is 3.0.

As has been pointed out previously, soil aeration depends on the transport of O_2 and CO_2 between the soil and the atmosphere.

Reduced Chemical Forms

The aeration status of a soil is related to its oxidation and reduction status. Therefore, the ionic status of several ions is indicative of soil aeration. If a soil is aerobic, oxidized states of ferric iron (Fe^{3+}), manganic manganese (Mn^{4+}), nitrate (NO_3^-), and sulfate (SO_4^{2-}) are dominant. In poorly drained, waterlogged soils, the reduced forms of ferrous iron (Fe^{2+}), manganous manganese (Mn^{2+}), ammonium (NH^{4+}), and sulfides (S^{2-}) are found. These reduced forms are so soluble in the soil solution that toxicities may develop. In addition, poorly aerated soils tend to contain a wide variety of only partially oxidized products such as alcohols, organic acids, and ethylene gas (C_2H_4), which can be toxic to higher plants.

Response of Plants to Poor Soil Aeration

Several factors affect plant response to poor aeration. These include the age of the plant, nutrition, temperature, light intensity, and the plant species or variety. Under field conditions, it is somewhat difficult to measure all of these effects because of time and depth complications. Plant species respond differently to the duration of poor oxygen regimes and to the time of occurrence in their life cycle, and they differ in their ability to develop specialized oxygen transport pathways.

In the case of clayey or poorly drained soils, there is a mosaic of aerobic and anoxic zones during wet periods of the year. Plant roots tend to grow through the larger pores (>100 μm diameter), where O_2 diffusion is rapid. Oxygen can also diffuse for limited distances through plant tissues from regions of high to low O_2 supply, an attribute best shown in rice. However, when heavy rain coincides with temperatures high enough to maintain rapid respiration rates, the root zone can become anaerobic, a condition that, even if sustained for only a day, has serious effects on the metabolism and growth of sensitive plants. Plant species intolerant of waterlogging experience an acceleration in the rate of glycolysis and in the accumulation of endogenously produced ethanol, cell membranes become leaky, and ion and water uptake is impaired. The characteristic whole-plant symptoms are wilting, yellowing of leaves, and the development of adventitious roots at the base of the stem. Stimulation of the production of ethylene (C_2H_4) may also cause growth abnormalities such as leaf epinasty, the downward curvature of a leaf axis. Thus, poor soil aeration typically results in reduced plant growth and development, reduced nutrient uptake and water absorption, and the accumulation of certain toxic inorganic elements and organic compounds.

Cited References

Campbell, G. S. 1985. *Soil Physics with Basic Transport Models for Soil-Plant Systems*. Elsevier Science, New York.

Glinski, J., and W. Stepniewski. 1985. *Soil Aeration and Its Role for Plants*. CRC Press. Boca Raton, FL.

Hillel, D. 1982. *Introduction to Soil Physics*. Academic Press. New York.

Koorevaar, P., G. Menelik, and C. Dirksen. 1983. *Elements of Soil Physics*. Elsevier. New York.

Sextone, A. J., N. P. Revsbech, T. B. Parkin, and J. M. Tiedje. 1985. Direct measurement of oxygen profiles and denitrification rates in soil aggregates. *Soil Science Society of America Journal* 49:645–651.

Stolzy, L. H. 1974. Soil atmosphere. In E. W. Carson (ed.), *The Plant Root and Its Environment*, pp. 335–361. University Press of Virginia. Charlottesville.

Wild, A. 1993. *Soils and the Environment: An Introduction*. Cambridge University Press. Cambridge, England.

Additional References

Campbell, G. S. 1985. *Soil Physics with Basic Transport Models for Soil-Plant Systems*. Elsevier Science Publishing Co. New York.

Kirkham, D., and W. L. Powers. 1972. *Advanced Soil Physics*. Wiley-Interscience. New York.

Payne, D., and P. J. Gregory. 1988. The soil atmosphere. In *Russell's Soil Conditions and Plant Growth*, pp. 298–314. John Wiley and Sons. New York.

Problems

1. If the diurnal soil temperature varies between 300 K and 280 K, calculate the volume and percentage change in aeration porosity at values of f_a of 0.5, 0.25, and 0.10 m^3/m^3. What happens when the volume of soil changes?

2. Show the influence of temperature in the range between 15 and 35°C and pressure between 80 and 100 kPa on the molecular diffusivity in water of O_2 and CO_2.

3. Calculate the influences of temperatures of 15, 25, and 35°C and pressures of 80, 90, and 100 kPa on the molecular diffusion coefficients of O_2 and CO_2 in air and compare with the same coefficients in water.

4. Calculate the relationship between the O_2 concentration in the gas phase and in the soil as a function of aeration porosity. Make similar calculations for the relationship between the fraction of the concentration of O_2 in the aqueous phase and that in the gaseous phase and volumetric soil water contents.

5. Show the influence of the magnitude of the diffusion coefficient on the O_2 concentration distribution in the soil profile to 0.5 m depth. Assume that the oxygen concentration at the soil surface is 0.20 kg/m^3.

6. A soil profile has a root zone of 1.4 m. At steady state, the oxygen consumption is 0.4 $mg/m^3 \cdot s$ and is uniform throughout the root zone. Consumption of O_2 below the root zone is negligible. Calculate the partial pressure of O_2 as a function of depth from the following data:

   air-filled porosity = 0.32 m^3/m^3
   oxygen diffusion coefficient in air = 2×10^{-5} m^2/s
   T = 20°C
   molecular weight of oxygen = 32 g/mol
   volume fraction of O_2 in the atmosphere = 0.20 (partial pressure = 20 kPa)

7. The equation that predicts the steady-state concentration of oxygen in the soil profile is

   $C(z) = C_O - (S/2D)(2/Lz - z^2)$

where C(z) is the oxygen concentration at depth z, C_o is the concentration at the soil surface, S is the oxygen respiration rate, L is the depth of the impermeable boundary, and D is the oxygen diffusion coefficient. At the lower boundary of the biologically active zone, the concentration is

$$C(L) = C_o - SL^2/2D$$

Using this equation, do the following:

a. Derive the equation for the respiration rate, S', when the critical level of oxygen concentration at the lower boundary is C'.

b. Derive the equation for the critical value of the diffusion coefficient, D', that ensures oxygen concentrations above C' in the profile down to depth L.

c. Derive the equations for S' and D' assuming that the value of C' is zero.

d. If the profile is not all biologically active, S is less than S'. Derive an equation that predicts the depth of the aerobic zone, L_o.

8. Calculate the amount of oxygen (in grams) in 5 m^2 of the root zone of a soil profile that has a soil water content of 0.3 kg/kg, bulk density of 1420 kg/m^3, rooting depth of 20 cm, and oxygen concentration of 0.15 kg/m^3. What is the oxygen status in the root zone?

9. Distinguish between the mechanisms of transport of gases in the soil. What are the conditions that cause gases to move from one location to another?

10. Derive the conservation equation for a gas moving in a unit volume of soil in the z direction only. Assume that the gas is insoluble in water.

11. Graph the relation between the daily average soil respiration (g/m^2d) and soil temperature (°C) assuming that the Q_{10} is equal to 3.0 and that for the first 6 months of the year S_o is equal to 1.2 and for the second 6 months of the year S_o is equal to 0.9. Why is there a difference in the two curves?

12. Assume that a soil profile has the following properties:

Property	Horizon	
	Ap	B21t
Bulk density (kg/m^3)	1350	1450
Water content (m^3/m^3)	0.45	0.40
Thickness (m)	0.15	0.15
O_2 concentration (vol%)	20.0	16.0

Determine the amount of O_2 present in each soil horizon and in the profile in units of

a. m^3 of O_2/m^3 of soil, and

b. g of O_2.

How would you characterize the O_2 status of this soil?

13. In the same soil given in problem 12, O_2 was consumed at a rate of 5 $g/m^2/d$. How many days' supply of O_2 are contained in the combined Ap and B21t horizons? What does this mean in terms of soil aeration?

CHAPTER EIGHT

Soil Water Principles

Introduction

Water is the most prevalent substance on the earth and covers about two-thirds of the surface. This water is primarily located in the oceans, seas, and lakes. Water is the only substance found in vast quantities in nature in three states: solid, liquid, and gas. Of the common liquids, water

is the most universal solvent,
has the highest surface tension (0.0727 N/m),
has the highest dielectric constant (approximately 80),
has the greatest heat of vaporization (2.25 MJ/kg), and
with the exception of ammonia, has the greatest heat of fusion (0.334 MJ/kg).

Unlike most substances, water expands when it freezes under low pressure. Knowledge of the basic physical properties of water is essential for an understanding of its behavior and function in nature, its interactions with soil particles, and its state and movement in the soil-plant-atmosphere system.

The distribution of the world supply of water is summarized in Table 8.1. Note that the percentage of the total supply of water in the soil relative to the total supply is low. The soil contains only about 0.05% of the total freshwater supply. Thus, the water in the soil must be carefully managed.

Soil water is of primary importance to plant growth and development. It also impacts many aspects of soil physical behavior. For example, the amount and strength of the films of water around soil particles affect the following soil properties and processes:

| plasticity | permeability | water availability | runoff |
| stickiness | aeration | solute transport | redistribution |

Table 8.1. The Source and Percentage Contribution of the
World Supply of Water

Source	% of Total Supply	% of Fresh-water Supply
World oceans	96.5	—
Ground waters	1.7	
Fresh groundwater	0.76	30.1
Soil water	0.001	0.05
Glaciers and snowpack	1.74	68.7
Ground ice	0.022	0.86
Water in lakes	0.013	
Freshwater lakes	0.007	0.26
Saltwater lakes	0.006	—
Marsh water	0.0008	0.03
Water in rivers	0.0002	0.006
Atmospheric water	0.001	0.04

Source: Adapted from Shiklomanov and Sokolov 1983 as presented by Dingman 1994.

Mean Annual Precipitation

Figure 8.1. Distribution of long-term annual precipitation in the United States.

Over the world, an average of 76 cm of rain falls per year. In the United States, however, the average annual rainfall is 91 cm. It is well known that rainfall is spatially and temporally distributed, which significantly contributes to the spatial and temporal distributions of soil water (Figure 8.1).

In this chapter, we examine the principles that affect the water status of a soil. We begin with a review of the water content and the types of energy with which the water is held by the soil. The water content of a soil is sometimes called the *capacity factor,* and the energy status is the *intensity factor* (similar to the terminology of thermal problems, where the capacity factor is the heat content and the thermal intensity factor is the temperature). Later in this chapter, we explore the energy relations of soil water, the distributions of total and component potential energies under various field conditions, the soil water retention curve, methods of determining soil water status, and soil water terms used in classification systems. Excellent presentations of the mechanics of fluids in porous media are given in Corey (1994) and in Kutilek and Nielsen (1994).

Review of Soil Water Content Expressions

Soil water content, θ, is usually expressed on an oven-dry weight basis. For most purposes, soil water is taken to be that which can be removed by drying to a constant weight at 110°C. This temperature standard was set because it is practically impossible to determine the absolute water content of soil. However, in most cases, the relative water content between two energy states is more important than absolute values. Thus, the problem of soil water content measurement becomes one of reproducibility more than one involving absolute accuracy.

Mathematically, soil water content on a dry-weight basis (θ_{dw}) can be calculated from

weight: θ_{dw} = water lost/oven-dry soil weight [1]

volume: $\theta_v = \theta_w(\rho_b/\rho_w)$ [2]

where ρ_b and ρ_w are the bulk densities of the soil and of water, respectively. In those situations where the volumetric soil water content is known, the equivalent depth at which water would stand if accumulated from a specified depth interval or the volume of water per unit area per indicated depth interval can be calculated. Mathematically, this equivalent depth of water, D_{eq}, is calculated from

$$D_{eq} = \theta_v \Delta z$$ [3]

where Δz is the depth interval. The unit of D_{eq} is reported as length, but the proper interpretation must always be a volume of water per unit area.

Soil water content can also be expressed on a wet-weight basis. In equation form, the water content on a wet-weight basis, θ_{ww}, is calculated from

$$\theta_{ww} = \frac{(\text{wet-soil weight} - \text{dry-soil weight}}{\text{wet-soil weight})}$$ [4]

Water expressed on a wet-weight basis can be converted to the water content on a dry-weight basis, θ_{dw}, by the following equation:

$$\theta_{dw} = \theta_{ww}/(1 - \theta_{ww})$$ [5]

Example

The soil water content is 0.40 kg/kg on a wet-weight basis. Calculate the water content on a dry-weight basis using equation [5].

$$\theta_{dw} = 0.40/(1.00 - 0.40) = 0.67 \text{ kg/kg}$$

Example

Show that the right-hand side of equation [5] is equivalent to θ_{dw}.

$$\theta_{dw} = \frac{(\text{wet weight} - \text{dry weight})/\text{wet weight}}{[\text{wet weight} - (\text{wet weight} - \text{dry weight})]/\text{wet weight}}$$

$$= (\text{wet weight} - \text{dry weight})/\text{dry weight}$$

In soil physics literature, two additional unitless soil water content expressions are defined. The relative water content, θ_r, is

$$\theta_r = \theta_v/\theta_{sat}$$ [6]

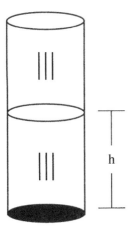

Figure 8.2. Pressure exerted on a cylinder of water. The term h represents the height of water in the cylinder.

where θ_{sat} is the volumetric soil water content at saturation. The reduced water content, Se, is defined as

$$Se = (\theta_v - \theta_{res})/(\theta_{sat} - \theta_{res}) \qquad [7]$$

where θ_{res} is the residual volumetric water content and is dimensionless.

Review of Pressure Concepts

Pressure is defined as "normal force (F) per unit area (A)," or F/A. Consider in Figure 8.2 the pressure exerted on the top of the water in the cylinder by the atmosphere and on the bottom of the cylinder itself by the water.

The absolute pressure P (N/m^2 or Pa) at the bottom of the static column of water exerted on the sealing plate is the sum of the pressure exerted by the water column and the atmospheric pressure:

$$P = F/A = \rho_w gh + p_0 \qquad [8]$$

where F is the force (N), A is the cross-sectional area (m^2), ρ_w is the density of water (kg/m^3), g is the acceleration of gravity (m^2/s), h is the height of the water column (m), and p_0 is the atmospheric pressure at the top of the column of water (Pa). The pressure of any fluid in the column may be calculated by equation [8] if the fluid density is known. Thus, the total pressure on the bottom of the cylinder shown in Figure 8.2 is the force per unit area of the fluid plus the pressure of the atmosphere. Frequently, the height of water in the column is called the *head* of the water.

Pressures, like temperatures, can be expressed by either absolute or relative scales. Whether relative or absolute pressure is measured in a pressure-measuring device

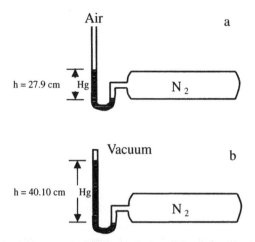

Figure 8.3. (a) Open-end manometer showing a pressure above atmospheric pressure. (b) Absolute-pressure manometer.

depends on the nature of the instrument used to make the measurements. For example, an open-end manometer measures a *relative pressure* (gauge pressure), since the reference for the open end is the pressure of the atmosphere at the open end of the manometer (Figure 8.3a).

Closing off the end of the manometer and creating a vacuum in the end results in a measurement against a vacuum, or against "no pressure"; therefore, p_0 in equation [8] is zero. This measurement is called *absolute pressure.* Since absolute pressure is based on a vacuum (a fixed reference point that is unchanged regardless of location, temperature, weather, or other factors), it is a precise, invariable value that can be readily identified. The zero point for an absolute-pressure scale corresponds to a perfect vacuum, whereas the zero point for a relative-pressure scale usually corresponds to the pressure of the air that surrounds us at all times, which varies slightly over time and location.

If a reading on a column of mercury, illustrated in Figure 8.4 with the bowl open to the atmosphere, is taken with the device called a barometer, the reading of pressure is termed barometric pressure.

In the pressure-measuring devices examined so far, the fluid (either water or mercury) is at equilibrium. That is, a state of hydrostatic balance is reached in which the manometer fluid is stabilized, and the pressure exerted at the bottom of the U-tube by the fluid in the part of the tube open to the atmosphere or vacuum exactly balances the pressure exerted at the bottom of the U-tube in the part of the tube connected to the tank of water or gas.

Another common pressure-measuring device is the Bourdon gauge, which normally (but not always) reads zero pressure when open to the atmosphere (Figure 8.5). The pressure-sensing device in the Bourdon gauge is a thin metal tube with an elliptical cross section and closed at one end that has been bent into an arc. As the pressure increases at the open end of the tube, the tube tries to straighten out. The movement of the tube is converted into a dial movement by gears and levers.

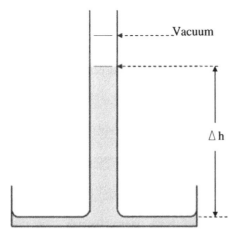

Figure 8.4. A column of mercury with the bowl open to the atmosphere, a configuration that serves as a barometer.

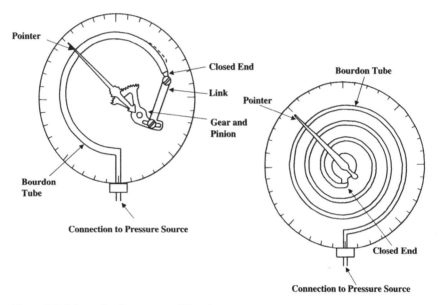

Figure 8.5. Schematic of two types of Bourdon gauges.

The reference point or zero point for the relative-pressure scales is not constant. The relationship between relative and absolute pressure is given as

gauge pressure + barometer pressure = absolute pressure

Figure 8.6 shows the four common systems of pressure measurement: pounds per square inch (psi or lb/in^2), inches of mercury (in Hg), mm of Hg (mm Hg), and kilopascals (kPa). Other frequently used systems are feet of water (ft H$_2$O), atmospheres (atm), and bars (100 kPa = 1 bar).

lb/in²	in Hg	mm Hg	kPa	
19.3 -------	39.3 ------	998 ----	133	A pressure above atmospheric
14.7 -------	29.92 ----	760 ----	101.3	Standard pressure
14.3 -------	29.1 -----	739 ----	98.5	Barometric presssure
0.0 ------	0 ------	0 -----	0.00	Perfect vacuum

Figure 8.6. Absolute-pressure comparisons when barometric pressure is 29.1 and 29.92 inches of Hg.

Another important point to remember about pressure is that the standard atmosphere should not be confused with atmospheric pressure. The *standard atmospheric pressure* is defined as the pressure equivalent to 1 atm or 760 mm Hg at 0°C or other equivalent value. *Atmospheric pressure* is a variable and must be obtained from a barometer each time you need it. The standard atmosphere is not equal to the barometric pressure in any part of the world except perhaps at sea level on certain days, but it is extremely useful in converting from one system of pressure measurement to another. Expressed in various units, the standard atmosphere is equal to

- 1.00 atmospheres
- 33.91 feet of water
- 14.7 pounds per square inch absolute
- 29.92 inches of Hg
- 760 mm of Hg
- 1.013×10^5 Pa, or 101.3 Pa

Forms of Energy

Energy is defined as the capacity for doing work. The units of energy are force times distance, i.e., the units of work. Energy is a force vector applied to a certain distance vector. Energy may be mechanical, thermal, chemical, electrical, etc. In this chapter, we will concentrate on the applications of mechanical energy in soil water.

The soil consists of three interactive phases: solid, liquid, and gas. Force fields exist at the solid-liquid, solid-gas, and liquid-gas interfaces as well as within the phases and strongly influence the behavior of soil water and thereby the soil physical properties. These forces can be categorized as follows:

- pressures and tensions due to curved water films
- hydrogen-bonding, covalent-bonding, and van der Waals–London forces, which cause cohesion between water molecules and adhesion between water and soil surfaces

- osmotic forces caused by salt or ion concentration differences across a membrane
- gravitational force
- pressure head in saturated conditions

Soil water, like other bodies in nature, contains energy in different quantities and forms. The complex nature of the pore space and the water held by soil makes it difficult to specify directly the force fields acting on soil water. The overall effect of the force fields may be obtained by measuring the work required to remove an increment of water from the soil at some equilibrium state. Knowledge of the relative energy state of soil water at each point within the soil can allow us to evaluate the forces acting on soil water in all directions and to determine how far the water in a soil system is from equilibrium.

Classical physics recognizes two principal forms of mechanical energy: kinetic and potential energies.

Kinetic Energy

Kinetic energy is due to motion. Mathematically, it is defined as

$$\text{kinetic energy} = KE = \tfrac{1}{2}mv^2 \tag{9}$$

where m is the mass of the body and v is the velocity. The units of kinetic energy (KE) are

$$KE = kg(m/s)^2 = Nm = J$$

Kinetic energy is considered to be negligible under most conditions in the soil because the velocity of soil water movement is low. However, KE may be important in preferential flow of water since the velocity of preferential flow is considerably more rapid than in normal conditions of soil water flow.

Potential Energy

Potential energy is energy due to position in a force field and is the primary form of energy in soil water. It is measured by the force required to move a body directly against the force field and is the product of the magnitude of the force (F) and the distance moved (Δz): $F \cdot \Delta z$. Work is done to move a body against the force field or work is done by the body if it is allowed to move within the force field. Use of potential energy concepts is restricted to those situations where the temperature is uniform throughout the system or where the temperature change may be regarded to have a negligible effect on the process under consideration. For practical purposes, we will assume that the effects of temperature on potential energy of soil water are negligible.

The potential energy of soil water varies over a wide range. Differences in the potential energy of soil water from one point to another give rise to the tendency of water to flow within the soil. The spontaneous and universal tendency of all matter in nature is to move from where the potential energy is higher to where the potential energy is lower, that is, for each parcel to equilibrate with its surroundings.

Soil water obeys the same universal pursuit of equilibrium. It moves constantly in the direction of decreasing potential energy. The rate of decrease of potential energy with distance is called the *hydraulic gradient* and is the primary moving force causing flow of water. As soil water is transferred from one equilibrium system to another, energy is either expended or acquired in each of the systems. Because work can be performed by water as it moves from one system to another, the water in an equilibrium system has the potential to do work; the energy is potential energy.

Total Soil Water Potential

Theoretically, a potential energy is associated with each force acting on water in soil. However, some of the separate potentials are combined for practical convenience.

Definition

The total soil water potential has been formally defined by the Soil Science Society of America (1997):

> The amount of work that must be done per unit of a specified quantity of pure water in order to transport reversibly and isothermally an infinitesimal quantity of water from a pool of pure water, at a specified elevation and at atmospheric pressure, to the soil water at the point under consideration.

Because of the complexities of soil systems, absolute potential energies, or potentials, are difficult to define. Therefore, each potential energy must be defined with respect to an arbitrary reference level.

The total soil water potential energy can also be defined mathematically as

$$\psi_t = \psi_g + \psi_m + \psi_p + \psi_o \tag{10}$$

where ψ_t is the total soil water potential energy, ψ_g is the gravitational potential energy, ψ_m is the matric potential due to capillary pressure and other soil and water surface forces, ψ_p is the pressure potential, and ψ_o is the osmotic potential due to salts.

Expressions of the Potential Energy per Unit Quantity of Water

The potential energy of soil water can be expressed in three ways, depending on the units chosen for the unit parcel of soil water.

Energy per Unit Mass

The fundamental expression of potential energy is per unit mass, ψ_{pm}. Physically, we know the relation

work = force $\times$ distance = energy

The units of the potential energy expressed on a mass basis are

work/mass = energy/mass = J/kg or J/mole

Energy per Unit Volume

The expression of water potential energy per unit volume, ψ_{pv}, is work divided by volume, which is defined as

force $\times$ distance/volume = force/area = pressure

The units of the potential energy on a volume basis are

work/volume $= J/m^3 = Pa$

Potential energy on a volume basis is obtained from potential energy on a mass basis by multiplying by the density of water, ρ_w.

Energy per Unit Weight

Work per unit weight of water, ψ_{pw}, is known as *hydraulic head*. Potential energy per unit weight can be obtained on a mass basis by remembering that weight equals mass times gravity. Therefore, we write

work/weight = force $\times$ distance/weight = distance or head
$N \cdot m \cdot s^2 / kg \cdot m = kg \cdot m \cdot s^2 / kg \cdot s^2 = m$

In addition, hydraulic head is defined as:

pressure/weight = (force/area) / (density $\times$ gravity) = head
$(kg \cdot m/m^2 \cdot s^2) / (kg \cdot m/m^3 \cdot s^2) = m$

This energy/weight method of expressing potential energy is simpler and more convenient because it only has units of length. Therefore, it is used mostly to characterize the state of soil water in terms of the total hydraulic head, the gravitational head, and the matric head. Head units are normally used to describe the potential energy of soil water in the field and can be considered as the height of a column of water corresponding to a given pressure.

Useful Conversions

The three methods of expressing soil water potential energy can be converted to one another by the following relations:

$\psi_{pw} = \psi_{pm}/g = \psi_{pv}/(\rho_w g)$
$\psi_{pv} = \psi_{pm}\rho_w$
$\psi_{pm} = \psi_{pw}g = \psi_{pv}/\rho_w$

where ψ_{pw} is the potential energy on a weight basis, ψ_{pm} is the potential energy on a mass basis, ψ_{pv} is the potential energy on a volume basis, ρ_w is the density of water, and g is the gravitational acceleration. Useful conversions of soil water potential units are given in Table 8.2.

Example

Convert the potential energy/mass of 100 J/kg to its equivalent in volume (Pa) and weight (m) units.

On a volume basis, we use $\psi_{pv} = \psi_{pm}\rho_w$:

$(100 \text{ J/kg})(1000 \text{ kg/m}^3) = 100{,}000 \text{ J/m}^3$
$= 100{,}000 \text{ Pa} = 100 \text{ kPa}$

On a weight basis, we use $\psi_{pw} = \psi_{pm}/g$:

$(100 \text{ J/kg})/(9.8 \text{ m/s}^2) = 10.2 \text{ m} = 1020 \text{ cm}$

Table 8.2. Useful Conversions of Soil Water Potential Units

Expression	J/kg	Pa	bar	psi	cm H_2O	cm Hg
Energy/mass						
1 J/kg	1	10^3	10^{-2}	0.145	10.2	7.5
Energy/volume						
1 Pa	10^{-3}	1	10^{-5}	1.45×10^{-4}	0.0102	0.0075
1 bar	10^2	10^5	1	14.5	1020	75
1 psi	6.89	6890	0.0689	1	70.2	51.2
1 atm	101.3	1.013×10^5	1.013	14.7	1033	76
Energy/weight						
1 cm H_2O	0.0981	98.1	9.81×10^{-4}	0.0142	1	0.735

Note: To convert from the unit of the expression, multiply by the factor to obtain the dimensions of the other factor.

Component Water Potentials

In this section, the definitions of each of the component potential energies of soil water are given, as well as the mathematics used to quantify their magnitudes in soil systems.

Gravitational Potential Energy

Every body on the earth's surface is attracted toward the earth's center by a gravitational force equal to the weight of the body. This gravitational force is

gravitational force = weight = mass $\times$ acceleration
$$= mg$$

The gravitational force acts on a body irrespective of whether the body is static or in motion.

Gravitational potential energy, ψ_g, can be derived from Newton's law of gravitation as follows:

$$F = GmM/z^2$$

where G is the universal gravitational constant, m and M are the masses of the two bodies, and z is the distance between their centers of mass (see also Appendix C).

To raise a body against this attraction, work must be expended. This work is in the form of the gravitational potential energy. Mathematically, this work can be expressed on a mass basis as

$$\psi_g = \int (F/m)dz = GMz/(R - z)R \cong GMz/R^2 \qquad [11]$$

where R is the radius of the earth and z is the depth in the soil profile. Note that R >> z. The work needed to lift the water from some depth in the profile (R − z) to the soil surface (R) is obtained by integrating over the distance moved.

By convention, the gravitational force acting on a body of mass M at the earth's surface is F = mg, where g is the gravitational acceleration. Therefore,

$$g = GM/R^2$$

where $R - z$ approaches R for a body near the earth's surface. Thus, the potential energy per unit mass is

$$\psi_{gm} = gz \qquad [12]$$

Equation [12] forms the definition of gravitational potential of soil water when expressed on a mass basis. It shows that ψ_g is linearly related to changes in elevation and is independent of the chemical or pressure conditions of soil water; that is, it is independent of soil properties. Note that the units of equation [12] are m^2/s^2, which are equivalent to J/kg.

In terms of gravitational potential energy per unit volume, equation [12] becomes

$$\psi_{gv} = \rho_w gz \qquad [13]$$

Note that since the density of water has dimensions of $1000 \ kg/m^3$, the magnitude of the volume expression is 1000 times larger than the mass expression. However, if the potential energy by volume is expressed in kilopascals, the dimensions are the same.

In terms of gravitational potential energy per unit weight, equation [11] becomes

$$\psi_{gw} = z \qquad [14]$$

where z is the distance measured in the vertical direction from some arbitrary chosen reference level. Once again, this equation shows that ψ_g depends only on the position of the point relative to the reference level, and not on any soil property or characteristic.

In equations [12], [13], and [14], positive values of z are measured when the soil water at the point of interest is above the reference level. This means that soil water is capable of doing work under the acceleration of gravity. Negative values of z are measured when the water is below the reference level, which means that work must be performed in moving the water to the reference level.

The Soil Science Society of America (1997) defines the gravitational potential as follows:

> The amount of work that must be done per unit of a specified quantity of pure water in order to transport reversibly and isothermally an infinitesimal quantity of water, from a pool of soil solution at a specified elevation and at atmospheric pressure, to a pool identical to the source pool but at the elevation of the point under consideration.

Example

Assume that a piezometer (an open-ended cylindrical tube) has been inserted into a soil profile and that this tube contains water at the 0.90 m depth (Figure 8.7). On a mass, volume, and weight basis, compute the gravitational potential at the water table assuming that the reference level is (1) located at the soil surface, (2) at the depth of the water table, and (3) at a drain that is located at a depth of 1.2 m, respectively. The solution is given in the table below for the three positions and three bases of expression.

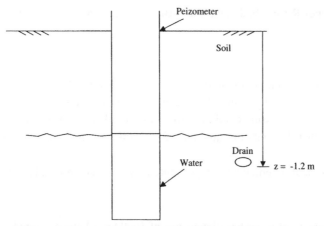

Figure 8.7. Schematic of a piezometer inserted into a soil profile containing a drain at the 1.2 m depth.

Position	Mass (J/kg)	Volume (Pa)	Weight (meters)
	Basis of expression		
Reference level at the soil surface			
Soil surface	0	0	0
Water table	−8.82	−8,820	−0.9
Drain	−11.76	−11,760	−1.2
Reference level at the water table			
Soil surface	8.82	8,820	0.9
Water table	0	0	0
Drain	−2.94	2,940	−0.3
Reference level at the drain			
Soil surface	11.76	11,760	1.2
Water table	2.94	2,940	0.3
Drain	0	0	0

These results show that varying the position of the reference level changes the absolute value of the gravitational potential. The position of the reference level can, however, be arbitrarily chosen since in quantifying the flow of soil water we are mainly interested in the difference in potential energy between two positions.

Example
Using the results of the example above, graph the gravitational potential distribution on a weight basis in the soil profile with the reference level at the three positions in the profile. The results are shown in Figure 8.8.

It should be noted that the lines in Figure 8.8 are parallel and differ only in their position; that is, the slopes of the lines are the same but the lines differ in the location of the intercept (i.e., the reference level) of the zero potential axis.

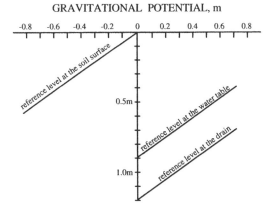

GRAVITATIONAL POTENTIAL, m

Figure 8.8. Relationships of the gravitational potential energy distribution at three positions of the reference level.

Pressure Potential Energy

The soil water pressure potential, ψ_p, is due to the weight of water at a point under consideration or to a gas pressure that differs from a gas pressure at a reference point. The reference for the pressure potential is by convention selected as atmospheric pressure. When soil water is at a hydrostatic pressure greater than atmospheric, its pressure potential is considered to be positive. Water under a free water surface has a positive pressure potential. The pressure potential at the free water surface is zero. In the field, the pressure potential applies mostly to saturated soil water below water tables or to "perched" water tables.

The formal definition of ψ_p as given by the Soil Science Society of America (1997) is as follows:

> The amount of work that must be done per unit of a specified quantity of pure water in order to transport reversibly and isothermally an infinitesimal quantity of water from a pool of soil solution at the elevation and external air pressure of the point under consideration to the soil water at the point under consideration (below the water table).

The positive pressure potential that occurs below the groundwater table has a hydrostatic pressure of water per unit mass with reference to the atmospheric pressure of

$$P = gh \qquad [15]$$

where h is the submergence depth below the free water surface. The potential energy of this water is then

$$E = P \, dV \qquad [16]$$

Therefore, the pressure potential energy per unit volume of soil water is

$$\psi_p = \rho_w gh \qquad [17]$$

and the potential energy per unit weight is

$$\psi_p = h \tag{18}$$

This indicates that on a weight basis, the pressure potential is the vertical distance from a point in question to the free water surface. The pressure potential results from a net pressure difference such as resulting from hydraulic pressure in a saturated soil, pneumatic pressure in a pressure membrane apparatus, and turgor pressure in plant cells. It does not include any effects that result from a curved air-water interface or adsorption to a solid, because where the system is open to the atmosphere, there is no overall pressure difference at the interface. Pressure potentials can be measured with devices that measure pressure. Under hydrostatic conditions in the field, ψ_p can most easily be measured by the depth of water in a piezometer.

Example

Calculate the pressure potential distribution in the profile of a soil with a water table at the 60 cm depth.

The results are shown in Figure 8.9. Above the water table, ψ_p is zero; below the water table, ψ_p increases directly with distance from the water table surface.

Under some conditions, the total soil water potential, ψ_t, is the sum of the gravitational and pressure potentials only. For example, calculate the water potential distributions of a 1-liter cylinder filled with water. Assume that the reference level is first located at the bottom of the cylinder and then at the top. Interpret your answers.

The distributions of the total, gravitational, and pressure potentials in the 1-liter cylinder are shown in Figure 8.10. These results show that the gravitational potential energy of the water in the cylinder is compensated by the pressure potential, and that there is no change in the distribution of ψ_t with position in the cylinder. The uniform distribution of ψ_t with depth indicates that there is no vertical movement of water in the cylinder.

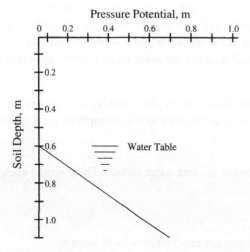

Figure 8.9. Profile distribution of the pressure potential in a soil with a water table at 0.6 m depth.

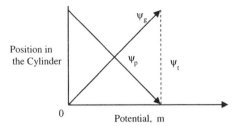

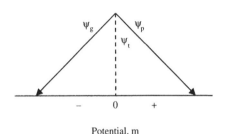

Figure 8.10. Distribution of the total, gravitational, and pressure potentials within the 1-liter cylinder filled with water.

Matric Potential Energy

Soils consist of a solid matrix with pores distributed between the particles. The pores have different sizes, shapes, and spatial distributions and provide space for storage and transport of water and gases. Storage or retention of water by soils is a result of attractive forces such as adhesion and cohesion between the solid and liquid phases, which are responsible for the binding of water in the pores or capillaries.

The formal definition of matric potential energy, ψ_m, as given by the Soil Science Society of America (1997) is as follows:

> The amount of work that must be done per unit of a specified quantity of pure water in order to transport reversibly and isothermally an infinitesimal quantity of water, from a pool of soil solution at the elevation and the external gas pressure of the point under consideration, to the soil water at the point under consideration (above the water table).

The general relationship between the forces of attraction of water by soil surfaces and water content is shown in Figure 8.11. This shows that the first layer of water is held with great forces of attraction. This is because water is a dipolar compound, which is attracted to the opposite-charge sites on the mineral soil. Generally, this process, which is called *sorption*, results in relatively immobile and fixed water. The layer close to the particle surface is referred to as the *Stern layer*. However, the magnitude of the forces of attraction for water decreases as the distance from the particle surface increases. The water that does not belong to the Stern layer is considered to belong to the *diffuse layer*,

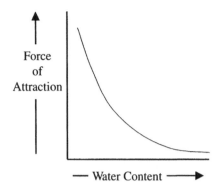

Figure 8.11. Relationship between the forces of attraction and the soil water content.

which is generally of undefined thickness. Almost all plant growth depends on a water film thickness of approximately 5–10 molecular layers thick.

In relatively dry soils, the attraction of the soil matrix for water involves sorption of water by soil surfaces and by ions sorbed to the exchange sites. These adhesion forces can be characterized by the ratio of the potential energy of water molecules near the solid surface to the area of the solid-water interface. Because the surfaces of soil solids are usually composed of a layer of either oxygens or hydroxyls, hydrogen bonding can develop, with oxygens atoms attracting the positive corners and hydroxyls attracting the negative corners of water molecules. The formation of hydrogen bonds alters the electron distribution from that in "normal" water, thus making it easier for bonded molecules to form additional bonds with molecules in the same and the next layer. Therefore, the "sorbed" water molecules are highly oriented in the first adsorbed layer, with the structure becoming less orderly and more like that of "free" water as the water layers or water content increase.

With 2:1 clays, water is strongly associated with several of the exchangeable cations and only weakly bonded, if at all, to the oxygen surfaces. The electric field of the cation orients the polar water molecules around the ion to form a hydration shell, which is an example of an ion-dipole interaction. The energy of hydration depends on the cationic radius and charge, the energies being in the order $Al^{3+} > Mg^{2+} > Ca^{2+} > Na^+ > K^+ > NH^{4+}$. The greater the hydration of the cation, the larger the reduction in energy of the water molecules in the hydration shell. Thus, soil water in the very dry state is mostly associated with the exchangeable cations.

The type of cation on the exchange site determines the geometry or structure of the water and the thickness of the oriented water layers around soil particles. Calcium, which usually is the most common adsorbed cation, tends to develop well-oriented water to a thickness of four molecular layers. Any additional water is unoriented; thus, there is a sharp break between liquid and nonliquid water. In an air-dried state, Ca^{2+} tends to develop two molecular layers of well-oriented water. Magnesium exerts a similar influence as calcium, but the maximum possible thickness of the well-oriented water is probably slightly less and the thickness of the water layer is slightly reduced. Sodium adsorbed to the exchange site favors the development of thick layers if water is available. The oriented water is not rigid, and there is no marked separation between oriented and liquid water. In an air-dried state, Na^+ favors the development of a single molecular layer of water. The ions K^+, H^+, Al^{3+}, and Fe^{3+} may be grouped together because they form a tight bond between soil particles and have a small potential for the growth of thick, oriented water layers. These oriented water structures act as a bonding force between soil particles and directly influence the organization of the water molecules and bonding forces between the particles.

The concentration of cations also increases as negatively charged clay surfaces are approached. Because of this increased cation concentration and the restriction on diffusion of ions from the vicinity of the surface as a result of electrostatic attraction, water molecules tend to diffuse toward the surface in order to equalize the cation concentration.

In the absence of solutes, the liquid and vapor phases in an unsaturated soil at equilibrium are related to the relative vapor pressures. For example, the ψ_m can be measured in equilibrium with various vapor pressures. On a mass basis, the relation is

$$\psi_m = (RT/M) \ln(e/e_0) \qquad [19]$$

where e is the vapor pressure of the soil water, which is lower than the vapor pressure of pure free water because of the attraction of the solid matrix; e_0 is the vapor pressure of pure water at the same temperature; R is the gas constant (8.314 J/K·mol); T is the Kelvin temperature; and M is the mass in kilograms of a mole of water (0.018015). The ratio e/e_0 is known as the *relative vapor pressure* (or relative humidity) and has values equal to or less than 1.

Example

Assume that the relative vapor pressure of water in soil at a temperature of 20°C is equal to 0.85. Calculate the matric potential of the soil water.

$$\psi_m = [(8.314 \times 293)/0.018] \times \ln(0.85)$$
$$= -21,989 \text{ J/kg} = -21,989 \text{ kPa} = -2244 \text{ m}$$

If the relative vapor pressure is 0.989, what is the matric potential?

$$\psi_m = [(8.314 \times 293)/0.018] \times \ln(0.989)$$
$$= -1496 \text{ J/kg} = -1496 \text{ kPa} = -152.7 \text{ m}$$

This last calculation shows that the range of interest in e/e_0 from a plant growth standpoint lies between 0.989 and 1.00, a very small range of relative humidity.

As soil water content increases, cohesive forces become more important. Inside the aqueous phase, the net force on the water molecule is zero. In contrast, water molecules at an air-water interface are attracted only toward the interior, because there are fewer water molecules on the air side of the air-water interface. Thus, the energy content of water molecules at the interface is higher than that of water molecules inside bulk water. This also is characterized by the surface tension, which is the interface potential energy divided by the interface area and is expressed in terms of force per unit length (N/m). The result is that water in contact with air will always attempt to reduce its surface area. The surface area for a given mass of water is called the *specific surface,* and the small-

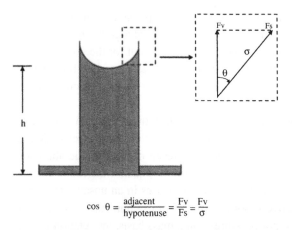

$$\cos\ \theta = \frac{\text{adjacent}}{\text{hypotenuse}} = \frac{Fv}{Fs} = \frac{Fv}{\sigma}$$

Figure 8.12. Diagram of the capillary tube in water, with an enlargement of the area around the point of contact. The symbols Fv and Fs represent the forces expressed vertically upward and to surface tension, respectively.

est specific surface is that of a sphere. This explains the formation of spherical droplets of rain, sprays, etc. Any other shape would represent a larger specific surface and, thus, a higher energy content. For the same reason, the meniscus of water in a capillary assumes the spherical shape, which is compatible with the angle of contact. As water layers increase, water begins to fill the smallest pores and interstices between mineral particles and forms curved air-water menisci.

The definition of ψ_m does not require that a distinction be made between the different forces responsible for attracting water to the soil matrix. Therefore, ψ_m is associated with the attraction of solid surfaces for water as well as with the attraction of water molecules for each other. It also includes the unbalanced forces across air-water interfaces that give rise to the phenomenon of surface tension. The matric potential is equivalent but opposite in sign to the soil water tension, soil suction, soil water suction, and matric suction.

Derivation of the Capillary Rise Equation

The forces active in soil pores can be illustrated by considering the soil as a system of parallel cylindrical glass capillaries gradually decreasing in diameter. When solid, water, and air phases come together at one point, as in a capillary tube, the shape of the interface, as characterized by the angle of contact θ, depends on whether $\cos \theta$ is positive, negative, or zero and on its magnitude with respect to the surface tension, σ. This can be shown by the enlargement of the area around the point of contact in Figure 8.12. The angle between the meniscus and the solid surface at the point is θ.

The attraction of water molecules for the glass walls of the capillary tube, along with the surface tension of the water, causes the water to curve upward in the meniscus. This upward curvature causes the pressure just below the meniscus to decrease below atmospheric pressure. The atmospheric pressure on the water outside the capillary tube forces water up into the tube until the difference of pressures is equilibrated by the hydrostatic pressure ($\rho_w gh$) at height h. Water is not pulled up by capillarity but is forced up by pressure differences.

Consider the water in the triangular volume of unit width perpendicular to the paper, for which the maximum distance to the solid surface is d. The distance d is arbitrary, except that it should be large compared to the distance over which the adhesive and cohesive forces are active, and small compared to the radius of curvature of the meniscus.

The column of water is held up by a force equal to $2\pi r \sigma \cos(\theta)$, where σ is the surface tension (N/m or kg/s^2), and θ is the angle of contact of water with the capillary tube. This expression arises from the force, $\sigma \cos(\theta)$, acting vertically upward per unit length of the circumference of the circle of length $2\pi r$ along which the air-water interface contacts the wall of the tube. For all the way around the cylinder we multiply by the line of contact between the liquid and the capillary tube, that is, the circumference of the tube. Thus, the upward forces can be expressed by

$$F_{up} = (2\pi r)\sigma \cos(\theta) \qquad [20]$$

The downward force is equal to the weight of water, $\pi r^2 h \rho_w g$, where ρ_w is the density of water, g is the gravitational acceleration, r is the radius of the capillary tube, and h is the equilibrium height above the free water surface. We neglect the mass of the air displaced by the column of water and the mass of the water lens of the meniscus. Thus, the downward forces are

$$\begin{aligned} F_{down} &= (\text{mass of water}) \times \text{gravity} \\ &= \text{volume} \times \text{density} \times \text{gravity} \\ &= \pi r^2 h \rho_w g \end{aligned} \qquad [21]$$

At equilibrium, the upward and downward forces are equal. Equating the magnitudes of the upward and downward forces, we have the following:

$$(2\pi r)\sigma \cos(\theta) = \pi r^2 h \rho_w g$$

or

$$h = 2\sigma \cos(\theta)/(r\rho_w g) \qquad [22]$$

where h is the height of water in the capillary tube (m). Setting P = pressure = $\rho_w g h$, we can write

$$P = 2\sigma \cos(\theta)/r \qquad [23]$$

where P is the pressure difference between water at the top of the column and the atmosphere. The pressure P is given negative values with reference to the atmosphere when the center of curvature is on the air side of the interface, indicating a suction or tension. Assuming that the surface is completely wet and the contact angle is zero, equation [23] becomes

$$P = 2\sigma/r \qquad [24]$$

where r is the radius of curvature of the water-air meniscus. Equation [24] shows that smaller values of r result in larger values of P. In terms of diameter of the capillary tube, equation [24] can be written as

$$P = 4\sigma/d \qquad [25]$$

Table 8.3. The Effect of Temperature on Surface Tension of Water

Temperature (°C):	0	10	15	20	25	30	35	40
Surface tension (mN/m):	75.7	74.2	73.4	72.8	71.9	71.2	70.3	69.6

Source: Hillel 1980, p. 49.

In terms of diameter of the capillary table, equation [22] becomes

$$h = 4\sigma \cos(\theta)/(d\rho_w g) \qquad [26]$$

where h is the equilibrium height of water. Substituting for the properties of water at 20°C and assuming a contact angle of 0 degrees gives

$$h \approx 0.15/r \qquad [27]$$

where h has units of centimeters.

The surface tension of water, σ, is the force per unit length needed for an increase in the surface area. It decreases with an increase in temperature. Surface tensions of water at several temperatures are given in Table 8.3.

Example

Calculate the equilibrium height of rise of water in a capillary tube of 10 μm diameter (silt size) and assume that the contact angle is zero and the temperature is 15°C.

$r = d/2 = 5 \mu m = 5 \times 10^{-6}$ m
$\cos \theta = \cos 0 = 1$

Since $P = \rho_w gh$, we solve for h using equation [22] to obtain

$$h = (2 \times 0.073)/9.8 \times (5 \times 10^{-6}) \times 1000$$

where $\sigma = 0.073$ kg/s^2; $\rho_w = 1000$ kg/m^3; and $g = 9.8$ m/s^2 = 9.8 N/kg. Thus, the equilibrium height of rise of water in the capillary is h = 2.98 m.

If the appropriate values are substituted into equation [26], we find that

$$h = 4\sigma \cos(0)/\rho_w gd = 0.291/1000 \times 9.8 \times d = 2.97 \times 10^{-5}/d$$
$$= (3 \times 10^{-5})/d \text{ (m)}$$
$$= 2.98 \text{ m}$$

Example

Calculate the equilibrium heights of capillary rise of water and mercury in clean glass cylindrical capillary tubes of the following diameters at 20°C.

For mercury, we have

$$h = (2 \times 0.43 \times \cos 180°)/13,600 \times 9.8 \times (1 \times 10^{-3})$$
$$= -6.45 \times 10^{-3} \text{ m}$$

The results for water as a function of diameter are

2.0 mm:	h = 0.0148 m
0.5 mm:	h = 0.0592 m
0.1 mm:	h = 0.296 m

The results for mercury are

2.0 mm:	$h = -6.45 \times 10^{-3}$ m
0.5 mm:	$h = -2.58 \times 10^{-2}$ m
0.1 mm:	$h = -1.29 \times 10^{-1}$ m

The surface tension of mercury is 0.43 kg/s^2, the contact angle is 180°, and the density is 13,600 kg/m^3. This example shows that, for a hydrophobic condition, the angle of contact is greater than 90°. Therefore, the height is negative, which indicates a capillary depression.

In soil, the air-water interface of a pore is not always spherical, and the pores are not always cylindrical. The curvature of the air-water interface is either convex or concave, depending on the phase. Therefore, the pressure difference cannot be described by a single radius of curvature. Two radii of curvature, r_1 and r_2, are in planes normal to one another and are required to describe the shape at any point on the interface. This can be demonstrated by considering the side view of a water film in the soil in three dimensions (Figure 8.13). The radius r_1 lies in a horizontal plane and has its center of curvature on the water side of the interface. The radius r_2 lies in a vertical plane and has its center of curvature on the air side of the interface. Thus, it is negative. The total curvature is equal to $1/r_1 + 1/r_2$ (Kirkham and Powers 1972).

If the contact angle is zero, the pressure difference across the interface at a point is given by

$$P_a - P_c = \Delta P = \sigma(1/r_1 - 1/r_2) \hspace{3cm} [28]$$

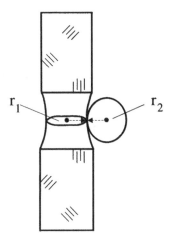

Figure 8.13. Three-dimensional diagram of soil particles and water films.

where P_a and P_c are the atmosphere (usually presumed to be zero) and soil water pressures, respectively, and r_1 and r_2 are taken as positive or negative according to whether the center of curvature lies on the water side or on the air side of the interface. Note that the pressure in equation [28] is negative if $r_2 < r_1$.

An infinite number of combinations of r_1 and r_2 correspond to a given soil water pressure. As the amount of water increases, r_1 approaches the radius of the sphere, r_s, as a maximum limit; r_2 becomes infinite (i.e., $1/r_2$ becomes small). As water is withdrawn by evaporation, by plant extraction, or through contact with drier soil, the total curvature of the water surface between particles becomes zero and then negative in the region between the particles. The two spheres vanish together as the amount of water in the soil decreases. Ultimately, a point is reached where the surface tension is no longer sufficient to maintain water between the particles in a configuration for which the curvature of the air- water interface is meaningfully associated with the pressure in the vapor phase. The water films are of molecular dimensions at this point, and the pressure depends on sorptive forces rather than on the curvature of the air-water interface. The geometry of the air-water interface throughout an unsaturated porous material with particles of irregular shapes and various sizes is too complex for precise quantitative treatment.

To determine the effect of pore size on the soil water pressure, we use equation [22] for calculation of an "equivalent radius," r_e, of the largest pore to remain full of water when a matric head, h, is used to drain the soil. Assuming the contact angle to be zero, the matric potential in terms of weight is given as

$$\psi_m = -2\sigma/\rho_w g r_e \qquad\qquad\qquad [29]$$

The volume of water withdrawn upon decreasing the matrix head from ψ_{m_1} to ψ_{m_2} in soil with a rigid matrix represents the volume of pores in soil having an equivalent radius between r_{e_1} and r_{e_2}. This determination of an equivalent radius for pores is a simplification and may be compared to that made for particles of irregular shape in determining particle size from sedimentation rates using Stokes's law.

Osmotic Potential Energy

The soil water potential caused by solutes in the soil or plant water system where a semipermeable membrane is present is designated as the osmotic potential, ψ_o. Thus, ψ_o is important where the salinity of the soil is high and there is a semipermeable membrane. A semipermeable membrane is a material that allows water to flow but does not permit solutes to pass through. Examples include plant roots (not perfect because some salts are absorbed by the roots) and water-air interfaces (near-perfect semipermeable membranes). The value of ψ_o is not important where liquid soil water flow occurs since there is no semipermeable membrane.

When only pure water and not dissolved solutes can move from one location to another, plants may wilt even though there is a high soil water content, and in the case of bare soils, salts may be transmitted to or near the soil surface and left behind when water is evaporated. This may result in reduced evaporation and infiltration of water and changes in color of the soil surface. In both cases, the osmotic effects must be taken into account because water can be transmitted across the interface but solutes remain behind in the liquid phase. Thus, where the salts are not allowed to move with the soil water because of the presence of a semipermeable membrane, there will be a contribution from ψ_o to ψ_t.

The formal definition of ψ_o given by the Soil Science Society of America (1997) is as follows:

> The amount of work that must be done per unit of a specified quantity of pure water in order to transport reversibly and isothermally an infinitesimal quantity of water from a pool of pure water, at a specified elevation and atmospheric pressure, to a pool identical to the source pool but containing soil solution.

Osmotic potential is the portion of the water potential that results from the combined effects of *all* solute species present in the solution. It results from the hydration of ions in the soil solution, which lowers the potential energy of the soil solution. Osmotic potential is similar to osmotic pressure, osmotic suction, and solute suction but is opposite in sign; that is, $\psi_o = -\Pi$, where Π is the osmotic pressure.

As with the other component potential energies, osmotic potential can be expressed in terms of mass, volume, or weight. The units of potential osmotic energy based on mass are joules per kilogram (J/kg), and the governing equation is

$$\psi_o = gh_0 \tag{30}$$

where h_0 is the height of a hanging water column with respect to the point of measurement in the solution. The quantity h_0 is always negative and can be considered as the gauge pressure.

The unit of potential osmotic energy based on volume is the pascal (Pa), and the governing equation is

$$\psi_o = \rho_w gh_0 \tag{31}$$

The SI unit of potential osmotic energy based on weight is the meter (m), and the governing equation is

$$\psi_o = h_0 \tag{32}$$

Values of ψ_o can be calculated if the concentrations and the interactions of all the separate solute species in solution are known. The van't Hoff equation can be used to calculate the osmotic potential from

$$\psi_o = -RTC_s \tag{33}$$

where R is the gas constant (8.314 J/K·mol), T is the Kelvin temperature, C_s is the solute concentration (mol/m^3), and ψ_o is the joules per kilogram or cubic meter of water, depending on whether the unit quantity of water is expressed on a mass or a volume basis. However, since ionization of the various salts varies with concentration and also in the presence of other salts, it is difficult to determine accurately the concentration of each solute species in soil solutions. Consequently, ψ_o of soils cannot be accurately determined by taking the product of the osmotic pressures and the number of species and summing over all species.

As shown above, the ψ_o of a solution varies with the chemical composition of the solution. Primarily, this variability is based on the number of ions produced per formula weight of salt and not the weight of dissolved solutes alone. Two approaches have been used to estimate the ψ_o of a soil solution. One is to use a linear regression between ψ_o and the total dissolved solids (TDS). The equation developed by the U.S. Salinity Laboratory Staff in 1954 (Bresler et al. 1982) and modified to SI units is

$$\psi_o = -0.056 \text{ TDS} \tag{34}$$

where TDS is measured in milligrams per liter (mg/L) and ψ_o is in kilopascals (kPa).

Another approach is to relate the electrical conductivity (EC) to the ψ_o. Electrical conductivity is based on the principle that the amount of electrical current transmitted by a salt solution under standardized conditions increases as the salt concentration of the solution increases. Values of EC are often used to infer values of total salt concentration. The resulting relationship between ψ_o and EC developed by the U.S. Salinity Laboratory Staff in 1954 (Bresler et al. 1982) and modified to SI units is

$$\psi_o = -36 \text{ EC} \tag{35}$$

where EC is measured in decisiemens per meter (dS/m) and ψ_o is in kilopascals (kPa).

In a soil-water system, it is more appropriate to estimate the combined effects of the solutes by special techniques that incorporate an air gap (a semipermeable membrane). Probably the most useful instrument is the vapor pressure psychrometer. Since the ψ_o of pure water is zero, the ψ_o of a solution at the same temperature is negative. The magnitude of ψ_o is dependent only on the concentration of the solution and on the properties of the solvent. When the unit quantity of water is expressed in terms of mass, ψ_o becomes

$$\psi_o = (RT/M) \ln(e/e_0) \tag{36}$$

where R is the gas constant (8.314 J/K·mol), T is the Kelvin temperature, M is the partial molal volume (kg/mol), e is the equilibrium vapor pressure for the solution, and e_0 is the equilibrium vapor pressure for the pure solvent. The ratio e/e_0 is less than 1, which makes $\ln(e/e_0)$ negative. The ratio is sometimes called the relative humidity.

When ψ_o is expressed as per unit volume of water, equation [36] becomes

$$\psi_o = (RT/V) \ln(e/e_0) \tag{37}$$

where V is the molar volume (the volume of 1 mole of water is 18 cm³/mol), e_0 is the vapor pressure of water, and e is the vapor pressure of the solution. Note that for water, ρ_w is equal to M/V.

A molal solution of an undissociated substance at 0°C has a theoretical osmotic pressure potential of –2270 kPa. Since the osmotic pressure is directly proportional to the number of solute molecules per solvent molecules, a 0.5 molal solution would have a theoretical osmotic pressure of –1135 kPa.

Example

Compute the value of ψ_t of a saturated soil at 20°C at a point C through which the reference level passes. The water table is 1.2 m above point C. Assume that at saturation, the volumetric water content of the soil is 0.5 m³/m³ and that 1 cm³ of soil at point C has 3×10^{-4} moles of solute.

We calculate the potentials on a volume basis. First, we write

$$\psi_t = \psi_g + \psi_p + \psi_m + \psi_o$$

In this particular case, $\psi_g = 0$; since the soil is saturated, $\psi_m = 0$; and the other potentials are calculated as follows:

$$\psi_o = -RTC_s = -8.31 \text{ J/K·mol} \times 293 \text{ K} \times (3 \times 10^{-4} \text{ mol}/5 \times 10^{-7} \text{ m}^3)$$
$$= -1.46 \times 10^3 \text{ kPa}$$
$$\psi p = 1.2 \times 9800 = 11.8 \text{ kPa}$$
$$\psi_t = -1.46 \times 10^3 + 11.8 \text{ kPa} = -1.448 \times 10^3 \text{ kPa}$$

Note the C_s is defined on a soil solution basis and not on a soil basis.

Summary of Soil Water Potential Concepts

Water at any position in a soil profile is under the influence of forces such as elevation or gravity, weight of water standing above a submerged soil, interaction with the matrix of solid particles such as in unsaturated soil, curved air-water interfaces, and the presence of solutes in soil water. Since these forces act in various directions, it may be difficult to find their combined effect on water movement. Thus, it is easier to work with ψ_t.

Signs of the component potentials are important. In tabular form, the signs of the potentials can be summarized as follows:

Potential		Sign	
Total	+	0	–
Gravitational	+	0	–
Pressure	+	0	
Matric		0	–
Osmotic		0	–

Soil water moves from a region of higher total potential energy to a region of lower total potential energy. The energy scale can be visualized as follows:

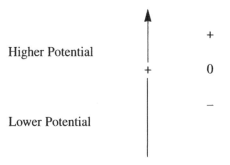

Hydrologic Horizons

For purposes of describing water movement in soils, it is useful to define a set of horizons or zones based on the normal range of soil water contents and potentials. The depths and thicknesses of the hydrologic horizons vary in both time and space, and one or more of them may be absent in a given situation in the field.

Groundwater Zone

The groundwater zone, which is sometimes called the *phreatic zone,* is saturated, and the pressure below the surface is positive. The word "phreatic" comes from the Greek word for "well." If there is no flow of groundwater, the pressure will be hydrostatic, thereby increasing linearly with depth according to

$$Hp = z - z_0 \qquad [38]$$

where Hp is the soil water pressure head (m), z is the vertical distance (m) and is positive upward, and z_0 is the depth of the water table or potentiometric surface (m). The top of the water table is at atmospheric pressure.

Most groundwater tables rise and fall in response to seasonal climatic variations and plant uptake and to recharge from individual storm events. In arid regions, the depth of groundwater may be tens of meters, if groundwater is present at all, whereas in humid regions, its depth may be shallow.

Vadose Zone

The vadose zone commonly refers to the entire variably saturated zone above the water table. The lowest portion of this zone is a region that is nearly saturated or satiated as a result of capillary rise. The pore spaces in the soil or porous medium act like capillary tubes, and surface tension forces draw the water into the region above the water table. The vertical extent of this tension-saturated zone, or capillary fringe, can be approximately calculated by equation [22], repeated below:

$$h = 2\sigma \cos(\theta)/\rho_w gr \qquad [39]$$

where h is the height above the water table, σ is the surface tension, θ is the angle of contact, r is the radius of the tube, and ρ_w is the density of water. The height of capillary rise may range from almost negligible in coarse materials to several meters in clay materials. The pressure head at the top of this zone is the air-entry pressure, since at greater tensions, the pores are partially filled with air.

Root Zone

The root zone is the zone in the soil profile from which plant roots can extract water during transpiration. Its upper boundary is the soil surface, and its lower boundary is indefinite and depends on the ability of the plants to explore the soil profile. This, in turn, depends on the type of plant, the phenological development of the plant roots, and soil physical and chemical properties. In the root zone, the soil water content can be both saturated and unsaturated, depending on the time of the year and depth. In general, plants tend to have greater root densities near the soil surface; thus, more soil water is extracted near the surface than deeper in the profile. Factors that influence plant root distribution and root activity in the soil profile also affect the thickness of the root zone.

Relation between Pedologic and Hydrologic Horizons

Because of the variability in development of pedologic and hydrologic horizons, only a few generalization can be made. Usually, bluish, grayish, and greenish mottling of subsoils, called gleying, indicates reduced aeration because the horizon is below the water table for a significant fraction of the year. Soil hydraulic properties may change relatively abruptly in successive pedologic horizons. In some situations, an impermeable or nearly impermeable layer or horizon called a "hardpan," tillage pan, or fragipan develops at or below the B horizon, and percolating water may accumulate above this layer, forming a perched water table above the general regional water table. Other soils may show a more or less gradual decrease in hydrologic properties with depth.

Potential Distributions of Soil Water under Field Conditions

Knowledge of the total water potential and its component potentials in the soil profile is required for determining the direction of the flow of soil water and the hydraulic gradient. Under isothermal conditions, soil water moves from locations where ψ_t is higher to locations where ψ_t is lower. For example, consider two points within the soil profile having the following total potentials:

point A $\psi_t = -10$ J/kg $= -10$ kPa $= -102$ cm
point B $\psi_t = -15$ J/kg $= -15$ kPa $= -153$ cm

Since the total potential energy is higher at point A than at point B, water will flow from A to B. The driving force to move this water is the difference in total potential energy divided by the distance between the points. This ratio is known as the *hydraulic gradient*.

Frequently, we are interested only in liquid flow of water within the soil profile. Under these conditions, the osmotic potential energy is zero. For liquid flow only, the total soil water potential can be expressed as

$$\psi_t = \psi_g + \psi_p + \psi_m \qquad [40]$$

This particular combination of potentials is commonly known as the *hydraulic potential*.

Under equilibrium conditions, any ψ_t is the same everywhere. Thus, the difference in ψ_t between two locations is zero and there is no flow of soil water in the system.

Example

A DeWitt silt loam has a perched water table above the clayey B horizon. Assuming that the B horizon is at the 40 cm depth, and the height of water ponded above this horizon is 8 cm, calculate the vertical distributions of the total soil water potential and the component potential energies under hydraulic equilibrium conditions to the 50 cm depth in 10 cm intervals. Show the effects on the potential energies of moving the reference level from the soil surface to the water table.

First, we assume that the reference level is placed at the soil surface. The potential energy distributions within the profile are tabulated below.

z (cm)	ψ_g	ψ_o	ψ_p (cm of water)	ψ_m	ψ_t
0	0	0	0	-32	-32
10	-10	0	0	-22	-32
20	-20	0	0	-12	-32
30	-30	0	0	-2	-32
40	-40	0	8	0	-32
50	-50	0	18	0	-32

Notice that values of ψ_t are the same at each depth within the profile. Thus, the flow of water in this profile is zero and the system is at hydraulic equilibrium.

Next, we assume that the reference level is placed at the water table and recalculate the potential energy distributions.

z (cm)	ψ_g	ψ_o	ψ_p (cm of water)	ψ_m	ψ_t
0	32	0	0	-32	0
10	22	0	0	-22	0
20	12	0	0	-12	0
30	2	0	0	-2	0
40	-8	0	8	0	0
50	-18	0	18	0	0

A comparison of the results in the two tables shows that the magnitudes of ψ_t and of one of the component potentials (ψ_g) changed depending on the placement of the reference level. The other potential energies did not change. However, the difference between ψ_t from one location to another did not vary in the profile regardless of the placement of the reference level.

Under most conditions in the field, equilibrium conditions do not occur. Thus, water moves within the profile in various directions and at various rates. Under these conditions, ψ_t is unequal within the soil profile, and nonequilibrium conditions exist.

Example

Assume that water is evaporating from an unsaturated DeWitt silt loam. Given the values of ψ_m in the table, find values of ψ_t and its component potentials to the 50 cm depth in 10 cm increments. Assume that the reference level is placed at the soil surface.

z (cm)	ψ_g	ψ_o	ψ_p (cm of water)	ψ_m	ψ_t
0	0	0	0	−1200	−1200
10	−10	0	0	−250	−260
20	−20	0	0	−165	−185
30	−30	0	0	−80	−120
40	−40	0	0	−50	−90
50	−50	0	0	−40	−90

In this example, values of ψ_t increased from the surface to the 40 cm depth; therefore, soil water was moving toward the soil surface. Hydraulic equilibrium conditions existed between the 40 and 50 cm depths.

Example
Assume that a 10 cm tile drain, which contained water at a height of 2 cm, was placed on top of the B horizon of the DeWitt silt loam and water moves to the drain. The clayey B horizon is located at the 40 cm depth. Find the values of ψ_t and its component potentials to the 50 cm depth in 10 cm increments. Assume that the reference level is at the soil surface.

z (cm)	ψ_g	ψ_o	ψ_p (cm of water)	ψ_m	ψ_t
0	0	0	0	−15	−15
10	−10	0	0	−12	−22
20	−20	0	0	−9	−29
30	−30	0	0	−4	−34
40	−40	0	2	0	−38
50	−50	0	12	0	−38

These data show that ψ_t decreased with depth to the tile drain, and thus, water was moving toward the drain. The height of water ponded in the drain was 2 cm.

Methods for Soil Water Determination

Over the years, numerous techniques have been developed for measuring the water status of a soil. Each technique has advantages and disadvantages in terms of use, cost, precision, and reliability. In this section, we discuss several methods used to measure the soil water status and list the advantages and disadvantages as given in the excellent reviews by Schmugge et al. (1980) and Kutilek and Nielsen (1994).

Water Content Methods

Methods used to determine soil water content include gravimetric techniques, nuclear techniques such as neutron probe and gamma-ray attenuation, and electromagnetic techniques such as time domain reflectance.

Gravimetric Techniques

The most widely used technique for soil water content determination is to take an in situ sample of soil and oven-dry it at 105°C in a forced-draft oven until a constant weight is obtained. Usually, this requires 10–12 hours of drying; however, for large samples and clayey soils, a longer drying time may be required. The gravimetric soil water content is calculated as the ratio of the water lost to the oven-dry soil weight. If volumetric water content is required, the gravimetric water content is multiplied by the ratio of the soil bulk density to the density of water.

There are several advantages and disadvantages of the gravimetric techniques. Some advantages include the following: (1) the samples of soil can be taken with an auger or sampling tube, (2) sample acquisition is inexpensive, and (3) soil water content is easily calculated. Some disadvantages include the following: (1) it is difficult to obtain representative soil water contents in a heterogeneous soil profile, (2) many samples may be required to monitor water movement or water content over time and space, and (3) the sampling is destructive.

Nuclear Techniques

Two nuclear techniques widely used for measuring soil water content involve neutron scattering and gamma-ray attenuation.

The neutron moderation method consists of two parts. The first part is a probe, which is lowered into an access tube, which is usually either an aluminum irrigation pipe or an iron pipe inserted vertically into the soil profile. The probe contains a source of fast neutrons, a detector of slow neutrons, a preamplifier, shielding material, and a cable. The detector is usually filled with BF_3 gas. When "thermalized," neutrons encounter a ^{10}B nucleus and are absorbed, and an alpha particle is emitted, creating an electrical pulse on a charged wire, which is counted by the scaler. The radiation source usually is a 2–5 mCi mixture of Ra and Be or a 100 mCi mixture of ^{241}Am and Be. ^{241}Am emits alpha, beta, and gamma radiation; Be is stable but emits neutrons when struck by alpha particles from Am. The alpha and beta particles are stopped by the walls of the source container. The penetrating power of gamma radiation depends on the density. The neutrons are extremely small, very dense particles that are electrically neutral and quite penetrating. The penetrating power of neutrons depends on the composition of the material and not on the density of the material. They are slowed most effectively by materials containing a similar mass, for example, H atoms such as water or polyethylene. The shield is a polyethylene cylinder composed of hydrogenous plastic, which acts as a water equivalent standard for the probe. The second part of the neutron moderation method is a rate meter, or scaler, which is usually battery powered and portable. It is used to monitor the flux of slow neutrons, which is proportional to the soil water content.

The fast neutrons (4.5–5.0 MeV or 1–12 MeV) are emitted radially into the soil. There they are "thermalized," or slowed, when they encounter and collide elastically with various atomic nuclei and lose their kinetic energy. In practice, the attenuation of fast neutrons in soil is proportional to the number of H atoms, or H content. The slowed neutrons (thermal to 0.025 eV) scatter randomly in the soil, forming a cloud around the probe. Some of these neutrons return to the probe, where they are counted by the gas-filled proportional detector of slow neutrons. Since the only significant source of H in most soils is water, the technique offers a convenient means of estimating the water content.

The count rates, adjusted for background and referenced to counts in tanks of pure water, are then calibrated against direct volumetric determinations of water content. A calibration line fits all soils except those with high clay content, abrupt changes in bulk density, and/or large amounts of Cl, Fe, and B. The equation has the form

$$\theta_v = (Rs/Rstd)(b - a) \tag{41}$$

where Rs is the count rate in the soil, Rstd is the count rate in the shield, and b and a are calibration factors. The count rate ratio is Rs/Rstd.

The neutron probe technique has several advantages and disadvantages. Advantages of this technique are that it (1) measures a large soil volume that is a function of the soil water content near the probe, (2) measures volumetric soil water content, (3) is nondestructive, (4) has no lag period while the soil water equilibrates with a sensing element, and (5) can be repeated at the same site. Disadvantages are that (1) it is a radiation instrument and so should not be used near the surface of dry soils and must be handled carefully, (2) a calibration curve is needed for each site, (3) the installation of an access tube disturbs the site, (4) the equipment is somewhat expensive, and (5) it does not accurately measure water content distributions near the soil surface.

Gamma-Ray Attenuation

The gamma-ray attenuation method is also a radioactive technique and can be used to determine soil water content within a 1–2 cm soil layer. This method assumes that scattering and absorption of gamma rays are related to the density of matter in their path and that the density of soil remains relatively constant as water content changes.

The gamma-ray source is surrounded by a collimator, a detector with a collimator, and a scaler. The source may be 25 mCi of ^{137}Cs with a lead collimator. The beam emerges from a circular hole 4.8 mm in diameter, and a scintillation counter, used as a detector, is shielded by a lead collimator having a 12.5 mm diameter hole. Gamma rays are collimated to a narrow beam, which permits a representative reading to be obtained at any position in the profile. The source and detector are lowered down parallel holes.

If N_0 is the count rate for gamma rays transmitted from the source through air to the detector, N_m is the count rate through wet soil, and s is the path length, then

$$N_m/N_0 = \exp[-s(\mu_s\rho_b + \mu_w\theta_v)] \tag{42}$$

where μ_s and μ_w are the mass attenuation coefficients for soil and water, respectively. Under absolutely dry conditions, the count rate is N_d and equation [42] becomes

$$N_d/N_0 = \exp(-s\mu_s\rho_b) \tag{43}$$

Substituting equation [43] into equation [42] and solving for the volumetric soil water content give

$$\theta_v = \ln(N_d/N_m)/s\mu_w \tag{44}$$

The gamma-ray attenuation technique has the same advantages and disadvantages as the neutron probe. An additional advantage is that water contents can be obtained over a small horizontal or vertical distance. Additional disadvantages are that (1) large variations in bulk density and water content can occur in highly stratified soils and limit spatial resolution, (2) two access holes, one for the emitter and one for the receiver, must be parallel, and (3) field instrumentation is costly and difficult to use.

Electromagnetic Techniques

Electromagnetic techniques include those methods that depend upon the effects of water content on the electrical properties of soil. Electromagnetic approaches exploit the water dependence of the dielectric properties of the soil, which varies with solid water content.

Dielectric properties of moist soil may be characterized by a frequency-dependent complex dielectric response function:

$$\xi(w) = \xi_r(w) + j\xi_i(w) \hspace{4cm} [45]$$

where $\xi_r(w)$ is the real part of ξ, $\xi_i(w)$ is the imaginary part of ξ, j is the square root of -1, and w is the angular frequency. The function $\xi_r(w)$ is almost constant from where the water content is zero to the neighborhood of the relaxation frequency w_r of dipoles in the medium. The real part of the dielectric response function is a measure of the energy stored by the dipoles aligned in an applied electromagnetic field. The function $\xi_i(w)$ is a measure of the energy dissipation rate in the medium. The behavior described is due to permanent dipoles in the soil medium.

In dry soil, values of ξ are typically between 3 and 5; in water, the value is about 80. Therefore, relatively small amounts of water in a soil will greatly affect its electromagnetic properties. At low water content, there is a slow increase of ξ with soil water content. This is due to the sorption by soil surfaces and the influence of water structure on the freedom of water molecules to become aligned. The dielectric properties of this water resemble that of ice, in which ξ is 3.5. As the water films become larger, the influence of binding to the particles decreases and the water molecules behave as they do in the liquid state and the slope of the curve sharply increases.

There are a variety of soil water sensors that depend on the dielectric properties of soil. In general, three types are available: soil resistivity, soil capacitance, and remote sensors.

Sensors that depend on the resistivity also depend on water content. It is possible either to measure the resistivity between electrodes in a soil or to measure the resistivity of a material in equilibrium with the soil. Sensors of either kind can be compact, and an array of them can be connected to standard data collection platforms. The difficulty with resistivity sensors is that the absolute value of soil resistivity depends on ion concentration as well as on water content. Therefore, careful and frequent calibration is required for this technique.

Capacitance sensors are sensitive to polarization of ξ_r, which is the electrical quantity that most directly indicates water content. When water held in the soil can be regarded as free (i.e., above a certain low soil water content), the relationship between ξ_r and water content is almost linear. Several types of sensors are designed for operation from less than 1 to 100 MHz.

The main advantages of resistivity and capacitance devices are that (1) with calibration, they are capable of providing absolute values for soil water content, (2) they can be implanted at any depth, (3) a wide variety of sensor configurations, from small to large, are possible, and (4) their precision is high. The main disadvantages are that (1) the moisture sensor must be implanted properly to minimize disturbances to the soil, (2) the long-term reliability and maintenance of the calibration curve may be questionable, and (3) the cost of readout devices and interfaces with remote collection platforms can be high.

One relatively new tool for measuring soil water content is time domain reflectometry (TDR). In TDR, the elapsed time of an electromagnetic pulse traveling along a waveguide of known length is measured, and the average pulse speed and subsequent dielectric constant are determined. Travel time is measured from the start of the waveguides to the endpoints by a microprocessor. The instrument launches a series of fast-rise step pulses down the cable into the waveguide. Voltage samples are taken at precise time intervals. A TDR waveform for the soil is produced by combining over many averaged sample points. The number of sample point intervals from waveguide start to endpoint (i.e., the end reflection) will determine lapse time, average pulse speed, and the dielectric constant of the soil. Conversion tables are used to convert the derived dielectric constant into the volumetric soil water content. The advantage of this method is that it is safe, quick, does not need to be calibrated (but probably should be calibrated), and measures volumetric soil water content. It can be influenced by salinity.

Potential Energy Methods

Methods used to determine the potential energy status of soil water include piezometers, tensiometers, and psychrometers.

Piezometer

If a cylindrical tube, open at both ends, is inserted into the soil to some depth under hydrostatic conditions, water will gradually fill this tube to the elevation equal to that of the free water level outside the tube. A piezometer can be used for measuring the pressure of soil water at the point where it is attached to the soil. The height of water in the piezometer is measured, and this column of water acts upon the bottom of the tube with pressure

$$P = \rho_w gh \qquad\qquad [46]$$

where ρ_w is the density of water, g is the acceleration of gravity, and h is the height of water in the water column. Flow of water along the piezometer biases the reading.

Tensiometer

A tensiometer is a widely used device for measuring the pressure potential of soil water. It is composed of a porous cup, usually ceramic, and a pressure-measuring device such as a closed manometer or a pressure gauge. If the wall material of a tensiometer cup does not restrict the passage of solute ions, the pressure measured by the tensiometer is equal to the value of the pressure potential of the water divided by the partial volume of the soil water. The matric pressure is calculated from

$$P = -\rho_w gh \qquad\qquad [47]$$

where h is the height of water in the manometer and can take on positive, zero, or negative values. Negative values indicate that the pressure on the soil water is lower relative to atmospheric pressure (i.e., a matric potential). Tensiometers are applicable to a pressure range of 0 to –85 kPa and a small range of positive values. At tensions exceeding these values, air comes out of solution, and the water column in the tensiometer breaks.

Advantages of tensiometers are that (1) the systems are easy to design and construct, are relatively inexpensive, and can be easily installed in the soil with minimum disturbance, (2) information can be obtained on pressure potential distributions under saturated and unsaturated conditions in near real time, (3) systems can operate over long time periods if properly maintained, (4) the response time is rapid in wet soil, and (5) they can operate under near-frozen conditions. Disadvantages are that (1) they can be broken easily during installation, (2) results can be determined only to –85 kPa water tension, (3) response time is slow in dry soil, and (4) field installations using pressure transducers frequently drift electronically.

Psychrometer

The potential of an aqueous solution at its reference water level equals the potential of water vapor immediately above the water level. For a plane water level of pure water, we take conventionally $\psi_m = 0$ and $\psi_o = 0$. The water vapor pressure above that water level is p_o. When the water level is curved as in a capillary tube, the surface pressure is decreased and the water vapor pressure also decreases. The ideal gas law can be written as

$$PV = nRT \qquad [48]$$

so that

$$P = nRT/V \qquad [49]$$

After some manipulation (Marshall et al. 1996), the matric pressure head, h, of the soil water is given by

$$h = (RT/Mg)\ln(p/p_o) \qquad [50]$$

For $T = 20°C$, $h = 1.46 \times 10^4 \ln(p/p_o)$. The units of h are meters.

Since water in the soil is never pure but has a particular solute concentration, the effects of osmotic potential ψ_o and matric potential ψ_m cannot be easily separated. However, the sum of the osmotic and matric potentials can be written as

$$\psi_o + \psi_m = (RT/M)\ln(p/p_o) \qquad [51]$$

Psychrometers have to be extremely accurate. Usually they are constructed on the principle of a thermocouple enclosed in a small cell provided with porous walls. The potential of water in the walls of the cell is in equilibrium with that of the surrounding soil water, and the water vapor pressure inside the cell is identical with that in the soil pores. Inside the cell, the temperature of the "dry bulb" of the thermometer and, using the Peltier effect, the temperature of the junction in the cell are reduced by an application of electromotive force. Water vapor condenses on this cool junction, and the "wet bulb" temperature is read. The accuracy of the procedure is usually on the order of $h = \pm 100$ cm.

In summary, the choice of instrument for measuring the water status of a soil depends upon whether water content or water potential is the better criterion for the intended use, whether the measurements are to be made in the field or in the laboratory, the precision required, the labor involved, and the cost, reliability, and availability of the measuring equipment (Marshall et al. 1996).

Soil Water Retention Curve

The relationship between water content and matric potential is a fundamental physical property of a soil and historically has been called by at least three terms: soil water characteristic, soil water release curve, and the soil water retention curve. The highly nonlinear function θ(h) when plotted is called a *water retention curve.* However, even though soil water content and potential energy are related, the relationship is not always the same for a given soil or even for a horizon within the same profile. The soil water retention curve is both spatially and temporally variable and should be determined experimentally for each horizon in each soil. It is the one measurement that gives the most information about the physical properties of a soil. It does not, however, give enough information to characterize the flow of soil water.

In a saturated soil at equilibrium with free water at the same elevation, the pressure is atmospheric. Therefore, the matric potential is zero (Figure 8.14). If a slight pressure is applied to the saturated soil in a pressure chamber, no outflow may occur until a certain critical pressure is exceeded, at which time the largest pore of entry begins to empty. This critical matric potential is called the *air-entry pressure* and is the matric potential (minus suction or minus tension) at which significant volumes of air begin to appear in soil pores. For coarse-textured soils, the air-entry value is close to saturation, and for fine-textured soils, it may be closer to 10 kPa. As the pressure in the chamber (minus suction or minus tension) is increased or the matric potential is decreased, water is drained from the soil, and more of the relatively large pores, which cannot retain water against the pressure applied, will empty. Thus, a gradual decrease in matric potential (or a gradual increase in applied pressure) will result in the emptying of progressively smaller pores and decreasing thickness of the water films covering the soil particles. Decreasing the matric potential is thus associated with decreasing soil water content.

The amount of water remaining in the soil at equilibrium is a function of the diameters and volumes of the water-filled pores and of the suction (tension) or matric potential with which the water is held by the soil. The pressure (i.e., tension) required to remove water from soil must overcome the various matric forces that retain the water. Thus, for removal of water from the soil, the pressure applied must be greater than the suction (tension) with which the water is held by the soil.

When the soil water content is low, most of the pores are filled with air, and the pressure necessary to remove further water must overcome these molecular forces of attraction of the individual water molecules to the solid particles and exchangeable cations. This matric potential is low, which means that rather high pressures are required to remove this water from the soil. When the water content is higher, the function of this pressure in nonshrinking soil is to maintain the curvature of the air-water interfaces

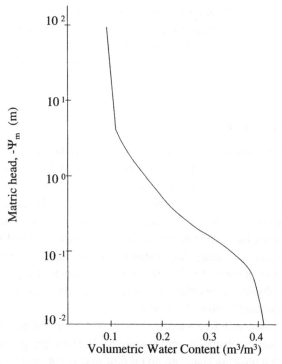

Figure 8.14. Water retention curve for a nonshrinking soil.

between the particles. In shrinking soil, the pressure is also required to overcome the mutual repulsions between the negative charges of the clay particles.

In nonshrinking soil, a quantitative interpretation of the pore size distribution can be obtained. Consider the situation represented by the point x on the drying curve AB in Figure 8.14. This indicates that at the matric suction P, the volume of water held in a unit volume of soil is V. Therefore, in accordance with the definition of the equivalent radius of curvature, this volume of soil water is held under a liquid-air interface whose effective radius of curvature is R, where R is calculated from

$$P = 2\sigma/R \tag{52}$$

where σ is the surface tension (N/m or kg/s^2). This equivalent radius of the pores (i.e., those channels on the point of allowing the entry of air and the loss of water) can just maintain a curvature as sharp as this. Thus, a gradual increase in pressure in the chamber will result in the emptying of progressively smaller pores. The shape of the curve AB provides quantitative evidence of the distributions of the pore space and pore sizes of a soil.

Example

Use equation [52] to calculate the equivalent diameter of the pores filled with water that can be maintained at a matric head of 100 cm at equilibrium and a soil temperature of 15°C.

$$d = 4\sigma/\rho_w gh \times 0.0734/(1000 \times 9.8 \times 1.0) = 2.996 \times 10^{-5} m = 30 \ \mu m$$

What changes in water surface curvature in the pores can be expected if the soil temperature is increased to 20°C? Temperature affects the magnitude of the surface tension. Thus, substituting the surface tension at 20°C gives

$$d = 4 \times 0.0728/(1000 \times 9.8 \times 1.0) = 2.971 \times 10^{-5}m = 29.7 \ \mu m$$

This shows that the 5°C increase in soil temperature results in a change in calculated pore diameter of about 1%.

As demonstrated by the example above, at a given soil temperature, several of the parameters in equation [52] are constant. Thus, at 20°C the equivalent radius is approximated by

$$R \approx 0.15/h \qquad\qquad\qquad [53]$$

where h is the matric head (cm).

The relationship between equivalent pore size and the soil water pressure (tension) required to empty the pores during drying is presented in Table 8.4. There is a wide range of pore sizes in soil. The larger pores are created by the activities of earthworms, insects, and plant roots, by shrinkage of clays due to changes in soil water status, by changes in soil structure due to the weather, and by management operations such as tillage.

Of particular importance to the shape of the water retention curve is the influence of texture (Figure 8.15). For a sandy soil, the water retained decreases rapidly from saturation with a small change in pressure. As ψ_m is decreased further, the slope of the curve decreases to low values, which reflects the small changes in water content with large changes in ψ_m. For clayey-textured soils, the curve intercepts the abscissa at a higher water content, reflecting the greater volumes of water retained at saturation. As ψ_m is decreased, the slope of the curve changes slowly. Therefore, only a gradual reduction of water is found. The curve for a silty-textured soil is intermediate between the sand and the clay.

At high matric potentials, i.e., near saturation, the water retention curve is strongly affected by soil structure. Compaction decreases total porosity, decreases the volume of

Table 8.4. Relationship between Equivalent Pore Diameter and the Soil Water Tension Required to Empty the Pores

Equivalent Pore Diameter (μm)	Critical Soil Water Tension (kPa)	Equivalent Matric Head (m)
20,000	0.015	0.002
4,000	0.075	0.008
300	1.0	0.10
30	10	1.0
2	150	15
0.2	1,500	150
0.003	100,000	10,000

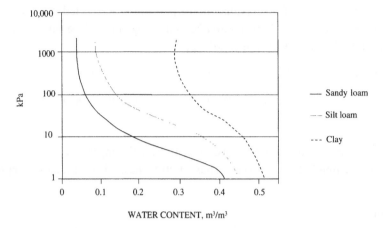

Figure 8.15. Water retention curves for three soil textures.

the large pores, and increases the intermediate-size pores. At low matric potential, the water retention curve is mostly affected by texture and specific surface.

The slope of the soil water retention curve is termed the *specific* or *differential water capacity*:

$$C(\theta) = d\theta/d\psi_m \qquad\qquad [54]$$

and in terms of matric suction head, equation [54] becomes

$$C(\theta) = -d\theta/dh \qquad\qquad [55]$$

Values of $C(\theta)$ are dependent on the slope of the water retention curve. High values of $C(\theta)$ are found when the slope of the water retention curve is high and indicate large amounts of water released per unit change in applied pressure (or matric potential). This is an important property of the soil in relation to water storage and availability to plants. The actual value of $C(\theta)$ depends on the soil water content, the soil texture, and the hysteresis effect (see next section).

Several mathematical models have been proposed to empirically describe the water retention characteristics of a soil. These expressions allow for more efficient representation and comparison of the hydraulic properties of different soils and soil horizons and also allow for interpolation between parts of the retention curve where little or no data are available. Some of these models are given in Table 8.5. An excellent discussion on the application of these models to field-determined water retention data is presented by Bruce and Luxmoore (1986). These authors stress the importance of characterization of errors and variability in determining water retention characteristics.

Hysteresis

The water characteristic curve in many coarse-textured soils is not unique but varies according to the history of wetting and drying. This means that the relation between

Table 8.5. Examples of Empirical Mathematical Models Used for the Soil
Water Retention Curves

Model	Comments
$h(\theta) = a(f - \theta)^n/\theta^m$	where h is the soil water tension (negative matric potential), f is the total porosity, θ is the water content, and a, n, and m are empirical constants.
$h(\theta) = a\theta^{-b}$	where a and b are empirical constants.
$h(\theta) = a\theta_s^{-b}$	where θ_s is the relative saturation, i.e., θ/θ_{sat}.
$h(\theta) = h_0 e^{-b\theta}$	where a, b, and h_0 are empirical constants.
$Se = (\theta - \theta_r/\theta_s - \theta_r) = [1 + (\alpha|h|)^n]^{-m}$	where Se is the effective degree of saturation; θ_r and θ_s are the residual and saturated volumetric water contents, respectively; α is a parameter inversely related to the air entry value (m^{-1}); and m and n are unitless parameters that affect the slope of the location of the inflection point on the retention curve (van Genuchten 1980).
$Se = (\theta - \theta_r/\theta_s - \theta_r) = (\alpha h)^{-\lambda}$ $\alpha h > 1$ $= 1$ $\alpha h < 1$	where Se, θ_r, θ_s, α are defined as above, and λ is a pore size distribution index, which ranges between less than 2 for well-structured soils and 5 for sands (Brooks and Corey 1964).

water content and pressure is not generally a single-valued one and is subject to hysteresis. Hysteresis results when the desorption curve differs from the sorption curve:

- desorption: an initially wet soil is dried by increasing amounts of suction or tension
- sorption: an initially dry soil is gradually wetted by reducing the suction or tension

Each process yields a continuous curve, but at a given soil water content, the desorption water content is greater than the sorption water content. This dependence of the equilibrium water content and state of soil water upon the direction of the process leading up to it is called hysteresis.

The water retention curves in Figure 8.16 show that within the closed loop of boundary wetting and drying curves are *primary* wetting and drying curves terminating at only one of the two boundary curves and *scanning* curves for which neither terminal is on a boundary curve.

The hysteresis effect may be attributed to several factors, including the following:

1. Geometric nonuniformity of the individual pores, i.e., ink bottle effect.
2. Contact angle effect: The angle of contact may be greater in an advancing meniscus than in a receding one. Surface roughness, impurities adsorbed on the solid surface, and the mechanism by which liquid molecules absorb or desorb when the interface is displaced, all contribute to contact angle hysteresis.
3. Entrapped air, which decreases the water content in recently wetted soils.
4. Swelling and shrinking, which change the structure of the soil.

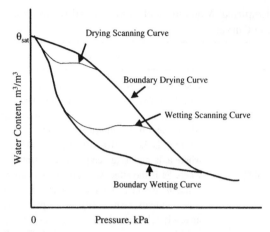

Figure 8.16. Water retention curve showing hysteresis.

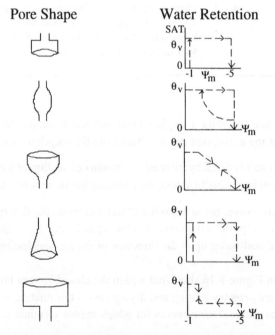

Figure 8.17. Examples of pore shapes and water retention curves.

Of particular interest is the influence of geometric nonuniformity of the pores (i.e., the changes in size and shape of the pores) on the shape of the water retention curves. Suppose we have the pore shapes given in Figure 8.17. If initially saturated, the pore will drain when the tension exceeds the ψ_r, where $\psi_r = -2\sigma/r$. For this pore to rewet, the tension must decrease below ψ_R (the radius of the large circle), where $\psi_R = 2\sigma/R$. Since

R > r, then $\psi_r > \psi_R$. Therefore, desorption depends on the narrow radii of the connecting channels, and sorption depends on the maximum diameter of the large pores. Another way of stating this is that if initially wet, the water content will not change with decreasing ψ_m as long as $\psi_m = -2\sigma/r$. When ψ_m becomes slightly less than $-2\sigma/R$, all of the pores will empty suddenly, and the water content will change abruptly to a new value and will remain constant at the new value until ψ_m is further decreased. Thus, during a drying cycle, pores with narrow necks on top tend to keep large pores filled with water. During wetting, after large pores have been drained, lower narrow necks tend to keep water from flowing out of the pore. Thus, the water content during drying will be greater than during wetting.

Several soil water retention examples for the several pore shapes are shown in Figure 8.17. The hysteresis effect is more pronounced in coarse-textured soils in the low suction range, where pores may empty at a much larger suction than that at which they fill.

Water Terms in the Soil Classification System

Soil classification recognizes the importance of soil water status in three ways (Soil Survey Division Staff 1993). Defined below are the water state classes, drainage classes, and soil water regimes.

Water State Classes

Three water state classes and eight subclasses have been defined for description of the wetness of individual horizons (Table 8.6). The water retention parameters of θ and ψ_m are considered in the definition of the water state classes; ψ_o is not considered.

In Table 8.6, MWR is the abbreviation for the midpoint water retention, which is halfway between the upper water retention and the retention at 1500 kPa; UWR is the upper water retention, which is the laboratory water retention at 5 kPa for coarse soil and 10 kPa for other textures.

Table 8.6. Water State Classes Used in Soil Survey

Class	Criteria
Dry	> 1500 kPa suction
Very dry	< 0.35 × 1500 kPa retention
Moderately dry	0.35–0.8 × 1500 kPa retention
Slightly dry	0.8–1.0 × 1500 kPa retention
Moist	< 1500 to > 1 kPa suction*
Slightly moist	1500 kPa suction to MWR
Moderately moist	MWR to UWR
Very moist	UWR to 1 or 0.5 kPa suction
Wet	< 1 to < 0.5 kPa
Nonsatiated	No free water
Satiated	Free water present

Source: Soil Survey Division Staff 1993.
*If coarse materials, to 0.5 kPa.

Note that the boundary between dry soil and moist soil is set at 1500 kPa suction. Wet is separated from moist at the condition where water films are readily apparent. The boundary between moist and wet is assumed to be about 0.5 kPa suction for coarse soils and 1 kPa suction for other soils.

Three conditions contribute to soil wetness. First, the soils may be wet because of an impermeable horizon or layer such as a fragipan or tillage pan. Second, the soil may be inundated by flooding or from runon from lands at a higher elevation. Third, the soils may have a high water table in the profile, which can have an extended capillary fringe.

Drainage Classes

Drainage refers to how rapidly excess water leaves the soil by runoff or internal drainage. The term generally describes a condition of the soil: how much time the soil is free of saturation. Seven natural drainage classes refer to the frequency and duration of wet periods under conditions similar to those under which a soil developed. These classes are listed in Table 8.7.

Table 8.7. Natural Drainage Classes of Soils

Drainage Class	Definition
Excessively	Water is removed very rapidly. The occurrence of internal free water is very rare or very deep. The soils are commonly coarse-textured and have very high hydraulic conductivity or are very shallow.
Somewhat excessively	Water is removed from the soil rapidly. Internal free water occurrence commonly is rare or very deep. The soils are coarse-textured and have high saturated hydraulic conductivity or are very shallow.
Well drained	Water is removed from the soil readily but not rapidly. Internal free water occurrence commonly is deep or very deep. Water is available to plants throughout the growing season in humid regions.
Moderately well	Water is removed from the soil somewhat slowly during some periods of the year. Internal free water occurrence commonly is moderately deep and transitory to permanent. The soils are wet for only a short time within the rooting depth during the growing season, but long enough that most mesophytic crops are affected. They commonly have a moderately low or lower saturated hydraulic conductivity in a layer within the upper 1 m of soil.
Somewhat poorly	Water is removed slowly so that the soil is wet at a shallow depth for significant periods during the growing season. The occurrence of internal free water commonly is shallow to moderately deep and transitory to permanent.
Poorly	Water is removed so slowly that the soil is wet at shallow depths periodically during the growing season or remains wet for long periods. The occurrence of internal free water is shallow or very shallow and common or persistent. Free water is commonly at or near the surface long enough during the growing season so that most mesophytic crops cannot be grown unless the soil is artificially drained.
Very poorly	Water is removed from the soil so slowly that free water remains at or very near the surface during much of the growing season.

Source: Soil Survey Division Staff 1993.

In well-drained soils, excess water leaves the soil profile quickly enough that roots do not suffer from oxygen deprivation. Poorly drained soils are wet soils that remain water-logged long enough to interfere with plant growth. Poor drainage interferes with tillage, planting, and spraying operations.

Soil Water Regimes

Five classes of soil water regimes have been defined in soil taxonomy. Important parameters in the classification system are the groundwater level, in terms of the seasonal presence or absence of water held at soil water suctions of less than 1500 kPa, and the location of the control section. The upper boundary of the control section is the depth to which a soil with a suction of more than 1500 kPa will be moistened by 2.5 cm of water within 24 hours. The lower boundary is the depth to which a dry soil will be moistened by 7.5 cm of water within 48 hours. The depth of the control section depends on texture. The classes of soil water regimes are defined in terms of water available for plant growth and are not affected by water management activities such as irrigation and fallowing. The definitions are listed in Table 8.8.

Table 8.8. Classes of Soil Water Regimes in Soil Taxonomy

Soil Water Regime	Definition
Aquic	This is a reducing regime in a soil that is virtually free of dissolved oxygen because it is saturated by groundwater or by water of the capillary fringe. It occurs when the soil temperature at the 50 cm depth is > 5°C.
Aridic	The moisture control section, in 6 out of 10 years, has no water available for plants for more than half the cumulative time that the soil temperature at 50 cm below the surface is > 5°C, and has no period as long as 90 consecutive days when there is water for plants while the soil temperature at 50 cm is continuously > 8°C. Soils that have an aridic moisture regime commonly occur in arid climates.
Udic	In 6 out of 10 years, the moisture control section is not dry in any part for as long as 90 cumulative days per year nor for as long as 60 consecutive days in the 90 days following the summer solstice at periods when the soil temperature at 50 cm below the surface is > 5°C. Soils that have udic moisture regimes commonly occur in humid climates that have a well-distributed rainfall distribution.
Ustic	This moisture regime is intermediate between the aridic and the udic regime. A limited amount of water is available for plants but occurs at times when the soil temperature is optimum for plant growth. Soils with an ustic moisture regime commonly occur in temperate subhumid or semiarid regions or in tropical and subtropical regions with a monsoon climate.
Xeric	This moisture regime is the typical moisture regime of Mediterranean climates, where winters are moist and cool and summers are warm and dry. A limited amount of water is present but does not occur at optimum periods for plant growth.

Sources: Soil Survey Division Staff 1994; Soil Science Society of America 1997.

Concluding Remarks

In this chapter, we have examined several physical principles involved in quantifying the aqueous phase of soil. These principles included (1) how the soil water status is defined, (2) the components of soil water potential, (3) the methods of determination of soil water status, (4) distributions of soil water potential in the field under equilibrium and transient conditions, (5) the soil water retention curve, (6) hysteresis in the soil water retention curve, and (7) soil water terms used in classification systems. With these concepts, the hydrologic status of the soil can be examined as it relates to transport of water through the soil profile, which will be discussed in the next chapters.

Cited References

Bresler, E., B. L. McNeal, and D. L. Carter. 1982. *Saline and Sodic Soils: Principles, Dynamics, Modeling.* Springer-Verlag. New York.

Brooks, R. H., and A. T. Corey. 1964. Hydraulic properties of porous media. Hydrology Paper 3. Colorado State University. Fort Collins.

Bruce, R. R., and R. J. Luxmoore. 1986. Water retention: Field methods. In A. Klute et al. (eds.), *Methods of Soil Analysis,* part 1, *Physical and Mineralogical Methods,* pp. 663–686. 2d ed. Agronomy Monograph 9. Soil Science Society of America. Madison.

Corey, A. T. 1994. *Mechanics of Immiscible Fluids in Porous Media.* Water Resources Publications. Highlands Ranch, CO.

Dingman, S. L. 1994. *Physical Hydrology.* Prentice-Hall. Englewood Cliffs, NJ.

Hillel, D. 1980. *Fundamentals of Soil Physics.* Academic Press. New York.

_____. 1982. *Introduction to Soil Physics.* Academic Press. New York.

Kirkham, D., and W. L. Powers. 1972. *Advanced Soil Physics.* Wiley-Interscience. New York.

Kutilek, M., and D. R. Nielsen. 1994. *Soil Hydrology.* Catena Verlag. Cremlingen-Destedt, Germany.

Marshall, T. J., J. W. Holmes, and C. W. Rose. 1979. *Soil Physics,* Third Edition. Cambridge University Press. New York.

Schmugge, T. J., T. J. Jackson, and H. L. McKim. 1980. Survey of methods for soil moisture determination. *Water Resources Research* 16:961–979.

Shiklomanov, I. A., and A. A. Sokolov. 1983. Methodological basis of world water balance investigation and computation. In *New Approaches in Water Balance Computations.* International Association for Hydrological Sciences Publication No. 148. (Proceedings of the Hamburg Symposium.)

Soil Science Society of America. 1997. *Glossary of Soil Science Terms* 1996. Madison.

Soil Survey Division Staff. 1993. *Soil Survey Manual.* USDA Agriculture Handbook No. 18. U.S. Government Printing Office. Washington, DC.

_____. 1994. *Keys to Soil Taxonomy.* Pocahontas Press. Blacksburg, VA.

Van Genuchten, M. T. 1980. A closed form equation for predicting the hydraulic conductivity of unsaturated soils. *Soil Science Society of America Journal* 44:892–898.

Additional References

Hanks, R. J., and G. L. Ashcroft. 1980. *Applied Soil Physics.* Springer-Verlag. New York.

Koorevaar, P., G. Menelik, and C. Dirksen. 1983. *Elements of Soil Physics.* Elsevier. New York, NY.

Problems

1. The A horizon of a soil has a clayey texture and a thickness of 23 cm. The B horizon of the same soil has a silt loam texture and a thickness of 45 cm. A static water table occurs at the 45 cm depth. Calculate in 10 cm increments the profile distributions of the total soil water potential and each component potential to a depth of 50 cm. Assume that equilibrium conditions exist.

2. A bare soil has a silt loam A horizon over a clayey B horizon. The thickness of the A horizon is 40 cm. A tile drain is installed on top of the B horizon. Show the profile distributions of the total soil water potential and each component potential for the situation in which water is simultaneously flowing to the drain and toward the soil surface, where evaporation is occurring.

3. A soil profile was initially dry. It received rain that wet the soil from the surface downward to partway through the profile. After a few days, the surface had dried and water moved upward in response to evaporation at the soil surface. At lower depths, however, the soil water moved downward into the soil that had not been wet by the rain. Estimate values of the total potential and the component potentials to a depth of 1.0 m at 10 cm intervals.

4. Using the same soil as above, graphically show the temporal changes after the rain in the total and matric potentials within the profile to 40 cm.

5. Draw a diagram of the relationship between pressure and the specific water capacity of the water retention curve. Determine the radius of the pores at applied pressures of 10 and 1500 kPa. What do you conclude from the pore sizes calculated at these two pressures?

6. Convert the potential energy per unit weight of 500 cm of water to its equivalent in potential energy per unit mass and volume.

7. Calculate the matric potential energy on a per unit weight basis if the relative humidity of the soil air is 0.9 and the temperature is 20°C.

8. Calculate the osmotic potential of a soil solution at 25°C if the solute concentration is 0.005 mol/m^3.

9. The following water retention data were obtained for a silt loam soil.

Applied Pressure (cm)	Water Content (kg/kg)
1	0.30
100	0.22
300	0.18
500	0.15
1,000	0.13
5,000	0.10
15,000	0.09

At each pressure, compute (a) the equivalent radius of the pores in cm and in μm and (b) the relative saturation. Graph the relationships between the following: (c) applied pressure and water content and (d) water content and equivalent pore radius.

10. Using linear and/or nonlinear regression techniques, model the water retention data in problem 9 with the following empirical models.

Model 1: $h = a\theta^{-b}$

Model 2: $h = a \exp(-b\theta)$

where a and b are regression coefficients.

11. Comment on the application of this statement to soils in the field: "Pore size is associated with particle size and particle size distribution. The smaller the particles, the smaller the pore size."

12. Describe the capillary pressures of unsaturated and flooded soils on the liquid and gas phases of soil.

13. In a field, it was observed that at two depths 14 cm apart vertically the soil water content did not change with time but the water contents were not identical. Describe the state of the system.

14. Bernoulli's equation is used to express the total mechanical energy. In head units, the mechanical energy, E_m, of soil water is

$$(P/\rho_w g) + z_0 + (\alpha v^2/g) = E_m/\rho_w g$$

where the first two terms on the left-hand side represent potential energy head and the third term represents the kinetic energy head. The pressure head is $P/\rho_w g$, where P is the gauge pressure; z_0 represents the gravitational head, positive upward; v is the average water velocity over a defined cross-sectional area; and α is a kinematic correction factor.

Assume that v = 1 mm/s, g = 980 cm/s^2, and α = 1. Calculate the contribution of the kinetic energy term. Under these conditions, what is the total mechanical energy head at a microscopic point?

CHAPTER NINE

Soil Water Flow

Introduction

The flow of water in soil is important for calculations of water balances and for the redistribution of water, solutes, and energy within the soil-plant-atmosphere system. Water flow affects the transport paths of solutes and gases in soil and serves as a storage medium for the transport of heat and solutes. In soils, water tends to be the dominant fluid for transport of solutes.

In this chapter, we develop the governing equations used to describe the flow of water in soils. Both saturated and unsaturated soil conditions are examined, and the similarities and dissimilarities noted. We begin the discussion by considering the movement of water in a model system, progress to the defining equation to quantify saturated flow in homogeneous and in composite soils, and then to unsaturated flow in field soils. Emphasis is placed on application of the principles of soil water flow to soil behavior under field conditions.

Flow of Water in Model Systems: Tubes and Pipes

Flow of water in a straight, narrow, cylindrical tube varies with several properties of water and characteristics of the tube. If we consider only nonturbulent flow conditions, the velocity distribution in the tube is near zero at the walls because of adhesion of water to the solid surface (Figure 9.1). The velocity is maximal on the axis and constant on cylindrical surfaces that are concentric about the axis. Adjacent cylindrical laminae moving at different velocities slide over each other. This parallel fluid motion is called *laminar flow*. The movement of the fluid is in response to a pressure gradient $\Delta P/L$ acting in the axial direction. The parameter L is the length of the tube. The change in pressure is

$$\Delta P = P_1 - P_2 \tag{1}$$

The aqueous fluid is accelerated by the pressure gradient and retarded by the frictional resistance.

The velocity or flux of a fluid in a straight tube can be calculated from Poiseuille's law, which is

$$Q/t = \pi r^4 \Delta P / 8\eta L \tag{2}$$

where Q/t is the volume per unit time known as the flux (m^3/s), r is the radius of the tube (m), η is the viscosity of the fluid (kg/m·s), and $\Delta P/L$ is the pressure gradient (N/m^3). Poiseuille's law indicates that the flux is proportional to the fourth power of the radius,

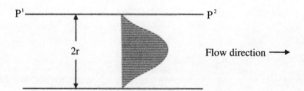

Figure 9.1. Cross section of the velocity distribution in a cylindrical tube. P^1 and P^2 are pressures at the inlet and outlet, respectively, and r is the radius of the tube.

a very strong dependence. Dividing the flux by the cross-sectional area of the tube gives the flux density:

$$q = Q/At = r^2 \Delta P/8\eta L \qquad [3]$$

where q is the flux density (m/s). This equation shows that the flux density is directly proportional to the square of the radius of the capillary tube.

When the adjacent cylindrical laminae, moving at different velocities, slide over each other, the flow is said to be *laminar*. In laminar flows, molecules of water follow smooth lines called *streamlines*. Laminar flow of water prevails only at relatively low flow velocities and in straight, narrow tubes. As flow velocity increases, the fluid gains kinetic energy, and a point is reached where the mean flow velocity is no longer proportional to the difference in pressure, and the flow is said to be *turbulent*.

The quantitative criterion to determine whether flow of a fluid is laminar or turbulent is the Reynolds number, Re. Values of Re are dimensionless and for a cylindrical tube can be calculated from

$$Re = qd\rho_f/\eta \qquad [4]$$

where q is the mean water flux density (m/s), d is the effective tube or pore diameter (m), ρ_f is the density of the fluid (kg/m^3), and η is the viscosity of the fluid (kg/m·s). In a straight tube, turbulence occurs when Re is greater than 1000–2000. In a curved tube, the value of Re at which the flow becomes turbulent is greatly reduced. For transport of water in soils, Re is usually less than 1, so we assume a linear relation between the hydraulic gradient and the flux density. Thus, laminar flow of water is normally found in soils, except possibly in macropore flow through large soil pores, large fractures, and shrinkage planes.

Example

Suppose that the soil water flux density through an undisturbed core is 1×10^{-4} m/s. If the average pore diameter is 5×10^{-5} m, and the soil temperature is 20°C, calculate the Reynolds number and conclude whether laminar or turbulent flow conditions exist.

$$Re = qd\rho_f/\eta$$
$$= (1 \times 10^{-4}\ m/s)(5 \times 10^{-5}\ m)(1000\ kg/m^3)/0.001\ kg/ms$$
$$= 0.005$$

Since RE < 1, we conclude that flow is laminar.

Example

Suppose that the water flux density through an earthworm burrow of diameter 0.5 cm is 1000 cm^3/h. Calculate the Reynolds number at a temperature of 20°C and conclude whether laminar or turbulent flow conditions exist.

$$area = A = \pi r^2 = \pi \times (0.0025)^2 = 1.96 \times 10^{-5}\ m^2$$
$$q = Q/At = (1000\ cm^3)(m/100\ cm)^3/[(1.96 \times 10^{-5}\ m^2)(1\ h)(3600\ s/h)]$$
$$= 0.0142\ m/s$$
$$Re = qd\rho_f/\eta = (0.0142\ m/s)(0.005\ m)(1000\ kg/m^3)/0.001\ kg/m \cdot s$$
$$= 70.9$$

In the earthworm burrow, the Reynolds number is considerably higher than in the soil matrix. Perhaps, a combination of laminar and turbulent flow conditions exists.

Darcy's Law

If the soil consisted of a bunch of straight and smooth cylindrical tubes, we could apply Poiseuille's law and assume the overall flow rate to be equal to the sum of the separate flow rates through the individual tubes. In this way, the total flow of water could be calculated through a bundle of tubes caused by a given pressure difference using Poiseuille's equation.

Unfortunately, soil pores do not represent uniform, smooth tubes but are highly irregular, tortuous, and interconnected. Flow through soil pores is limited by numerous constrictions, or necks, and occasional dead-end spaces. Hence, the actual geometry and flow pattern of a typical soil are much too complicated to be described in any microscopic detail. For this reason, the flow rate of water through complex porous media such as soil is described in terms of a macroscopic flow velocity, which is the overall average of the microscopic velocities determined over the total soil volume.

Darcy's Discovery

In 1856, the French civil engineer Henri Darcy developed an empirical equation that described the seepage of effluent through saturated sand filters. The law that bears his name is one of the most important developments in hydrology and serves as the basis for quantifying fluid flow in porous media. The general relationship developed by Darcy was that the volume discharge from a one-dimensional column of soil is proportional to the column cross-sectional area, the length of the column, and the total hydraulic head loss:

Q/t is proportional to $A\Delta H/L$

where Q is the quantity of water known as the discharge volume (m^3), t is the time (s), A is the cross-sectional area (m^2) of the sand column, ΔH is the hydraulic head difference (m), and L is the column length (m). The ratio $\Delta H/L$ is known as the *hydraulic gradient* (m/m) and is the driving force for moving water through soil. The specific discharge rate is known as the flux density and for water is symbolized by q with units of $m^3/m^2 \cdot s$. This equation, which is known as Darcy's law, is written in incremental form as

$$q = Q/At = K(\Delta H/L) \qquad [5]$$

where K is the proportionality coefficient known as the hydraulic conductivity (m/s).

The flux density, q, is the volume of water passing through a unit cross-sectional area per unit time. It can be considered as the amount of water, Q, passing through a plane perpendicular to the direction of flow during a specified time interval, divided by the area, A, of that plane and the magnitude of the time interval, t. The amount of water can be expressed as a volume, mass, or weight. The flux density of water flowing in soil, which is usually expressed as a volume per unit area per unit time, is

$$q = Q/At \qquad [6]$$

since water is considered to be incompressible, unlike air. It is an average flux density since it is based on a gradient of average hydraulic head.

The flow rate of water in the soil pores can also be calculated by dividing the flux density by the volumetric soil water content. Mathematically, this becomes

$$q/\theta_v = v \qquad [7]$$

and is known as the *pore water velocity*. The magnitude of v will be greater than q since θ_v is less than 1.0. The pore water velocity also is an average velocity.

The proportionality factor K is the transfer coefficient and is generally called the *hydraulic conductivity*. The hydraulic conductivity is the rate of discharge per unit area when the potential gradient is unity. Hydraulic conductivity is a measurement of the ability of the soil to conduct water and depends upon the permeability of the soil to water. For steady-state saturated flow of water in columns of soil, K is defined by Darcy's law as

$$K = qL/\Delta H \qquad [8]$$

The units of K are normally given as length per unit time (in the SI system, m/s). Values of K are usually determined experimentally and include all the factors pertaining to (1) variations in velocity over the flow section, (2) tortuosity and size of the channels, and (3) properties of the flowing media such as viscosity and density.

The relation between the hydraulic conductivity and the properties of the fluid media is

$$K = k(\rho_f g/\eta) \qquad [9]$$

where K is the hydraulic conductivity (m/s), ρ_f is the fluid density (kg/m^3), g is the gravitational acceleration (m/s^2), η is the viscosity ($kg/m \cdot s$), and k is the intrinsic permeability (m^2). The ratio of the viscosity to the density is known as the *kinematic viscosity* (m^2/s). The ratio ($\rho_f g/\eta$) characterizes the properties of the fluid and is known as the *fluidity*. The fluidity (f') varies with the composition of the fluid and with temperature. Thus, we can rewrite equation [9] as

$$K = kf' \qquad [10]$$

The intrinsic permeability is an exclusive property of a porous medium and, thus, should be the same for all fluids. It is related to the particle sizes and to the tortuosity of the path through which the fluid moves.

In saturated, uniform soil, K is considered to be constant over time and space. Under saturated conditions K is often represented as Ksat. Values of Ksat range from 10^{-4} to 10^{-5} m/s in sandy soils and from 10^{-6} to 10^{-9} m/s in clayey soils.

According to Darcy's law, the saturated hydraulic conductivity is the ratio of the flux density to the hydraulic gradient. Graphically, this relation is shown in Figure 9.2. The slope of the lines will be constant as long as flow is laminar. Therefore, the hydraulic conductivity is the proportionality factor in Darcy's law as applied to the viscous flow of water in soil, that is, the flux density of water per unit gradient of hydraulic potential.

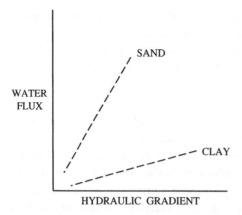

Figure 9.2. The relationship between soil water flux and hydraulic gradient for two saturated soil textures.

Although the Darcy equation is almost always used in practice, there are three distortions of the law, which are not often encountered (de Marsily 1986). These include (1) where the hydraulic gradients have low values, such as in compacted clay, (2) where the hydraulic gradients have high values at wetting fronts, and (3) in the transient state in fractured media and porous media where an additional transient term appears. The wetting front is the boundary between the wetted region and the dry region of soil during infiltration.

An interesting commentary on the character, life, and work of Henri Darcy was published by Philip (1995). He reported that the central square, town gardens, and other commercial and public buildings in the city of Dijon are named after Darcy. Apparently, however, the importance of his work has been forgotten by many citizens of the city, and his name persists only as a ubiquitous geographical label in a city perhaps known more for mustard.

Units of the Components of the Darcy Equation

The units of flux density, hydraulic conductivity, and hydraulic gradient depend on the units used in expressing the unit quantity of water in the potential energy. Table 9.1 gives three examples of the SI units of each component of Darcy's equation.

Table 9.1. Units of the Components of Darcy's Equation

Unit	Flux Density	Hydraulic Conductivity	Hydraulic Gradient
Potential energy/mass	kg/m²·s	kg·s/m	J/kg·m
Potential energy/volume	m/s	m³·s/kg	Pa/m
Potential energy/weight	m/s	m/s	m/m

It is helpful to remember that potential energy is expressed in joules per unit quantity of water, that a joule is a unit of work (i.e., N·m), and that a newton can also be expressed as kg·m/s^2. A quick inspection of Table 9.1 indicates that it is considerably easier to work with units of potential energy per unit weight, which is known as hydraulic head. The symbol that we will use for total hydraulic head is H.

Transport of Soil Water under Saturated Conditions

Darcy's work, which was conducted on the flow of water through saturated columns of sand, showed the importance of several soil and column factors. We will examine these in greater detail below.

Transport of Water in Saturated, Homogeneous Soil Columns

For saturated flow of soil water, the total hydraulic head is composed of the sum of the pressure and gravitational heads. Mathematically, this becomes

$$H = H_g + H_p \qquad [11]$$

where the gravitational head H_g is determined at any point by the height of the point relative to an arbitrarily placed reference level, and H_p is determined by the height of the water column above that point. Since water flow occurs in response to spatial differences in the total hydraulic head gradient, we must consider the contributions of all of the component potentials at the inflow and outflow ends of the soil column. Darcy's law becomes

$$q = Q/At = K[(H_i - H_o)/L] \qquad [12]$$

where H_i is the total hydraulic head at the inflow (m), and H_o is the total hydraulic head at the outflow (m).

Now, we consider the vertical flow of water in a column of homogeneous soil similar to the one studied by Darcy. A diagram of the system is shown in Figure 9.3. In order to calculate the flux of water through the column, we must know either the hydraulic conductivity and the hydraulic head gradient, which is the ratio of the hydraulic head drop between the inflow and outflow boundaries to the column length, or the volume of water moved through the soil core in a given time. We will use an example to illustrate the concept.

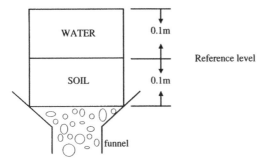

Figure 9.3. Diagram of a vertical column of saturated soil with water ponded on top. $H_i = 0 + 0.1 = 0.1$ m; $H_o = -0.1 + 0 = -0.1$ m. Water is draining into a beaker.

Example

Water is ponded on the surface of a saturated soil core to a constant depth of 10 cm. Assuming that the core has a cross-sectional area of 28.3 cm² and a height of 10 cm, and that 425 cm³ of water is collected in the beaker in 4 hours, calculate the saturated hydraulic conductivity (Ksat) and flux density of the soil. Place the reference level at the top of the core.

H inflow = $H_g + H_p = 0 + 10 = 10$ cm
H outflow = $H_g + H_p = -10 + 0 = -10$ cm
$K = QL/\Delta HAt = (425 \text{ cm}^3 \times 10 \text{ cm})/(20 \text{ cm} \times 28.3 \text{ cm}^2 \times 4 \text{ h})$
    = 1.877 cm/h = 0.01877 m/h = 5.21×10^{-6} m/s
$q = Q/At = 425 \text{ cm}^3/(28.3 \text{ cm}^2 \times 4h \times 3600 \text{ s/h})$
    = 1.04×10^{-5} m/s
$\Delta H/L$ = 20 cm/10 cm = 2 m/m

The magnitude of this Ksat indicates some restricted flow of water under saturated conditions. Possible reasons for this include a clayey texture, tillage pan, fragipan, or crust.

The spatial distributions of water pressure for the soil column described in the previous example are shown in Figure 9.4 for two different positions of the reference level. Note the different distributions of H_g and H depending on the position of the reference level. The distribution of H_p is not influenced by the position of the reference level. Values of the hydraulic gradient and pressure head are the same regardless of the position of the reference level.

In Figure 9.4a, the reference level is arbitrarily placed at the bottom of the column. The lower boundary shows that the pressure of the water at the outlet is atmospheric, and therefore, the pressure head at the bottom is zero. Considering the difference in total head at the inlet and outlet, ΔH for this system is equal to 0.2 m and $\Delta H/L$ is 2 m/m, which is the same as in the example above.

In Figure 9.4b, the reference level is arbitrarily placed at the top of the 10 cm column of water. The total head at the water and soil interface is zero. At the soil surface, the pressure head is equal to the height of water ponded above the soil, that is, 10 cm; the gravitational head at the surface is –10 cm. At the outflow, the pressure head at the outflow is zero due to the exposure of the column to atmospheric pressure. The total head is negative since the gravitational head is –20 cm and the pressure head is zero. Note that $\Delta H/L$ is 2m/m.

The soil water pressure distributions within the columns can be determined from boundary conditions and the fact that water flow occurs under steady-state conditions. If we assume that Darcy flow occurs, then the flux density of water is constant. Since under saturated conditions K is also constant, we have in differential form

$$dH/dz = \text{constant} \qquad [13]$$

Upon integration, we obtain

$$H(z) = bz + a \qquad [14]$$

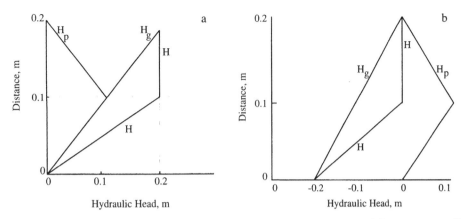

Figure 9.4. Hydraulic potential distributions in a saturated, constant-head, homogeneous soil column at two positions of the reference level: (a) at the bottom of the 10 cm soil column and (b) at the top of the water column.

where b is the slope and a is the intercept of a graph of H versus z. This shows that H varies linearly with position in the column. If values of H are known at two positions in the column, the slope, b, becomes the hydraulic gradient, and we have

$$b = (H_2 - H_1)/(z_2 - z_1)$$ [15]

and therefore, the intercept is

$$a = (H_1z_2 - H_2z_1)/(z_2 - z_1)$$ [16]

For saturated soils, $H = H_g + H_p = z + h_p$, where z is the gravitational head and h_p is the pressure head.

Example

Calculate the total and component potential heads under the constant-head ponding conditions shown in Figure 9.3 when the reference level is placed at the interface between the soil and water.

Column Position	H_g	H_p	H
Top of water	10 cm	0	10 cm
Water-soil interface	0	10	10 cm
Bottom	–10 cm	0	–10 cm

A comparison of the effect of the positioning of the reference level in the column shows that the potential distributions vary, but the hydraulic gradient between the inflow and outflow ends remains the same.

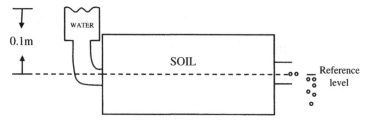

Figure 9.5. Diagram of a horizontal column of saturated soil.

Several questions are frequently asked about the flow of soil water under steady-state, saturated-flow conditions. These include the following:

1. What happens to K when the hydraulic gradient remains constant but q increases? According to Darcy's law, K varies directly with q. Therefore, as q increases, K increases.

2. What happens when the ponding depth increases but q remains constant? As the ponding depth increases, $\Delta H/L$ increases, and since $K = qL/\Delta H$, then K should decrease.

3. What is the significance of the magnitude of K? It is an indicator of the ease of flow, or conductance, of water through a soil.

For horizontal, saturated flow, there is no difference in the gravitational potential between the inflow and outflow ends of the soil column. Since the exact elevation of the reference level is unimportant, the absolute values of the pressure heads are immaterial, and only their differences from one point to another in the soil affect flow.

Now, let's consider a horizontal column of saturated soil and calculate the hydraulic gradient and hydraulic conductivity. Assume that 425 cm^3 of water was collected in 4 hours from a horizontal column of soil having a length of 10 cm and cross-sectional area of 28.3 cm^2. The system is shown in Figure 9.5.

The potential head distributions and calculations of K and q for this system are developed as follows:

$H_i = H_g + H_p = z + h = 0 + 0.1 = 0.1$ m
$H_o = H_g + H_p = 0 + 0 = 0$ m
$K = QL/\Delta HAt = (425 \times 10)/(10 \times 28.3 \times 4)$
    $= 3.75$ cm/h $= 0.0375$ m/h $= 1.04 \times 10^{-5}$ m/s

$\Delta H/L = 0.1/0.1 = 1.0$ m/m. Therefore,

$q = (425/28.3 \times 4)(1/3.6 \times 10^5) = 0.0133$ m/h $= 1.04 \times 10^{-5}$ m/s

In comparing the vertical and horizontal columns, q is the same in both columns, but in the horizontal column, ΔH is lower by a factor of 2, and thus, K is twice as high.

For the same saturated soil column, these results show that (1) the rate of flow of water in a vertical column is greater by the magnitude of the Ksat, and (2) if the hydraulic head at the inflow is zero, the flux density is equal to Ksat. This latter point is due to the fact that in the absence of a pressure gradient, the only driving force is the gravitational head gradient.

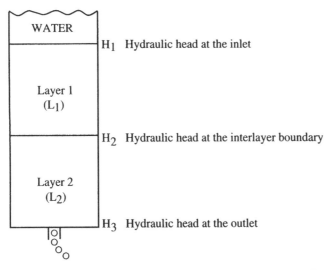

Figure 9.6. Diagram of the vertical flow of water through a saturated composite column of soil under constant ponding head.

Transport of Water in Saturated, Composite Soil Columns

A somewhat more difficult problem occurs when a soil consists of two or more hydraulically distinct layers. Each layer or horizon is assumed to be homogeneous within itself and characterized by its own thickness and saturated hydraulic conductivity.

First, we assume that a saturated soil has two horizons, each with differing thicknesses and saturated hydraulic conductivity. These are known as composite soils and occur frequently in older soils in humid regions and in alluvial soils deposited in layers of material of different hydraulic conductivity. Graphically, this system is shown in Figure 9.6.

For vertical flow of water at steady state and disregarding any possible contact resistance between the horizons, the flux density of water through both horizons must be equal.

$$q = K_1[(H_1 - H_2)/L_1] = K_2[(H_2 - H_3)/L_2] \qquad [17]$$

or

$$q = (K_1/L_1)(H_1 - H_2) = (K_2/L_2)(H_2 - H_3)$$

where the subscripts indicate the horizon or boundary. The value of H_2 at the interface between the horizons should be the same in both equations. Therefore, solving for H_2 gives

$$H_2 = H_1 - (qL_1/K_1)$$
$$H_2 = H_3 + (qL_2/K_2)$$

Therefore, we write

$$q(L_2/K_2 + L_1/K_1) = H_1 - H_3 = \Delta H$$
$$q = (H_1 - H_3)/[(L_1/K_1) + (L_2/K_2)]$$

The quantity $1/K$ has been called the *resistivity,* and the ratio L/K has been called the *hydraulic resistance per unit area.*

According to Darcy's law for the combined flow of water in a composite soil, the flux density is

$$q = \Delta H / (R_1 + R_2) \tag{18}$$

where ΔH is the total hydraulic head drop across the entire system (i.e., $H_1 - H_3$), and R_i is the hydraulic resistance (L_i / K_i). The effective hydraulic conductivity of the column or profile, K_{eff}, can be written as

$$K_{eff} = \sum_{i=1}^{m} L_i / \sum_{i=1}^{m} (L_i / K_i) \tag{19}$$

where m is the number of saturated layers. This equation, which is used to calculate the effective saturated hydraulic conductivity in series layers of composite media, has the same form as previously used to calculate the effective thermal conductivity in series layers of composite media.

Example

A 1 m long soil column is composed of 0.6 m of sandy loam soil that has a saturated hydraulic conductivity of 1.0×10^{-5} m/s and 0.4 m of silt loam soil that has a saturated hydraulic conductivity of 1.5×10^{-7} m/s. If 0.1 m of water is ponded on the soil surface and the bottom of the column is exposed to the atmosphere, calculate the soil water flux under saturated conditions. Assume the reference level is at the soil surface. Also, calculate the effective Ksat. A graph of this composite column is shown in Figure 9.7.

Substituting in equation [18] we obtain

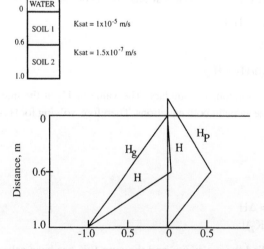

Figure 9.7. Diagram of a composite column and the hydraulic potential distributions within the column. The reference level is placed at the soil surface.

$q = (H_1 - H_3)/[(L_1/K_1 + L_2/K_2)]$
$H_1 = H_{g1} + H_{p1} = 0 + 0.1 = 0.1$
$H_3 = H_{g3} + H_{p3} = -1.0 + 0 = -1.0$
$\Delta H = 0.1 - (-1.0) = 1.1$ m
$q = 1.1/[(0.6/1.0 \times 10^{-5}) + (0.4/1.5 \times 10^{-7})]$
$\quad = 4.034 \times 10^{-7}$ m/s

The effective Ksat of the column is calculated from equation [19] as

$K_{eff} = (L_1 + L_2)[(L_1/K_1) + (L_2/K_2)]$
$\quad\quad = 1.0/[(0.6/1 \times 10^{-5}) + (0.4/1.5 \times 10^{-7})] = 3.67 \times 10^{-7}$ m/s

This shows that the magnitude of the effective Ksat is between the Ksat values of the two horizons, and that the more restrictive hydraulic horizon or layer (i.e., silt loam) in the column had the most influence on the Ksat for the soil column.

A comparison of the calculated value of q with the Ksats of the two layers in the previous example shows that in stratified soils the flux density of water is mostly influenced by the layer with the lowest Ksat. In this case, the lower horizon dominates the flux density of water.

The hydraulic potential distributions in the column are also shown in Figure 9.7. The hydraulic potentials at the 0.6 m depth are calculated from the following:

$\Delta H_1 = H_1 - H_2; \Delta H_2 = H_2 - H_3$
$\Delta H_1 + \Delta H_2 = \Delta H = 1.1$ m
$K_1 \Delta H_1/L_1 = K_2 \Delta H_2/L_2$
$\Delta H_1 = [(0.6 \times 1.5 \times 10^{-7})/(0.4 \times 1.0 \times 10^{-5})]\Delta H_2 = 0.0225\Delta H_2$
$\Delta H_2(0.0225 + 1.0) = 1.1$
$\Delta H_2 = 1.0758$ m
$\Delta H_1 = 0.0242$ m

This shows that the largest ΔH (or driving force) occurs in soil horizon 2, that is, the soil with the lowest Ksat.

The hydraulic potentials at the interface of the two horizons are calculated by substituting into the equation

$\Delta H_1 = (H_{g1} - H_{g2}) + (H_{p1} - H_{p2})$

Substituting the appropriate values for this column and solving for the pressure head at the interface gives

$H_{p2} = 0.6 + 0.1 - 0.0242$
$\quad\quad = 0.676$ m

If the reference level is placed at the bottom of the column of soil, the calculations are performed in a similar way:

$H_1 = H_{g1} + H_{p1} = 1.0 + 0.1 = 1.1$ m
$H_3 = H_{g3} + H_{p3} = 0 + 0 = 0$ m
$\Delta H = 1.1 - 0.0 = 1.1$ m

Therefore, we have the following:

$\Delta H_1 + \Delta H_2 = 1.1$ m

$\Delta H_1 = 0.0225 \Delta H_2$

$\Delta H_2 (0.0225 + 1.0) = 1.1$

$\Delta H_2 = 1.0758$ m

$\Delta H_1 = 0.0242$ m

$\quad = (H_{g1} + H_{p1}) - (H_{g2} + H_{p2})$

$0.0242 = 1.0 + 0.1 - 0.4 + H_{p2}$

$H_{p2} = -0.7 + 0.0242 = 0.676$ m

This again shows that changing the position of the reference level has no effect on the distribution of H_p but does affect the distributions of H_g and H. Note, however, that ΔH remains the same.

In the previous composite column of soil, the magnitude of K_1 was greater than K_2, and the lower layer of soil controlled the flow of water. As a result, the entire column of soil was saturated, and positive pressures occurred throughout the column. This is not always the case, however. Recall the following relationship from Darcy's law:

$$K_1 \left(\frac{H_1 - H_2}{L_1} \right) = K_2 \left(\frac{H_2 - H_3}{L_2} \right)$$

Solving for H_2, we have

$$H_2 = \frac{K_1 H_1 L_2 + K_2 H_3 L_1}{K_2 L_1 + K_1 L_2}$$

If we assume that the reference level is at the bottom of the column, then H_2 is defined as

$$H_2 = H_{p2} + L_2$$

Then, the pressure head is calculated from

$$H_{p2} = \frac{L_2 (K_1 L_1 + K_1 H_{p1} - K_2 L_1) + K_2 H_{p3} L_1}{K_2 L_1 + K_1 L_2}$$

When $H_{p3} = 0$, we obtain

$$H_{p2} = \frac{L_2 (K_1 L_1 + K_1 h_{p1} - K_2 L_1)}{K_2 L_1 + K_1 L_2}$$

If $(K_1 L_1 + K_1 h_{p1} - K_2 L_1) < 0$, then H_{p2} becomes negative and the lower layer is unsaturated. Transforming the inequality, we obtain

$$K_2 > K_1 \left(\frac{h_{p1} + L_1}{L_1} \right)$$

This condition indicates that negative soil water pressures (i.e., matric potentials) occur in the lower portion of the soil column. Thus, in this case, K is not constant but is a function of the soil water status. The hydraulic potential distributions for vertical flow in the composite column should be contrasted with those determined in the homogeneous column.

When flow of water is horizontal (i.e., parallel to the horizonation), the total flux density is the sum of the flux densities of each horizon. For a column of silt loam soil containing two horizons, Darcy's law is written as

$$q = [(K_1z_1 + K_2z_2)/(z_1 + z_2)]\Delta H/L \tag{20}$$

where K_1 and K_2 are the hydraulic conductivities of the two horizons, z_1 and z_2 are the thicknesses of the two horizons, and L is the length. In this case, the parameters within the brackets are known as the *apparent hydraulic conductivity,* which is calculated from

$$K_{app} = \Sigma K_i z_i / \Sigma z_i \tag{21}$$

General Saturated-Flow Equations

Under field conditions, we frequently use the Cartesian spatial coordinate system to indicate the direction of flow of water. In incremental form, Darcy's law for the z direction becomes

$$q = Q/At = -K(\Delta H/\Delta z) = -K[(H_2 - H_1)/(z_2 - z_1)] \tag{22}$$

where z is the spatial coordinate, H_1 is the total hydraulic head at z_1, and H_2 is the total hydraulic head at z_2. The length of the column of soil, which was equal to L in the former system of analysis, now is $z_2 - z_1$. Note that the subscripts of the numerator and denominator in equation [22] are in the same order.

A positive value of q or Q/At indicates flow in the positive z direction, which we will assume to be upward. Therefore, a negative value of Q/At indicates flow of water in the negative z direction, or downward.

The quantity $[(H_2 - H_1)/(z_2 - z_1)]$, or $\Delta H/\Delta z$, is called the hydraulic gradient and is the driving force for the movement of water in soil. The units of the hydraulic gradient depend on the units of H (see Table 9.1), which in turn depend on the units of the unit quantity of water. On a weight basis, the units of $\Delta H/\Delta z$ are expressed as meters per meter.

The quantity Q/At is called the flux per unit cross-sectional area, or flux density, and has units of $m^3/m^2 \cdot s$ or m/s. The flux density is interpreted as the volume of water flowing per unit area per unit time.

Darcy's law can also be written in differential form as

$$q = -K\partial H/\partial z \tag{23}$$

where ∂H is the change in total hydraulic head along an interval dz. Since q and K have both magnitude and direction, they are vectors. Thus, Darcy's law can be written as

$$q_x = -K_x\partial H/\partial x \tag{24}$$

$$q_y = -K_y\partial H/\partial y \tag{25}$$

$$q_z = -K_z\partial H/\partial z \tag{26}$$

The subscripts x, y, and z do not appear on H because potential energy has only magnitude and not direction; that is, H is a scalar quantity. The partial derivatives were used rather than the total derivatives to show that two directions are held constant while H varies with the direction of interest.

In most cases, soil water flow rates vary in both space and time. Those cases where the magnitude and direction of the flux density as well as the hydraulic gradients vary with time are called transient-state conditions. To incorporate these changes into Darcy's law, we use the law of conservation of mass.

For one-directional flow of water without plant root extraction, the mass conservation equation can be written as

$$\partial\theta_v/\partial t = -\partial q/\partial z \tag{27}$$

where z is the direction of interest. Incorporation of equation [23] into equation [27] results in the transient-state flow equation in the z direction:

$$\partial\theta_v/\partial t = -\partial q/\partial z = \partial/\partial z(K\partial H/\partial z) \tag{28}$$

In saturated soil, $\partial\theta_v/\partial t = 0$, and if Ksat is assumed to be constant, equation [28] becomes

$$Ksat(\partial^2 H/\partial z^2) = 0 \tag{29}$$

For three-dimensional flow, the flow equation becomes

$$K_x\partial^2 H/\partial x^2 + K_y\partial^2 H/\partial y^2 + K_z\partial^2 H/\partial z^2 = 0 \tag{30}$$

where K_x, K_y, and K_z are the saturated hydraulic conductivities in the x, y, and z directions, respectively. When the Ksats have the same magnitude regardless of direction, then equation [30] reduces to

$$\partial^2 H/\partial x^2 + \partial^2 H/\partial y^2 + \partial^2 H/\partial z^2 = \Delta^2 H = 0 \tag{31}$$

which is known as the *Laplace equation*. This equation can be solved analytically for certain simple boundary conditions. In most cases of interest, however, numerical techniques are needed to solve equation [31].

Transport of Soil Water under Unsaturated Conditions

Most of the processes involving soil water flow in the field occur while the soil is in an unsaturated state. Unsaturated-flow processes are, in general, much more complicated and difficult to describe quantitatively and mathematically than saturated flow because they often entail changes in the state and content of soil water during flow. These changes involve complex relationships among the water content, potential energy, and hydraulic conductivity. These relations are affected by time and position in the soil profile and landscape.

Water in unsaturated soil is subject to a subatmospheric pressure. Therefore, negative pressures or negative potential energies operate. The matric potential is due to the physical affinity of water to the soil particle surfaces and to curved air-water interfaces in the pores. Soil water tends to move from locations where the total soil water potential is higher to locations where it is lower. The gradient of the total potential energy constitutes the moving force for water in soil. The moving force is greatest at the wetting-front zone of entry into an initially dry soil. In this zone, the hydraulic gradient may be several meters per meter (or kPa/m). Therefore, such a hydraulic gradient constitutes a moving force several times greater than the gravitational force.

Darcy's Law for Unsaturated Flow

For unsaturated flow, Darcy's law is written as

$$q = -K(\theta_v)\partial H/\partial z \qquad [32]$$

where, expressed on a weight basis, q is the soil water flux density (m/s), H is the total soil water potential (m), z is the spatial coordinate (m), and $K(\theta_v)$ is the hydraulic conductivity (m/s), which is dependent on the volumetric soil water content (θ_v) and matric potential (h). The negative sign indicates that the flux density of water is positive in the positive z direction. This equation governs steady-state water flow processes (i.e., those processes where the flux density remains constant along the conducting system).

Assuming z is positive upward, the total hydraulic potential, or head, under unsaturated conditions can be defined as

$$H = H_g + H_m = z + h \qquad [33]$$

where H_g is the gravitational potential or head (m), and H_m is the matric potential or head (m). When the potential energies are expressed on a weight basis, z is used for H_g and h is used for H_m.

The total hydraulic gradient is defined as in the past. Specifically, for unsaturated conditions, we write the average hydraulic gradient within a depth interval as

$$\partial H/\partial z \approx \Delta H/\Delta z = [(h_{n+1} + z_{n+1}) - (h_n + z_n)]/(z_{n+1} - z_n) \qquad [34]$$

where the shallow location is designated as position n and the deeper location as position n + 1. Evaluation of $\partial H/\partial z$ is accomplished by selecting values for h and z in the same sequence. The reference level is usually set at the soil surface.

Two methods are available for determining the direction of water flow in soil. The first method uses the total soil water potential, H, which is calculated for each soil depth. Water flows from a region of higher total potential to a region of lower total potential, and therefore, flow is in the direction away from the higher H. The second method uses the sign of the hydraulic gradient. If the hydraulic gradient is positive, then water moves downward. If the hydraulic gradient is negative, then flow of water is upward. Use of the first method is illustrated in the following example.

Example
The data set given below contains values of the matric head, h, as a function of z. Compute values of H and determine the direction of flow of water.

Depth (cm)	z (cm)	h (cm)	H (cm)	Flow Direction
15	−15	−69	− 84	down
30	−30	−114	−144	down
45	−45	−143	−188	up
61	−61	−98	−159	up
76	−76	−34	−110	

The total hydraulic head is calculated for each depth in the profile. This is accomplished by summing the values of gravitational and matric potential heads for each depth. Next, the direction of flow within a depth interval is determined by comparing H across depth in the profile. Soil water moves from locations of higher H to locations of lower H.

The second method of calculating the hydraulic gradient is by using equation [34]. Note the following for vertical, unsaturated flow of soil water:

$$H = z + h$$
$$\partial H/\partial z = \partial z/\partial z + \partial h/\partial z = 1 + \partial h/\partial z \tag{35}$$

Therefore, Darcy's law in the z direction can be written as

$$q = -K(1 + \partial h/\partial z) \tag{36}$$

Example

In the previous example, the total potential and hydraulic gradients were as given below. Calculate the average hydraulic gradient using equation [34] and determine the direction of flow of water using the sign of the hydraulic gradient as the indicator.

Interval (cm)	Δz (cm)	H_{z_2} (cm)	H_{z_1} (cm)	ΔH (cm)	$\Delta H/\Delta z$ (cm/cm)	Flow Direction
15 to 30	−15	−144	−84	−60	4.00	down
30 to 45	−15	−188	−144	−44	2.93	down
45 to 61	−16	−159	−188	29	−1.81	up
61 to 76	−15	−110	−159	49	−3.27	up

In general, the following guidelines should be adopted:

1. Choose the reference level at the soil surface.
2. Choose upward as positive.
3. Positive gradients indicate downward flow of water.
4. Negative gradients indicate upward flow of water.

Unsaturated Hydraulic Conductivity

The most important difference between unsaturated and saturated flow of soil water is the magnitude of the hydraulic conductivity. When the soil is saturated, all of the pores are filled with water and are conducting water. Therefore, Ksat is the transport coefficient for water. At saturation, the most conductive soils are those with large and continuous pores. The least conductive soils are those in which the pore volume consists of numerous micropores. Therefore, at saturation, a sandy soil conducts water much more rapidly than a clayey soil (Figure 9.8).

When soils become unsaturated, some of the pores contain air, and the liquid conduc-

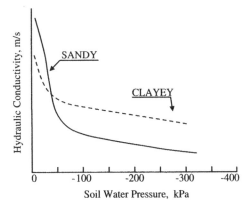

Figure 9.8. Relation of hydraulic conductivity and matric potential for two soils of differing texture.

tive portion of the soil's cross-sectional area decreases correspondingly. Also, as the matric potential decreases, the first pores to empty have the largest radii, which are the most conductive pores according to Poiseuille's law. Thus, under unsaturated conditions, water flows mostly in the smaller pores. As a result, the hydraulic conductivity decreases rapidly with a decrease in water content and matric potential. We represent these relationships as $K(\theta_v)$ and $K(h)$. Experimental measurements show that K is not linearly related to θ_v but can be modeled more adequately with a logarithmic or power-related function of θ_v. For these reasons, the transition from saturation to unsaturation results in a steep decrease in the magnitude of K, perhaps by several orders of magnitude between a matric potential of zero and –10 kPa. At the lower matric potentials, values of K may be so low that very large hydraulic gradients, or very long times, are required for any appreciable flow of soil water to occur.

As a sandy soil becomes unsaturated, the large pores empty quickly and become nonconductive as the matric potential decreases (Figure 9.8). Thus, its initially high K is decreased sharply. In contrast, in a soil with small pores, many of the pores remain full of water and conductive even at low matric potentials, so the value of K does not decrease as steeply and may actually be greater than that of a soil with large pores subjected to the same matric potential. Since in the field the soil is unsaturated most of the time, it may be that under dry conditions flow of water is faster and persists longer in a clayey than in a sandy soil. Therefore, the occurrence of a layer of sand in the fine-textured profile may actually impede unsaturated water movement until water accumulates above the sand and the matric potential of the sand increases sufficiently for water to enter the larger pores.

Examples of Unsaturated Flow in the Field

It is helpful to work through practical examples where the units of the component potentials are varied. The principles required to calculate hydraulic gradients are the same, however.

Example

1. Draw both linear and log-log graphs and comment on the shapes of the following relationships: (a) K versus θ_v; (b) h versus θ_v; (c) K versus h. The data are in the table below.
2. What is the estimated saturated soil water content?
3. What is the estimated saturated hydraulic conductivity?
4. What is the "field capacity" of this soil?
5. What is the "plant extractable" water content?
6. What is the texture of the soil?
7. What is the specific water capacity at -100 cm matric head?

Water Content (m^3/m^3)	Hydraulic Conductivity $(m/s \times 10^5)$	Matric Head (cm)
0.45	6.57	-1
0.44	4.40	-5
0.42	1.98	-22
0.40	0.98	-36
0.38	0.399	-46
0.36	0.179	-58
0.34	0.081	-72
0.32	0.036	-100
0.30	0.016	-146
0.28	0.0073	-200
0.26	0.0033	-292
0.24	0.0015	-415
0.22	0.00066	-540
0.20	0.00030	-690
0.18	0.00013	$-1,090$
0.16	0.000060	$-2,010$
0.14	0.000027	$-5,040$
0.12	0.000012	$-9,000$
0.10	0.000005	$-15,000$

8. Assume that tensiometers were used to measure the matric potentials at 15 cm (matric head $= -75$ cm) and 30 cm (matric head $= -125$ cm). At each depth, calculate the total hydraulic head, hydraulic head gradient, hydraulic conductivity, soil water flux density, and average pore water velocity. In which direction is water flowing? What is the equivalent depth of water retained between these two depths?

Example

In a silt loam soil, the matric potential was determined to be -75 and -50 cm of water at depths of 150 and 200 cm, respectively. Assume that the relationship between soil water content by volume, θ_v, and the matric potential, h, can be described by the relationship

$$\theta_v = a|h|^{-b} \qquad h < -45 \text{ cm}$$

where a is 1000 and b is 2.0. Also, the relationship between θ_v and the hydraulic conductivity, K, is

$$K = me^{n\theta_v}$$

where the parameter m is equal to 1×10^{-12} and n is 45. The units of K are meters per second. Calculate the soil water flux density in centimeters per day. What are the units of m and n? In which direction is water flowing?

The average hydraulic gradient is calculated as

$$\Delta H/\Delta z = 1 + (\Delta h/\Delta z) = 1 + \{[-50 - (-75)]/[-200 - (-150)]\} = 0.5 \text{ m/m}$$

One estimated K is obtained from

$$\log \theta_v = \log a - b \log h$$

or

$$\theta = 1000|h|^{-2.0}$$

where $|h| = (75 + 50)/2 = 62.5$ cm. Therefore, $\theta_v = 0.256$ m^3/m^3.

$$K = (1 \times 10^{-12})\exp(0.256 \times 45) = 1.007 \times 10^{-7} \text{ m/s} = 0.0087 \text{ m/d}$$

The estimated q is

$$q = -K\partial H/\partial z = -(0.0087)(0.5) = -5.035 \times 10^{-8} \text{ m/s}$$
$$= -0.004351 \text{ m/d}$$

The direction of flow of soil water is downward.

Soil Water Diffusivity

To simplify the mathematical and experimental treatment of unsaturated-flow processes, it is often advantageous to change the flow equation into a form analogous to the equations of diffusion of solutes and conduction of heat. The advantage is that analytical solutions to the differential equations describing transport are available in selected cases involving boundary conditions applicable to soil-water flow processes.

With use of the chain rule, we can write the hydraulic gradient as

$$\partial h/\partial z = (\partial h/\partial \theta_v)(\partial \theta_v/\partial z)$$

where the *specific water capacity* is defined as $C(\theta_v)$ or $C(h) = \partial \theta_v/\partial h$ and has units of m^{-1}. For horizontal flow of soil water in the y direction,

$$q = -K(\theta_v)\partial H/\partial y = -[K(\theta_v)/C(\theta_v)]\partial \theta_v/\partial y$$

We define the soil water diffusivity as $D(\theta_v)$, where

$$D(\theta_v) = K(\theta_v)/C(\theta_v) \qquad [37]$$

which has SI units of m^2/s. The diffusivity form of the soil water flow equation becomes

$$q = -D(\theta_v)\partial \theta_v/\partial y \qquad [38]$$

which is mathematically identical to Fick's first law of diffusion. This equation does not function properly near saturation because both $C(\theta_v)$ and $\partial\theta_v/\partial z \to 0$. The soil water diffusivity can be contrasted with the molecular diffusivity of solutes and the thermal diffusivity, which were similarly developed. However, care must be taken to be consistent with the units and to derive the various diffusivity parameters correctly.

Transient-State Unsaturated-Flow Equations

Darcy's law, originally conceived for saturated flow in porous media, was extended by L. A. Richards in 1931 to unsaturated flow in soil, with the provision that the hydraulic conductivity is now a function of the matric potential and/or water content and is expressed as $K(h)$ or $K(\theta_v)$. Under steady-state conditions (i.e., where q, θ_v, and h are constant over time), the steady-state flux density can be written as

$$q = -K\partial H/\partial z \tag{39}$$

Since these conditions are rarely achieved in the field, the equation of continuity for one-dimensional flow, without extraction, can be written as

$$\partial\theta_v/\partial t = -\partial q/\partial z \tag{40}$$

and upon substitution of equation [39] into equation [40], we obtain the transient-state form of the flow equation:

$$\partial\theta_v/\partial t = \partial/\partial z[K(h)\partial H/\partial z] \tag{41}$$

This equation is frequently referred to as the Richards equation.

For horizontal flow of soil water in the y direction, the flow equation has the following two forms:

Conductivity form: $\partial\theta_v/\partial t = \partial/\partial y[K(h)\partial h/\partial y] = C(h)(\partial h/\partial t)$ $\tag{42}$

Diffusivity form: $\partial\theta_v/\partial t = \partial/\partial y[D(\theta_v)\partial\theta_v/\partial y]$ $\tag{43}$

For vertical flow of soil water in the z direction, the flow equation becomes

$$\partial\theta_v/\partial t = \partial/\partial z[K(h)\partial h/\partial z] + \partial K(h)/\partial z \tag{44}$$

which is also known as the conductivity form for vertical flow of soil water. Similarly, the diffusivity form for vertical flow of soil water is written as

$$\partial\theta_v/\partial t = \partial/\partial z[D(\theta_v)\partial\theta/\partial z] + \partial K(\theta_v)/\partial z \tag{45}$$

This form of the flow equation can be used to describe transient water movement in uniform soil profiles.

Characteristic Flow Conditions of Water in Bare Soil

Three characteristic flow conditions of water in soil without vegetation have been distinguished. Discussion of these conditions follows.

Equilibrium

Equilibrium, or hydrostatic, conditions occur when soil water does not move. This results in $q(z, t) = 0$ at any depth and at any time. In a homogeneous soil, we get from Darcy's law

$$q(z, t) = -K(\theta_v)(\partial h/\partial z + 1) = 0 \qquad [46]$$

which reduces to

$$\partial h/\partial z + 1 = 0 \quad \text{or} \quad \partial h = -\partial z \qquad [47]$$

Therefore, equilibrium conditions occur in the soil profile where h changes by the same amount as the depth coordinate. Apparent equilibrium conditions may briefly occur in humid climates in the early spring before intensive evaporation occurs and a few days after a soaking rainfall. This distribution of h and θ_v in the profile during the early spring in humid regions is a reasonable approximation of field capacity for the particular soil profile.

Steady State

Steady-state conditions imply that the flux density of water is constant during a given time and depth interval. Mathematically, $q(z, t)$ is constant in a portion of the soil profile or in the whole soil profile. Darcy's law becomes

$$q(z, t) = -K(\theta_v)(\partial h/\partial z + 1) = \text{constant} \qquad [48]$$

and the equation of continuity becomes

$$\partial \theta_v/\partial t = -\partial q/\partial z = 0 \qquad [49]$$

This indicates that the soil water content profiles $\theta_v(z, t)$ do not change with time. These water content conditions are frequently found in the lower portion of the soil profile, which tends to be buffered by the more dynamic surface horizons, and below a hydraulically restrictive horizon where flow is due to gravity only.

Transient State

Transient-state conditions occur when the flux density is not constant, that is, $q(z, t) \neq q(z + \Delta z, t) \neq q(z, t + \Delta t)$. These are the predominant flow conditions in the field. The general case of transient flow of water in soil is described by one of the vertical, one-dimensional flow expressions:

$$\partial \theta_v/\partial t = \partial/\partial z[K(h, z)(\partial h/\partial z + 1)] \qquad [50]$$

$$= \partial/\partial z[D(\theta_v, z)(\partial \theta_v/\partial z) + K(\theta_v, z)] \qquad [51]$$

$$C(h, z)\partial h/\partial t = \partial/\partial z[K(h, z)(\partial h/\partial z + 1)] \qquad [52]$$

For homogeneous soils, the flow equations are simplified by writing K(h) instead of K(h, z) and $D(\theta_v)$ instead of $D(\theta_v, z)$. Under the usual nonhomogeneous conditions h and θ_v are both dependent on the depth z.

Determination of Hydraulic Conductivity

Quantitative prediction of soil water transport requires known values of the hydraulic conductivity, a highly variable soil parameter. Measured values of K often vary by 10-fold or more for a particular soil series and even with those samples taken close to each other. In addition, measured K values for a soil may vary dramatically with the method used for measurement and with volume of the core. Laboratory-determined values rarely agree with field measurements, the differences often being on the order of 100-fold or more. Field methods generally are more reliable than laboratory methods due to the closer approximation to natural conditions.

Over the years, numerous methods have been developed to measure and estimate the K of soils. These methods are arbitrarily divided into methods of measurement and methods of estimation.

Methods of Measurement

The first group we consider are the methods of measurement of Ksat. For the most part, these methods take advantage of the Darcy equation and assume constant temperature. Methods for measuring K can be divided into two groups: steady state and transient state. In steady-state methods, the flux density and hydraulic gradient are constant. In transient-state methods, the flux density and hydraulic gradient vary with time and must be measured. These transient-state methods are generally faster and considered to be less accurate.

For soils with high flow rates, the differences in hydraulic head generated by gravity are used to measure K. The constant-head permeameter is recommended for values of Ksat greater than 10^{-5} m/s. This method was used in the experiments of Darcy, and the setup is shown in Figure 9.9. With this method, Ksat is determined from

$$K = QL/A\Delta Ht \qquad [53]$$

where Q is the volume of effluent, L is the length of the column, ΔH is the change in total head through the column, and t is the equilibration time. Advantages of the constant-head method are that many samples can be run simultaneously, inexpensively, and rapidly under controlled laboratory conditions. The disadvantages are that the hydraulic properties of soil may change because of handling or by running water through the soil to establish steady-state conditions. Many samples are required because the small samples of soil often display considerable variability.

For soils with lower flow rates or those that have Ksats less than 10^{-5} m/s, the falling-head permeameter method is recommended (Figure 9.10). This is a transient-state method. The resulting equation for calculation of Ksat is

$$K = [aL/A(t - t_0)]\ln(h_0/h) \qquad [54]$$

where a is the cross-sectional area of the small tube (i.e., standpipe), L is the length of the sample, A is the cross-sectional area in the core sample, and h_0 and h are the hydraulic heads at t_0 (the initial time) and t (the final time), respectively. A plot of $\ln(h_0/h)$ versus time results in a straight line and the slope is proportional to K.

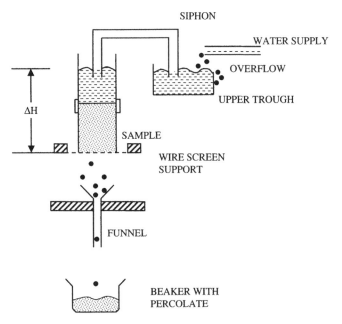

Figure 9.9. Diagram of the constant-head system for measuring saturated hydraulic conductivity. (Redrawn from Klute 1965.)

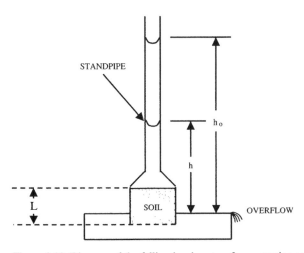

Figure 9.10. Diagram of the falling-head system for measuring saturated hydraulic conductivity. (Redrawn from Klute 1965.)

The core method can also be used to measure $K(\theta_v)$ or $K(h)$ within a range of 0 to -20 kPa. A soil core is placed in a specially designed pressure chamber between two porous plates. A constant pressure head of water is applied to each end of the core. Lower pressure at the lower end creates a hydraulic gradient across the sample. The water content of the soil is regulated by using an external gas pressure system. Values of K are calculated from the steady-state rate of water flow through the core and from the hydraulic gradient as measured with tensiometers placed in the side of the core. The advantage of this method is that many samples can be run simultaneously under controlled laboratory conditions. Disadvantages are that the K values may not adequately reflect actual soil conditions because of the small sample and possible disturbance of soil when collecting the sample.

The double-tube method can also be used to measure Ksat. Two concentric 1 m long tubes are placed in the soil to a given depth, and water flow is manipulated to move from the inner tube into the outer tube at a high and changing hydraulic head. Readings are transformed to Ksat values by using tables and graphs.

The auger hole method requires a hole that extends well below the water table. The sides of the hole must not be smeared. Water is rapidly removed from the hole and the resulting rate of water flow into the hole is determined from the rate of water rise in the hole. Ksat is calculated from the rate of water rise and the geometry of the system according to

$$Ksat = C \; dz/dt \eqno{[55]}$$

where z is the depth of the water level in the hole measured from the top of the water table, and C is the shape coefficient dependent upon the geometry of the test hole.

The percolation test requires a hole dug or augured into the soil to the depth required. A few centimeters of gravel are placed in the hole, and water is ponded to a height of 30 cm for at least 4 hours to saturate the soil. In clayey soils, measurements are made the next day. In sandy soils, measurements are made immediately after the 4-hour saturation period. The vertical velocity of the water is determined and reported in minutes per inch or, correctly but rarely, in seconds per meter. The value measured after 4 hours is considered representative. Usually three holes or more are needed per test. If presoaking is required, the test takes 2 days.

The instantaneous profile method can be used to measure $K(\theta_v)$ or $K(h)$ in the field between saturation and about -30 kPa, depending on the soil drainage. This is probably the most accurate method because the plots are relatively large. A relatively large area is saturated, then covered with plastic and insulating mulch material to prevent evaporation. Drainage is monitored with tensiometers, and the volumetric water content is determined with a neutron moisture probe, time domain reflectometry (TDR), or gravimetric sampling, along with measurements of bulk density and a water retention curve by depth in the profile. The plastic and mulch establish a plane of zero water flux at the soil surface; therefore, all changes in water content are due to redistribution of water within the profile. Water is assumed to move only vertically. The range in volumetric water content above any depth during a time interval is the flux density across that depth. Tensiometers measure the matric potential at each depth. The magnitude of K is calculated using Darcy's law to give curves of $K(\theta_v)$ or $K(h)$ for all depths. The working equation is

$$K(\theta_m) = -(\Delta W/\Delta t)(\Delta H/\Delta z)^{-1} \eqno{[56]}$$

where W is the equivalent depth of water stored in the profile between the soil surface and depth z, and θ_m is the mean volumetric water content related to the mean h. Initially, readings are taken hourly, and as the redistribution of water slows, once a day or longer is sufficient.

The advantages of the instantaneous profile method are the minimum disturbance and large plots, which reflect natural soil conditions better than other methods. Also, all horizons can be measured simultaneously. Disadvantages are that the sites must be fairly level, no lateral flow of water is allowed, it may be difficult to install tensiometers and neutron probe access tubes in soils containing coarse fragments, K can only be determined between a given range of soil water contents, and long times may be required for K measurement at greater depths and on slowly permeable soils.

Methods of Estimation

Modeling water flow dynamics in soils requires that the soil hydraulic properties be known. Considerable effort has been given to estimating $K(\theta_v)$ relationships because (1) the experimental measurements are costly and time-consuming, (2) the soil variability is high in the field and the number of data required to represent the hydraulic properties accurately is enormous, and (3) the K values of some soils may vary by several orders of magnitude within the water content range of interest, and most lab and field methods cannot efficiently cover such a wide range.

Empirical functions have been applied when measured data are unavailable. These functions (1) allow for a closed-form analytical solution for some unsaturated-flow problems, (2) simplify the computational procedures in numerical solutions, (3) save time and perhaps improve accuracy, (4) systematically extrapolate the measured curve, and (5) minimize the measurements required for statistical representation of K in the field (Mualem 1986). Examples of these models are presented in Mualem (1986) and are summarized in Table 9.2.

Statistical models have been used to compute the relative hydraulic conductivity, K/Ksat, based upon the shape of the water retention curve. This method assumes that the soil is composed of a set of randomly distributed interconnected pores. Using the capillary rise equation, the pore radius is related to the hydraulic head at which the pore is filled and drained, and the water content is a function of the pore radius. The Poiseuille equation is used to estimate the K of each pore. The total K for the soil is determined by integration over the contributions of the filled pores.

Table 9.2. Empirical Functions for $K(\theta_v)$ and $K(h)$ Based on h

Models	Condition	Source
$K = \alpha|h|^{-n}$		Wind 1955
$K = a/(|h|^m + b)$		Gardner 1958
$K = Ksat$	$h > h_{cr}$	Brooks and Corey 1964
$K/Ksat = (h/h_{cr})^{-n}$	$h < h_{cr}$	
$K = Ksat$	$h > h_{cr}$	Gardner and Mayhugh 1958
$K = \exp[\alpha(h - h_{cr})]$	$h_1 < h < h_{cr}$	
$K = K_1(h/h_1)^{-n}$	$h < h_1$	

Note: α, b, and m are empirical constants, and h_{cr} is the air-entry matric head. The coefficients to the models are determined by fitting the model to the experimental data using statistical regression techniques.

Another estimation method that describes the hydraulic relationships is a pedotransfer function. A pedotransfer function is a mathematical function that has as arguments basic data used in morphologically describing the soil (e.g., particle size distribution, bulk density, and organic carbon content). The $K(\theta)$ function and/or the $h(\theta)$ function are obtained by regression analyses. One major advantage of this method is the ability to extend pedon size data to larger land areas.

One interesting application of pedotransfer functions is the use of quantitative measures of soil morphology. Lin (1995) developed hydraulic pedotransfer functions to convert descriptive morphological classes into numerical values that were linked to soil hydraulic features. The five basic soil properties evaluated in the system included texture, pedality (grade, size, and shape of peds), macroporosity (quantity, size, and type of visible pores), initial water content, and density of living roots. For each morphological feature considered, descriptive classes were scaled in terms of their impacts on soil permeability. Optimal points that provided best fits to measured infiltration data were obtained for each morphological class using optimization techniques. The points assigned to each morphological property were then divided by the maximum value of that morphology to obtain an index system from 0 to 1. These values were called morphometric indices. Examples of hydraulic pedotransfer functions obtained by Lin (1995) through this approach were

$$Qmac = 4.7 + 97.2(MIs) - 219.6(MIs)^2 + 474.1(MIp) - 460.3(MIp)^2 +$$
$$673.6(MIs)(MIp) + 241.4(MIs)(MIr) - 769.8(MIp)(MIr) \qquad [57]$$

$$Qmic = -0.3 + 1.9(MIt) + 0.9(MIp) + 2.5(MIm) + 1.5(MIt)^2 - 1.2(MIm)^2 +$$
$$1.0(MIr)^2 - 22.9(MIt)(MIp) - 2.1(MIt)(MIm) \qquad [58]$$

where Qmac and Qmic are the macropore flow rate and micropore flow rate, respectively, and MIt, MIs, MIp, MIm, and MIr are the morphometric indices for texture, pedality, macroporosity, initial moisture condition, and density of living roots in soil, respectively. These equations can be used as input equations to estimate transport of soil water on a large scale.

Concluding Remarks

The transport of water in soil plays a very important role in the hydrologic cycle and, therefore, in applications to agricultural and environmental problems. Water is the dominant fluid in most soils and is the medium in which solutes are transported to biological surfaces such as roots and microorganisms and in which solutes are redistributed within and through the soil profile.

Under steady-state conditions, quantification of the transport of soil water is described by Darcy's law. Under transient-state conditions, Darcy's law is combined with the continuity equation to form the Richards equation.

Transport of soil water under saturated conditions occurs when all pores are filled with water. Transport of soil water under unsaturated conditions occurs when the larger pores are completely or partially filled with air. Thus, water is conducted more rapidly in saturated soils than in unsaturated soils. The impact that a soil has on the transport of water is expressed in the magnitude of the hydraulic conductivity. For the same hydraulic gradient, the soils with higher values of K have higher volumes of water moved in a given time interval and area.

Cited References

Brooks, R. H., and A. T. Corey. 1964. Properties of porous media affecting fluid flow. *Journal Irrigation and Drainage Division, American Society of Civil Engineering* 92(IR2):61–88.

de Marsily, G. 1986. *Quantitative Hydrogeology: Groundwater Hydrology for Engineers.* Academic Press. New York.

Gardner, W. R. 1958. Some steady state solutions of the unsaturated moisture flow equation with application to evaporation from a water table. *Soil Science* 85:228–232.

Gardner, W. R., and M. S. Mayhugh. 1958. Solution and tests of the diffusion equation for the movement of water in soil. *Soil Science Society of America Proceedings* 22:197–201.

Hanks, R. J., and G. L. Ashcroft. 1980. *Applied Soil Physics.* Springer-Verlag. New York.

Hillel, D. 1982. *Introduction to Soil Physics.* Academic Press. New York.

Iwata, S., T. Tabuchi, and B. P. Warkentin. 1988. *Soil-Water Interactions: Mechanisms and Applications.* Marcel Dekker. New York.

Klute, A. 1965. Laboratory measurement of hydraulic conductivity of unsaturated soil. In C. A. Black et al. (eds.), *Methods of Soil Analysis,* part 1, *Physical and Mineralogical Properties Including Statistics of Measurement and Sampling,* pp. 210–221. American Society of Agronomy Monograph 9. Madison.

Lin, H. S. 1995. Hydraulic properties and macropore flow of water in relation to soil morphology. Ph.D. diss. Texas A and M University, College Station.

Marshall, T. J., and J. W. Holmes. 1979. *Soil Physics.* Cambridge University Press. New York.

Mualem, Y. 1986. Hydraulic conductivity of unsaturated soils: Prediction and formulas. In A. Klute et al. (eds.), *Methods of Soil Analysis,* part 1, *Physical and Mineralogical Methods,* pp. 799–823. 2d ed. Agronomy Monograph 9. Soil Science Society of America. Madison.

Philip, J. R. 1995. Desperately seeking Darcy in Dijon. *Soil Science Society of America Journal* 59:319–324.

Wind, G. P. 1955. Field experiment concerning capillary rise of moisture in heavy clay soil. *Netherlands Journal of Agricultural Science* 3:60–69.

Problems

1. A DeWitt soil has a silt loam A horizon with a thickness of 37 cm over a clayey B horizon. Frequently, during the spring a perched water table occurs over the B horizon. Graphically, show the general hydraulic potential distributions in the profile to 50 cm (in 10 cm increments) under conditions in which evaporation is occurring at the surface.

2. The hydraulic conductivity of two horizons was modeled with the following equations:

$$K_1 = 5.4 \times 10^{-12} \exp(54\theta_v)$$
$$K_2 = 0.8 \times 10^{-8} \exp(22\theta_v)$$

Plot the $K - \theta_v$ relationships and find the volumetric water content where the K values are the same.

3. The soil water pressures (matric potentials) in a profile were determined in the field during early spring. Some of these values are given below (in mbars). Calculate the total soil water potentials, the hydraulic gradients, and the direction of flow in the profile for the 2 days. Plot the potentials as a function of soil depth.

Time	Soil Depth (cm)							
	15	30	45	60	76	92	106	122
Day 1	−534	−413	−332	−177	−98	−76	−70	−62
Day 2	−576	−501	−388	−254	−108	−80	−71	−62

4. A silt loam soil has an A2g horizon that is 15 cm thick. Water is moving into the horizon at a rate of 0.8 cm/h and out of the horizon at a rate of 0.26 cm/h. Calculate the approximate time rate of change of volumetric soil water content in the horizon. Is the horizon gaining or losing water?

5. Summarize the effects of changing the location of the reference level and boundary conditions on H, H_p, H_g, and ΔH for columns of homogeneous soil and for composite soil columns.

6. Assume that a composite soil column of 60 cm height is composed of two horizons. The top horizon has a thickness of 40 cm and a Ksat of 1.2×10^{-5} m/s. The bottom horizon has a thickness of 20 cm and a Ksat of 1.9×10^{-8} m/s. Calculate the effective saturated hydraulic conductivity and steady-state flux density. Assume that a constant head of 10 cm of water was placed on the soil surface.

7. Suppose that you had a column of composite soil with a clay horizon overlaying a sandy horizon. Under saturated steady-state flow conditions in the clay, is the sandy horizon saturated? Explain.

8. Calculate the effective saturated hydraulic conductivity in the vertical direction for the following soil profile:

Horizon	Depth (cm)	Ksat (m/s)
Ap	0 – 15	1×10^{-5}
B	15 – 25	1×10^{-6}
Bt	25 – 50	1×10^{-7}

9. A soil has an A horizon that is 17 cm thick. Water is added to the soil by irrigation at a rate of 0.35 cm/h. The B horizon has a high, expandable clay content, and water movement through it is essentially zero relative to the A horizon. On average, θ_v and f in the A horizon are 0.12 m^3/m^3 and 0.45 m^3/m^3, respectively. Approximate the time that it will take the A horizon to become saturated.

10. In the table below are soil water pressures (in mbars) measured at five depths in each of six plots. Calculate the following:

a. arithmetic mean and standard deviation of the pressures (in cm) by depth and by plot
b. the mean total soil water potential (cm) across all plots and by depths
c. the direction of flow of water in the profile

Discuss the results by emphasizing the means and variabilities of the soil water pressures measured in the field.

Soil depth (cm)	Plot					
	1	2	3	4	5	6
30	−23	−180	−135	−90	−67	−193
60	−29	−18	−18	−19	−47	−222
90	−46	−16	−43	−4	−39	−188
120	−18	−25	−16	−13	−35	−122
200	−14	−13	−28	−8	−18	−35

11. Contrast the differences between saturated and unsaturated flow in the following categories:
    a. flow equation
    b. transport coefficient
    c. potential energy
    d. flux density

12. Assume the following quantities are known or measured. Calculate the soil water flux density (in cm/h), and determine the direction of flow of water in the soil.

Depth (cm)	h (cm)	K(h) (m/s)
30	−500	1.0×10^{-7}
45	−445	1.5×10^{-7}

CHAPTER TEN

Soil Water Flow Processes in the Field

Introduction

Several important processes determine the fate and transport of soil water in the field. In this chapter, we examine the movement of water through the physical processes of infiltration, redistribution, runoff, and evaporation. These are very important dynamic processes in the hydrologic cycle and are components of the soil water balance equation. As in previous chapters, emphasis will be given to the characterization of soil water transport in the field with relatively simple mathematical models.

Overview of the Water Flow Processes

Assume that a rainfall or irrigation event occurs at a constant rate for a finite time on a landscape containing actively growing vegetation. The total precipitation that reaches the ground is partitioned into infiltration, plant interception and depressional storage, and runoff (Figure 10.1). A series of rate versus time curves can be visualized for each of these dynamic physical processes. Depending on the rainfall intensity and the potential soil infiltration capacity, the infiltration rate curve typically increases quickly and then decreases exponentially with time. The rate of plant interception also decreases rapidly with time. Once interception storage of the vegetation is filled, raindrops begin falling from the leaves, stems, and grasses, where water stored on these surfaces eventually becomes depleted through evaporation into the atmosphere. The rate at which depression storage is filled rapidly declines after the initiation of the precipitation event. Water stored in depressions on the soil surface will eventually infiltrate or

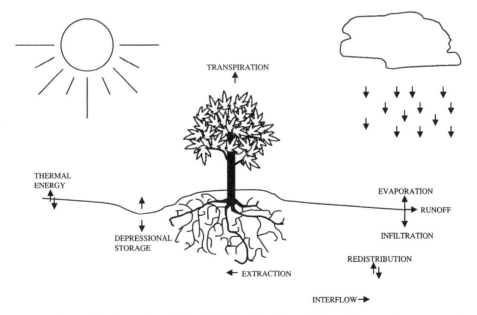

Figure 10.1. Schematic of the hydrologic cycle with emphasis on the water flow processes in the soil.

evaporate. A portion of the infiltrated water will redistribute within the profile, where it increases the soil water content and moves downward toward the groundwater or intercepts a relatively impervious layer and flows approximately parallel to the surface as interflow. Other portions of the infiltrated water will replenish the soil water in the upper portion of the profile. The component of the precipitation that exceeds the local infiltration rate will create a film of water on the surface until runoff begins. Eventually, extraction and evapotranspiration removes water from the soil profile, and the hydrologic cycle begins again.

Infiltration

Infiltration is the downward entry of water into the soil from rainfall, irrigation, or snowmelt. The infiltration process is dynamic and complex and varies with several soil, vegetative, and climatic parameters. It is one of the more important processes in the soil phase of the hydrologic cycle since infiltration determines the amount of runoff as well as the resupply of water to the soil profile. The two important parameters used to characterize infiltration of water into the soil profile are the rate and the cumulative amount.

Infiltration Rate

The infiltration rate, i, is defined as the flux density of water passing through the soil surface and flowing into the soil profile. Since it is a flux density term, mathematically the infiltration rate is defined as

$$i = Q/At \qquad [1]$$

where Q is the quantity or volume of water absorbed or infiltrating by unit area A of soil surface per unit time t. The SI units of i are $m^3/m^2 \cdot s$, which results in units such as m/s. Other frequently used units are cm/h and mm/h. The shape of the i versus t curve depends upon whether infiltration is supply controlled, profile controlled, or surface controlled. Three cases will be examined.

If infiltration is *supply* controlled, then i occurs at a constant rate as long as the rate of water delivery to the surface is smaller than the soil's ability to receive the water and to transmit the water away from the soil surface. Therefore, under supply-controlled conditions, the precipitated water infiltrates the soil surface as fast as it arrives, and the rainfall (or irrigation) rate is less than Ksat. The shape of the i versus t curve under these conditions is shown in Figure 10.2a.

The equation of continuity for one-dimensional flow of water through the soil surface is rewritten as

$$\partial\theta_v/\partial t = -\partial i/\partial z \qquad [2]$$

We see that if the infiltration rate does not change with space or time under supply-controlled conditions, $\partial\theta_v/\partial t$ equals zero at the soil surface. This also implies that $\partial h/\partial t$ also equals zero at the soil surface. Supply-control conditions occur in the field when the rainfall or irrigation rate is less than Ksat.

The second and most frequently observed relationship between i and t occurs when the

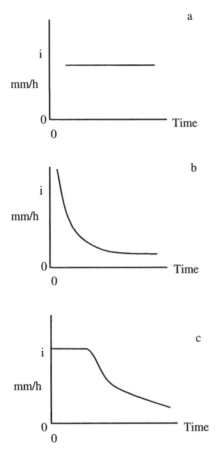

Figure 10.2. Relationship between the infiltration rate i and time for three situations: (a) supply controlled, (b) profile controlled, and (c) surface controlled.

soil *profile* controls the infiltration rate. Under these conditions, the delivery rate of rainfall exceeds the ability of the soil to receive water and to transmit the water away from the surface. The infiltration curve has the shape shown in Figure 10.2b. Values of i decrease rapidly with time after initiation of the infiltration process and approach Ksat as t approaches infinity, when the water table is deep because the potential gradient approaches unity. The rate of decrease in infiltration is rapid initially, but at longer times, the rate of decrease is much lower. The time that i approaches a constant rate varies with soil texture, structure, Ksat, and initial water content.

The third case, known as *surface*-controlled conditions, is a combination of supply- and profile-controlled conditions (Figure 10.2c). Initially, the infiltration rate is constant, and then it declines with time. This occurs when the rainfall exceeds Ksat but is less than the rate at which the soil can store water. Surface-controlled conditions may be observed during the early stages of rainfall when the soil water content at the surface varies with rainfall intensity.

Cumulative Infiltration

The second important infiltration parameter is the cumulative infiltration, I, which has SI units of length (e.g., m; it also may have units of cm or mm) and represents the equivalent depth of water infiltrated. The cumulative infiltration can be computed by two methods. In the first method, the time integral of the infiltration rate is calculated:

$$I = \int_{t_1}^{t_2} i(t)dt \qquad\qquad [3]$$

where t_1 is the initial time of infiltration and t_2 is the final time of infiltration. This shows that if the functional relationship between i and t is known between t_1 and t_2, then integration of the relation results in I.

The second method of calculating I is by integrating the change in volumetric soil water content distribution in the profile between the surface (z = 0) and the depth to the wetted front (z = z_f). Mathematically, this is represented by

$$I = \int_0^{z_f} \Delta\theta_v \, dz \qquad\qquad [4]$$

where z_f is the depth of the wetting front, and $\Delta\theta_v$ is the difference between the initial volumetric water content, $\theta_v(z, t_1)$, and the final volumetric water content, $\theta_v(z,t_2)$. The wetting front is the boundary between the wetted region and the dry region of soil during infiltration. This is the same method used to calculate the equivalent depth of water. In practice, equations [3] and [4] are calculated with summation approximations.

Graphically, the relationship between I and t usually has the smooth, monotonically increasing shape shown in Figure 10.3. The decrease in slope of the I versus t curve at the longer times is due to the decrease in i as shown in Figures 10.2b and 10.2c.

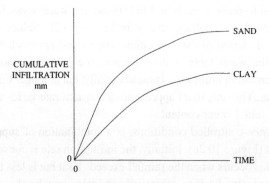

Figure 10.3. Relationship between the cumulative infiltration of water and time for two soil textures.

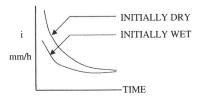

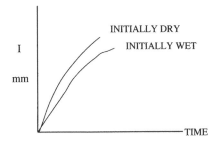

Figure 10.4. Relationships between the infiltration rate, cumulative infiltration, and time under two initial soil water contents at the soil surface. The symbols i and I represent the infiltration rate and cumulative infiltration, respectively.

Factors Affecting Infiltration

Infiltration of water depends on several soil physical parameters and on other factors associated with water application characteristics. Most of these parameters operate at or near the soil surface.

Initial Wetness and Matric Potential

The greater the initial water content of the soil, the lower the initial infiltration rate and the quicker the attainment of a constant infiltration rate. The drier the soil, the greater the matric potential gradients, which persist because of the slower advance of the wetting front since more water is required to fill the soil pores nearly to saturation. The influences of the initial water content at the soil surface on i and I with time are shown in Figure 10.4. Cumulative infiltration is greater in dry soil than in soil that is initially wet because the dryer soil has a higher storage capacity.

Texture

The saturated hydraulic conductivity is higher in coarse-textured soils than in fine-textured soils. Similarly, the infiltration of water is higher in coarse-textured soils than in fine-textured soils. A general relationship between soil texture and the final value of i is presented in Table 10.1. Soils with high percentages of coarse fragments tend to have high infiltration rates.

Nonuniformity of texture in the soil profile (horizonation of the profile) is also important. Horizons that impede flow reduce the infiltration rate. If a less permeable horizon is located near the soil surface, it restricts the wetting front and reduces the infiltration

Table 10.1. Relationship between Soil Texture and
Approximate Final Infiltration Rate

Soil Texture	Final Infiltration Rate	
	m/s $\times 10^{-7}$	mm/h
Sands	> 5.5	> 2.0
Sandy and silty soils	2.8–5.5	1.0–2.0
Loams	1.4–2.8	0.5–1.0
Clayey soils	0.3–1.4	0.1–0.5

rate. This restricting layer may be a shallow rock formation, a clay layer, a plow pan, or a fragipan. A light rainfall is easily absorbed, but heavier rains tend to saturate the surface layer. Since saturated soils have low infiltration capacity, this results in more runoff.

Structure

In general, differences in soil structure and aggregate size influence water movement in the same way as texture. The stability of aggregates at the soil surface and the rearrangement of the soil particles by the moving water affect soil structure. Stable soil properties imply that the $K(\theta_v)$ and $\theta_v(h)$ relationships are constant over time. Wetting and drying and freezing and thawing also affect soil structure, especially near the soil surface and make the soil properties unstable. Small rodent and earthworm burrows and old plant root channels at the soil surface significantly increase the infiltration rate.

Tillage

Tillage such as moldboard plowing and chiseling tends to increase soil porosity at the soil surface by 10–20%, depending on soil texture. This leads to an increase of i over nontilled soils. The increased i may not, however, be permanent because the porosity decreases during an intense rainfall.

Type of Clay

Soils containing montmorillonitic clays crack upon drying, thus providing a high initial infiltration rate. These soils may also swell quickly upon wetting, thereby sharply reducing the infiltration rate.

Vegetative Cover

Bare soils tend to have lower infiltration rates than soils protected by a vegetative cover. On bare soils, the impacting raindrops tend to puddle the soil, which results from the decomposition of aggregates, leading to the formation of a surface crust. A crust impedes infiltration. The net result is a lowering of i. Crop species and cropping patterns also affect the infiltration rate. In general, the presence of plants and plant roots tends to increase i compared with bare soil.

Rainfall Intensity

Rainfall intensity affects the infiltration rate in two ways. In high-intensity rains, the raindrops tend to be larger and have more energy when they strike the soil. Thus, high-intensity rains are more effective in sealing the soil surface than low-intensity rains. In the absence of any ponded water or runoff, the maximum possible i is the lesser of the rainfall rate or the soil's infiltration capacity. In the presence of ponded water or runoff, i equals the infiltration capacity until this surface supply of water is exhausted.

Slope

On a steep slope, the precipitated water tends to run off rapidly and, thus, has less opportunity to infiltrate than on a milder slope. Also, the soils on steeper slopes tend to have different physical characteristics (particularly, higher clay contents), in part due to erosion in previous years.

Channel Characteristics

The shape of the irrigation channel does not normally affect i significantly, as the entire border is covered with water in border or flood irrigation. The shape does influence i in furrow irrigation because the hydraulic conditions vary with shape, roughness, slope, and flow rate at any point in the furrow. The ponding depth influences the hydraulic gradients. This effect may be negligible except during the initial infiltration stages.

Water Temperature

As the temperature of the soil profile increases, the rate of water infiltration also increases. Temperature affects the viscosity and density of water (i.e., the fluidity) and its surface tension, which in turn affect the hydraulic conductivity. Temperature gradients can also cause thermally induced flow of water. The impact of freezing depends on the soil water content. High water contents decrease infiltration rates, but low water contents increase i.

Air Entrapment

The entrapment of air within the profile reduces the infiltration rate. As the wetting front moves downward in the profile, air is displaced, giving rise to a pore air pressure ahead of the front that increases above atmospheric air pressure. If the increased air pressure does not lead to a bubbling of air through the wetted topsoil, the infiltration rate is decreased (Kutílek and Nielsen 1994).

Soil Salinity

High-saline soils have low infiltration rates. Also, swelling and dispersion of clay in response to electrolyte composition and concentration in the flowing water change the size of the conducting pores.

Horizontal Infiltration

The simplest case of infiltration is horizontal imbibition in homogeneous soil where the influence of gravity is zero or negligible. Thus, water is drawn into the soil by matric

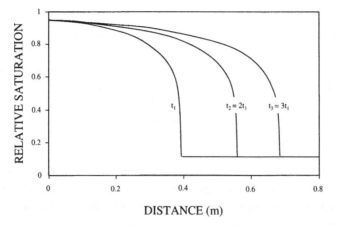

DISTANCE (m)

Figure 10.5. Horizontal infiltration into a clay loam. S represents the relative saturation. Values of the other parameters used were $S_r = 0.106$, Smax $= 0.98$, $S(x, t = 0) = 0.11$, $S(x = 0, t) = 0.95$, and van Genuchten pore size distribution index, m $= 0.283$.

forces only. In the last chapter, we saw that we could write the transient-state equation for horizontal flow of water as

$$\partial\theta_v/\partial t = \partial/\partial x(K\partial h/\partial x) = \partial/\partial x[K(\partial h/\partial\theta_v)(\partial\theta_v/\partial x)] \qquad [5]$$

Using the chain rule, we let

$$\partial h/\partial x = (\partial h/\partial\theta_v)(\partial\theta_v/\partial x) \qquad [6]$$

and set

$$D = K(\partial h/\partial\theta_v) \qquad [7]$$

where D is the soil water diffusivity. Substituting this into equation [5] gives

$$\partial\theta_v/\partial t = \partial/\partial x(D\partial\theta_v/\partial x) \qquad [8]$$

Equation [8] has been called the diffusion equation for horizontal flow of water in soil. It has been solved for the special initial and boundary conditions of flow of water into an infinitely long column of uniform wetness where the plane at x = 0 is instantaneously brought to and maintained at a water content of θ_0. These initial and boundary conditions can be written as

$$\theta_v(x, 0) = \theta_i \qquad x > 0, t = 0$$
$$\theta_v(0, t) = \theta_0 \qquad x = 0, t > 0$$

Solutions of the diffusion equation for these initial and boundary conditions for several times give curves with the shapes shown in Figure 10.5. At each time, the soil water content changes only slightly with distance along the soil column until at some distance from the source, it changes abruptly. This sharp decrease in water content is called the *wetting front* and results from a dramatic decrease in K and D with corresponding large changes in θ_v.

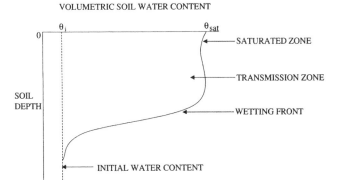

Figure 10.6. Relationship between soil water content and depth during vertical infiltration of water.

Vertical Infiltration

Downward infiltration of water into an initially dry, unsaturated soil generally occurs under the combined influences of gradients of matric potential and gravity. Since the soil at the surface normally is dry, matric forces dominate during the early stages of water entry into the soil.

The changes in soil water content during vertical infiltration into a uniformly textured soil profile can be visualized as shown in Figure 10.6. This shows that there are two wet zones: the saturated zone, which exists at and near the soil surface, and the transmission zone, which extends down into the profile. In the transmission zone, the water content is nearly uniform, and therefore, the hydraulic conductivity is uniform and the hydraulic gradient is nearly uniform. As with horizontal infiltration, there is a sharp boundary between the wet and dry soil. The distributions of hydraulic head and degree of saturation in the profile can be visualized as shown in Figure 10.7.

The cumulative infiltration of water into an air-dry soil profile can be approximated by a piston flow analogy. The depth of the wetting front is multiplied by the mean volumetric soil water content behind the wetting front, θ_m. If the initial profile water content is zero, I can be expressed as

$$I = z_f \theta_m \qquad [9]$$

where z_f is the depth of the wetting front. However, if the initial volumetric water content is constant with depth and equal to θ_i, then I can be approximated from

$$I = (\theta_m - \theta_i)z_f = \Delta\theta \cdot z_f \qquad [10]$$

The equivalent depth of water contained in the soil from the surface to any depth z is represented as

$$\int \theta_v(z)dz \approx \sum \Delta\theta_{vi}\Delta z_i \qquad [11]$$

where the subscript i is the depth interval (m). If θ_v is a known function of z, that is, $\theta_v(z)$, then equation [11] can be integrated using standard calculus or numerical techniques (Appendix B).

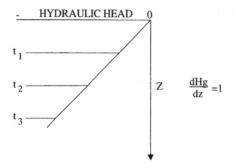

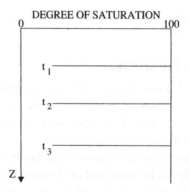

Figure 10.7. Temporal relationships of the hydraulic head and degree of saturation and depth in the profile.

The volume of water entering the soil (m³) to z_f can be determined from

$$Q = A\Delta\theta_v z_f \qquad [12]$$

where Q is the volume of water and A is the cross-sectional area. The infiltration rate can be calculated from

$$i = Q/(At) = (\Delta\theta_v/t)z_f = \Delta\theta_v V_f \qquad [13]$$

where t is the elapsed time and V_f is the velocity of advance of the wetting front.

Vertical Infiltration Equations

When water infiltrates a uniform soil profile under *supply-controlled* conditions (i.e., the supply of water is lower than the ability of the soil to receive or absorb the water), then the infiltration rate is constant and the downward velocity of the wetting front becomes constant. Thus, the water content of the transmission zone is constant, and the infiltration rate becomes

$$i = K + m \qquad [14]$$

where m is a constant. The parameter K varies with the intensity of the supply of water up to a value near Ksat, which is the hydraulic conductivity of soil with water ponded on the surface.

Several equations have been developed to approximate the vertical infiltration of water in soil under *profile-controlled* conditions. These infiltration equations can be classified into two groups: empirical and exact. The discussion that follows details several of the more frequently used infiltration equations.

Empirical equations are those equations that can be used to fit the experimental data. Essentially these equations contain curve fit parameters that, along with the use of statistical techniques such as nonlinear least squares, can be used to characterize the infiltration data. The regression parameters may or may not have physical significance.

Kostiakov Equation

The simplest of the empirical infiltration equations is the Kostiakov equation. Mathematically, this power equation can be written as

$$i(t) = kt^{-n} \tag{15}$$

where k and n are constants obtained in infiltration trials. Since in this equation i goes to 0 as t goes to infinity, the modified Kostiakov equation is usually used. This is

$$i(t) = kt^{-n} + C \tag{16}$$

where C is the infiltration rate at large t. This modified form of the Kostiakov equation usually fits the experimental infiltration data quite well, particularly for time periods of less than a few hours. Equations [15] and [16] can be integrated to obtain the cumulative infiltration at any time t:

$$I(t) = At^B \tag{17}$$

or

$$I(t) = At^B + Ct \tag{18}$$

where A is equal to $(k/-n + 1)$, and B is equal to $(-n + 1)$. Parameters A and B have no obvious physical meaning and are determined from the statistical fit of the model to the experimental data.

Horton Equation

Another well-known empirical equation is the Horton equation. Mathematically, this equation is

$$i(t) = i_f + (i_0 - i_f)\exp(-Bt) \tag{19}$$

where i_0 is the initial infiltration rate (cm/min); i_f is the final infiltration rate, which is achieved at long times (cm/min); and B is a soil parameter that describes the rate of decrease of infiltration (min^{-1}). The Horton equation indicates that if the rainfall exceeds infiltration capacity, infiltration tends to decrease in an exponential manner. The area under the curve represents the cumulative depth of water infiltrated beneath the soil surface.

The cumulative infiltration is determined by integrating equation [19] with respect to time to obtain

$$I(t) = i_f t + [(i_0 - i_f)/B][1 - \exp(-Bt)] \tag{20}$$

where I has SI units of meters or may be expressed as centimeters or millimeters.

Horton thought that the reduction in infiltration rate with time after the initiation of infiltration was largely controlled by factors operating at the soil surface. These factors included swelling of the colloids and the closing of small cracks, which progressively seal the soil surface. Later evidence showed that a crust may be formed at the soil surface, which reduces the hydraulic conductivity.

Example

Calculate the infiltration rate after 30 minutes using the Horton equation and assuming the following: initial infiltration rate = 15 cm/h, final infiltration rate = 6.0 cm/h, and B = 2.0 h^{-1}.

$$i(t) = i_f + (i_o - i_f)\exp(-Bt)$$
$$= 6 + (15 - 6)\exp(-2 \times 0.5) = 9.31 \text{ cm/h}$$

Philip Equation

Exact solutions of the flow equation for infiltration require analytical procedures. For a homogeneous, infinitely deep soil profile at a uniform initial water content, exact solutions were derived by Philip (1957) and Green and Ampt (1911).

The Philip solution describes the time dependence of cumulative infiltration in terms of a power series. For vertical infiltration, the cumulative infiltration is

$$I(t) = S_{or}t^{1/2} + A_1 t + A_2 t^{3/2} + A_3 t^{5/2} + \dots \qquad [21]$$

where the constant S_{or} is called the *sorptivity* (mm/min$^{0.5}$) and for short times is equal to $I/t^{1/2}$; and the parameters A_1, A_2, and A_3 are characteristics of the soil. The parameter S_{or} has meaning only in relation to an initial hydraulic state of the soil and an imposed boundary condition. It is the capacity to absorb or release water during the early times and is calculated from the slope of the line of a graph of I versus $t^{0.5}$.

In practice, it is generally sufficient to approximate infiltration by the first two terms of equation [21] as

$$I(t) = S_{or}t^{1/2} + At \qquad [22]$$

$$i(t) = 0.5S_{or}t^{-1/2} + A \qquad [23]$$

where A is a constant infiltration rate after long times and is closely related to Ksat. These Philip equations have the advantage of being based on physical theory, and the constants S_{or} and A have physical significance. Use of these equations, however, is limited to uniform, homogeneous soils. The sorptivity term, S_{or}, is the dominant parameter governing the early stages of infiltration. Therefore, management practices that affect infiltration affect the magnitude of S_{or}. As time increases, the parameter A becomes progressively more important and eventually becomes the only significant parameter in predicting infiltration.

Green-Ampt Equation

A simple, approximate but useful model that describes the steplike infiltration of water in soil was developed by Green and Ampt in 1911 (Haan et al. 1994; Viessman et al. 1989). In their model, water penetrates into the profile like a piston and proceeds with time to

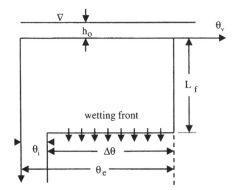

Figure 10.8. A schematic of the Green-Ampt infiltration model parameters.

greater depths. Below the abrupt horizontal wetting front, the soil remains at its initial water content. Their approach has been found to apply to infiltration into initially dry, coarse-textured soils (Figure 10.8). The following assumptions were used in the derivation:

1. Darcy's law governs the vertical flow of water.
2. Piston flow creates a distinct wetting front in the initially uniformly dry soil profile.
3. The suction head at the wetting front remains constant, regardless of time and position.
4. Behind the wetting front, the soil is uniformly wet and K is constant.
5. A constant ponded depth, h_0, is applied at the soil surface at $t = 0$.

For vertical infiltration, the infiltration rate equation is

$$i(t) = dI/dt = K\{[h_0 + h_f + z_f(t)]/z_f(t)\} \qquad [24]$$

where K is the hydraulic conductivity of the wetted part of the soil profile, $z_f(t)$ is the time-dependent depth of the wetting front, h_f is the effective suction head for wetting at the wetting front, and h_0 is the depth of ponding on the soil surface. Where the infiltration does not pond or saturate the soil, h_0 is small and is often neglected for practical purposes. The change in soil water content across the wetting front, $\Delta\theta_v$ (m³/m³), depends on the initial soil water content, θ_i (m³/m³), and the effective soil water content behind the wetting front, θ_e (m³/m³). Mathematically, $\Delta\theta_v$ can be estimated from

$$\Delta\theta_v = \theta_e - \theta_i \qquad [25]$$

For piston flow, the cumulative infiltration at time t is $I(t) = z_f(t)\Delta\theta_v$, then $z_f(t) = I(t)/\Delta\theta_v$, and taking h_0 as zero, the infiltration rate is

$$i(t) = K\{[h_f\Delta\theta_v/I(t)] + 1\} \qquad [26]$$

The cumulative infiltration as a function of time, I(t), is found by integrating equation [26] and substituting the condition that $I(0) = 0$ to obtain

$$I(t) = Kt + (h_f\Delta\theta_v)\ln\{1 + [I(t)/h_f\Delta\theta_v]\} \qquad [27]$$

If the ponding depth is not negligible, h_f should be replaced with $h_f + h_0$ or Δh. Since equation [27] cannot be solved explicitly for I(t), iteration is required. One way is to substitute a trial value for I(t), for example, Kt, into the right-hand side of equation [27], which is then compared to the left-hand side. This process is repeated until agreement between the two values is obtained. The infiltration time is

$$t = (\Delta\theta_v/K)\{z_f - (\Delta h)\ln[1 + (z_f/\Delta h)]\} \qquad [28]$$

The Green-Ampt solution gives no information on the details of the soil water profiles during infiltration but does offer estimates of i and I as a function of time. Rawls et al. (1983) tabulated values of θ_e, h_f, and K by textural class for use in the Green-Ampt infiltration equation.

Example

Assume that water is ponded on a silt loam soil to a depth of 2 cm. If the effective porosity is 0.486 m^3/m^3, the initial soil water content is 0.146 m^3/m^3, the hydraulic conductivity is 0.65 cm/h, and the suction head at the wetting front is 17 cm, calculate the infiltration rate for a cumulative infiltration of 1 cm.

$$\Delta\theta_v = \theta_e - \theta_i = 0.486 - 0.146 = 0.340 \text{ m}^3/\text{m}^3$$

Solving for the time, we rearrange equation [28] to get

$$t = \{I(t) - h_f\Delta\theta_v \ln[1 + I(t)/h_f\Delta\theta_v]\}/K$$
$$= \{1 - 17(0.34)\ln[1 + 1/17(0.34)]\}/0.65 = 0.12 \text{ h}$$

Using equation [26], the infiltration rate is

$$i(t) = 0.65\{[17(0.34)/1] + 1\} = 4.41 \text{ cm/h}$$

This calculation process can be repeated at other values of I to obtain the infiltration rate curve for this soil.

Example

Suppose that a double-ring infiltrometer (a device consisting of two concentric rings of different diameter) is used to pond a constant head of 5 cm of water on a clay loam soil, which has a Ksat of 0.54 cm/h, h_f = 106 cm, θ_r = 0.106, f = 0.469 cm^3/cm^3, and $\Delta\theta_v$ = 0.363 cm^3/cm^3. If the soil obeys the Green-Ampt theory, it will produce an infiltration curve like the ones shown in Figures 10.9, 10.10, and 10.11.

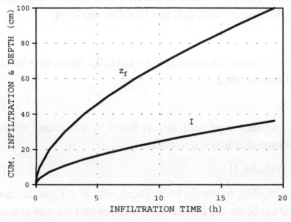

Figure 10.9. Cumulative infiltration, I, and depth of wetting front in clay loam as a function of time.

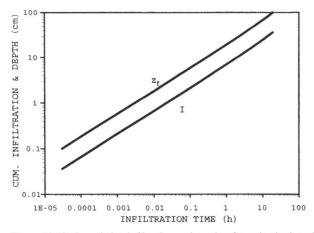

Figure 10.10. Cumulative infiltration and wetting front depth plotted on a log-log scale.

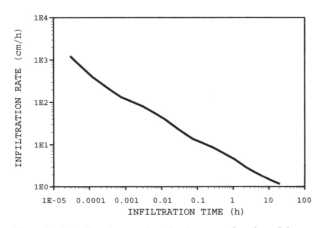

Figure 10.11. Infiltration rate in a clay loam as a function of time.

Richards Equation

Vertical infiltration of water into unsaturated soil occurs under the combined influences of gravity and matric gradients. Darcy's law for vertical flow can be written as

$$q = -K\partial H/\partial z = -K\partial(h + z)/\partial z \qquad [29]$$

At the soil surface, $q = i$ and $z = 0$; therefore, we can write

$$q = i = -K(\partial h/\partial z) - K \qquad [30]$$

Combining this with the equation of continuity gives the Richards equation:

$$\partial\theta_v/\partial t = \partial/\partial z(K\partial H/\partial z) \qquad [31]$$

Since H = h + z, some authors write equation [31] as

$$\partial\theta_v/\partial t = \partial/\partial z(K\partial h/\partial z) + \partial K/\partial z \tag{32}$$

to separate the flow into gravity and capillary components, although it is not necessary. The difficulty in separating the flow components this way occurs when the Richards equation is converted to a numerical method. There is a tendency to use separate values for K in the two flow components, and this produces a non-Darcian flow.

Since θ_v and h are related, we can write, using the chain rule,

$$\partial\theta_v/\partial t = (\partial\theta_v/\partial h)(\partial h/\partial t) = C(h)\partial h/\partial t \tag{33}$$

and

$$C(h)\partial h/\partial t = \partial/\partial z(K\partial h/\partial z) + \partial K/\partial z \tag{34}$$

When this form of the Richards equation is converted to a numerical method, it suffers from errors in mass balance (Celia et al. 1990). Alternatively, we could write

$$\partial\theta_v/\partial t = \partial/\partial z(D\partial\theta_v/\partial z) - \partial K/\partial z \tag{35}$$

This form fails in numerical models near saturation in field soils since $K(\theta_v)/C(\theta_v) = D$, which is indeterminant as both $C(\theta_v)$ and $\partial\theta_v/\partial z$ approach zero, creating larger numerical errors.

Infiltration into Layered Soils

Soil profiles are seldom uniform with depth and seldom have uniform soil water contents at the initiation of infiltration. These effects tend to reduce the infiltration rate more rapidly than would be predicted theoretically. In nonhomogeneous soils, the soil profiles are normally divided into horizons or layers that are assumed to be homogeneous. The matric potential and total soil water potential must be continuous throughout the profile regardless of the layering sequence even though the water content and hydraulic conductivity may exhibit discontinuities at horizon boundaries.

A layer or horizon having a different texture at some depth in the profile tends to reduce the infiltration rate regardless of whether it is coarser or finer than the soil surface. If the texture of the lower horizon is coarser, the reduction in infiltration is due to the fact that the water potential must increase to a point to permit entry of water into the relatively large pores of the coarse layer. We consider two general textural configurations of soil profiles.

Coarse Texture over Fine Texture

In this configuration, Ksat in the coarse layer is greater than Ksat in the fine-textured layer. Initially, i is controlled by the coarse layer (Figure 10.12a). When the wetting front reaches and penetrates into the finer-textured layer, i decreases to that of the finer texture alone. Thus, in the long term, the horizon with the lower hydraulic conductivity controls i, and the infiltration rate is similar to that of the uniform soil. If i continues for long times, then positive pressure heads (a perched water table) can develop in the coarse soil, just above the boundary with the impeding finer layer. After infiltration ends, a point is

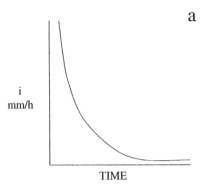

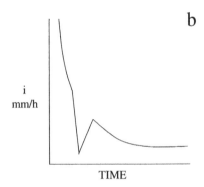

Figure 10.12. Infiltration rates as a function of time for (a) coarse texture over fine texture and (b) fine texture over coarse texture.

reached eventually when the water moves more rapidly in the finer-textured soil than in the overlying, unsaturated, coarse-textured soil.

Fine Texture over Coarse Texture

In this configuration, the Ksat in the second layer is greater than the Ksat in the first layer. The initial i is determined by the upper, fine-textured layer (Figure 10.12b). As the wetting front reaches the interface between the fine and coarse layers, i decreases immediately. However, when values of h at the wetting front increase somewhat with increasing water content of the soil above the coarse texture, more pores fill and the permeability increases and i increases. The water content at the wetting front is unsaturated, and the matric potential is too low to permit entry of water into the relatively large pores of the coarse soil. Infiltration at the soil surface does not stop. Note that the rate does not increase to that which would exist in the absence of the coarse layer. The coarse-textured horizon does not become saturated, since the restricted flow through the less permeable horizon is less than saturated flow (Ksat) of the coarse layer. The h and θ_v remain constant throughout a considerable depth, and i becomes constant over time.

Runoff

Runoff occurs when the rate of rainfall exceeds the rate of infiltration and surface retention by soils and plants. Runoff is the overland flow of water resulting from precipitation events. Some of this flow, which may occur during and immediately following the precipitation event, is known as storm water runoff, or *stormflow.* Flow occurring between precipitation events is generally supported by seepage and groundwater discharge and is known as *baseflow.* There is no definable point or time at which stormflow ceases and baseflow begins. One process gradually grades into the other. Generally, on small watersheds, the magnitude of baseflow is quite small and is often neglected in the computation of stormflow.

Soils vary greatly in the rates at which they can absorb water. Little or no runoff will occur on soils having permeabilities as high or higher than the rate of precipitation. For example, a rainfall rate of 2 cm/h would not produce runoff if the soils on which it fell had infiltration rates of 2 cm/h or greater. However, if the same 2 cm/h rain intensity occurred for an hour on a soil with an infiltration rate of 1 cm/h, there would be between 0.8 and 1.0 cm of runoff. The difference in runoff is due to storage by vegetation and to depressional storage areas at the soil surface.

Runoff can be reduced if the infiltration rate of the soil can be increased or if the water can be held on the soil surface for a longer period of time. Both of these conditions permit more water to enter the soil and less to run off. A granular soil structure, a rough soil surface, and numerous earthworm burrows and plant root channels promote infiltration and reduce runoff.

The effects of tillage on runoff vary greatly with the type of tillage, soil, and topography and with time. Soil covered by a thick stand of vegetation usually has a minimum of runoff. Crop residues give considerable protection from runoff if they are left on the soil surface. Plowing the soil and burying the residues usually produces a rough surface, thereby providing small depressions that can temporarily store water until it infiltrates or evaporates. This helps to reduce runoff. Raindrop impact on bare soil can produce a crust and reduce infiltration rate after a time. Further cultivation may temporarily increase infiltration but soon leads to reduced infiltration and increased runoff. Maximum runoff from sloping land will occur where the residues are plowed under and the surface is smoothed and left without further cultivation.

Plowing and planting around a hill (contour furrows) rather than up and down the slope forms small ridges across the slope. Water is stored behind the ridges, and therefore, more water infiltrates and less runs off. Terraces also control runoff due to increased infiltration behind the terraces. Even on land with low slopes, terraces can be effective in conserving water.

Next, we discuss two well-known methods of approximating runoff from agricultural plots and watersheds. In general, these methods are used at the macroscale level.

Water Balance Method

Recall that the water balance equation included terms for the inputs of water and the outputs of water (Chapter 5). The difference between the inputs and the outputs is the change in water storage. Mathematically, this water balance was written as

$$P + I \pm R - ET \pm D = \Delta W \tag{36}$$

where ΔW is the change in water, P and I are the inputs from precipitation and irrigation, and R, ET, and D are the outputs from runoff (or runon), evapotranspiration, and internal drainage (or capillary rise). The positive sign for R and D is used if water is added to the profile. Therefore, the negative sign is used for runoff. If values of all other parameters in equation [36] are known, runoff can be estimated.

NRCS Runoff Curve Number Procedure

The National Resources Conservation Service (NRCS) has developed a widely used curve number procedure for estimating runoff (SCS 1972). The effects of land use and treatment, and thus infiltration, are embodied in the procedure. The procedure was empirically developed from studies of small agricultural watersheds (<800 ha) across the United States.

The relationship used in computing cumulative runoff is

$$Q = (P - 0.2S)^2/(P + 0.8S) \qquad P > 0.2S \tag{37}$$

where Q is the accumulated runoff volume or rainfall excess (mm), P is the accumulated precipitation (mm), and S is a parameter, sometimes called a *maximum soil water retention parameter*. In equation [37], P must exceed 0.2S before any runoff is generated. Thus, a rainfall volume of 0.2S must fall before runoff is initiated.

When the parameters Q, P, and S are expressed in millimeters, the equation for S is

$$S = (25400/CN) - 254 \tag{38}$$

In equation [38], CN is the curve number that ranges between 30 and 100. Curve numbers indicate the runoff potential of an area and are available by soil series from the local NRCS office. Impervious areas and water surfaces are assigned a CN of 98–100. Curve numbers are functions of land use, antecedent runoff conditions, and soil hydrologic group and for selected land uses are given in Table 10.2. The hydrologic groups of some Arkansas soils of interest are Calloway (C), Captina (C), DeWitt (D), Sharkey (D), Memphis (B), and Crevasse (A). Selected characteristics of the profiles of these soils were given in Chapter 3. Curve numbers and hydrologic groups for other soils are given in Haan et al. (1994).

The procedure for estimating runoff consists of selecting a rainfall event and computing the direct runoff by the use of curves founded on field studies of the amount of measured runoff from numerous soils and land cover combinations. A runoff curve number (CN) is extracted from Table 10.2. Selection of the runoff curve number is dependent on the types of land cover, hydrologic soil group, and antecedent moisture conditions (AMC). Soils are classified in hydrologic soil groups as either A, B, C, or D according to the criteria in Table 10.3.

The curve numbers are applicable to average antecedent moisture conditions: AMC(II). Other antecedent moisture conditions are given in Table 10.4. Haan et al. (1994) provide an equation first given by Chow (1964) to convert CN(II) to those for dry conditions CN(I) and wet conditions CN(III):

$$CN(I) = 4.2 \times CN(II)/\{10 - [0.058 \times CN(II)]\} \tag{39}$$

$$CN(III) = 23 \times CN(II)/\{10 + [0.13 \times CN(II)]\} \tag{40}$$

Table 10.2. Runoff Curve Numbers (CN II) for Selected Land Uses

Land Use Description	Hydrologic Soil Group			
	A	B	C	D
Cultivated land				
Without conservation treatment	72	81	88	91
With conservation treatment	62	71	78	81
Pasture and rangeland				
Poor condition	68	79	86	89
Good condition	39	61	74	80
Meadow				
Good condition	30	58	71	78
Wood or forestland				
Thin stand, poor cover, no mulch	45	66	77	83
Good cover	25	55	70	77

Table 10.3. Description of Hydrologic Group Used to Determine Runoff

A: Low runoff potential and high infiltration rates even if thoroughly wetted. Chiefly deep, well to excessively drained sands or gravels. High rate of water transmission (> 0.3 in/h or 0.76 cm/h) from the soil surface.

B: Moderate infiltration rates if thoroughly wetted. Chiefly moderately deep to deep, moderately well to well drained soils with moderately fine to moderately coarse textures. Moderate rate of water transmission (0.15–0.3 in/h or 0.38–0.76 cm/h).

C: Slow infiltration rates if thoroughly wetted. Chiefly soils with a layer that impedes the downward movement of water or soils with moderately fine to fine textures. Slow rate of water transmission (0.05–0.15 in/h or 0.13–0.38 cm/h).

D: High runoff potential and slow infiltration rates if thoroughly wetted. Chiefly clay soils with a high swelling potential, soils with a permanent high water table, soils with a claypan or clay layer at or near the surface, and shallow soils over nearly impervious material. Very slow rate of water transmission (0–0.05 in/h or 0.13 cm/h).

Example

Determine the effects of antecedent moisture conditions for a soil having a CN of 80 for average conditions.

For AMC(I), we calculate using equations [39] and [38] the following:

$$CN(I) = (4.2 \times 80)/[10 - (0.058 \times 80)] = 62.7$$
$$S = (25,400/62.7) - 254 = 151 \text{ mm}$$

For AMC(III), we calculate using equations [40] and [38] the following:

$$CN(III) = (23 \times 80)/[10 + (0.13 \times 80)] = 90.2$$
$$S = (25,400/90.2) - 254 = 28.2 \text{ mm}$$

Table 10.4. Antecedent Moisture Conditions That Affect the CN

AMC(I): A condition in which the soils are dry but not to the wilting point, and when satisfactory plowing or cultivation takes place. Used when there has been little rainfall preceding the rainfall in question.

AMC(II): The average case for annual floods, i.e., an average of the conditions that have preceded the occurrence of the maximum annual flood on numerous watersheds.

AMC(III): If heavy rainfall or light rainfall and low temperatures have occurred during the 5 days previous to the given storm and the soil is nearly saturated. Used where there has been considerable rainfall prior to the rain in question.

Thus, we conclude that under dry soil conditions, that is, AMC(I), the revised CN is lower than the average CN value, but the water retention term S increases, which leads to a decrease in the estimated runoff from a given precipitation event. Under wet soil conditions, that is, AMC(III), the revised CN is increased above the average value and the water retention term is lower, which according to equation [38] leads to an increase in the estimated runoff.

For a large area such as a watershed where the soils and land cover vary, a composite CN may be computed by weighting each CN according to areal extent. If, for example, 80% of a watershed has a CN of 75 and the remaining 20% is impervious (CN = 100), then the weighted CN is calculated from CN = (0.8 × 75) + (0.2 × 100) = 80.

Example

Determine the influence of pasture quality on runoff from a 7 cm rainfall on a Captina silt loam.

The Captina soil is in hydrologic group C. For a rainfall of 70 mm, the CN of a good pasture is 74. Therefore, the estimated runoff is

$S = (25,400/74) - 254 = 89.24$

$Q = [70 - 0.2(89.24)]^2/[70 + 0.8(89.24)] = 19.23$ mm

For a pasture of poor quality, the CN is 86. Therefore,

$S = (25,400/86) - 254 = 41.35$

$Q = [70 - 0.2(41.35)]^2/[70 + 0.8(41.35)] = 37$ mm

These calculations show that the quality of the pasture affects the amount of runoff. The higher the quality of the pasture, the lower the curve number and the lower the runoff.

Example

Assume that a rainfall of 10 cm falls on a watershed that is 30% bare in hydrologic group C and has 30% of its soils in hydrologic group B under grass and 40% in hydrologic group D under forest. Determine the runoff from this rainfall event over the watershed.

The integrated CN for the watershed is calculated as

$(0.3)(88) + (0.3)(58) + (0.4)(83) = 77$

The estimated runoff from the watershed is

$S = (25,400/77) - 254 = 75.87$

$Q = [100 - 0.2(75.87)]^2/[100 + 0.8(75.87)] = 44.78$ mm

This calculation shows that the estimated runoff is an area-weighted composite of the various soil runoff characteristics, land uses, and land covers in the small watershed.

Internal Drainage and Redistribution

When rain or irrigation ceases and no more water is ponded on the soil surface, infiltration comes to an end. Downward movement of water within the soil continues as soil water redistributes within the profile. The water moves down within the soil profile under the combined influences of gravity and matric potential gradients. This downward movement of water in the profile is called *internal drainage* or *redistribution*. Its effects are to redistribute soil water to lower depths of the profile, thereby increasing the water content at the lower depths. In most cases, the redistribution rate of water within a depth interval decreases rapidly over a period of several days or weeks. In some cases, the hydraulic gradients may become small, at which time the flux density of water approaches zero.

The redistribution of water in the soil profile after rainfall or irrigation depends on the amount of water added and whether it is sufficient so that the wetting front reaches the wet soil below. In humid areas, the wetting front often will advance into wet soil during infiltration or shortly thereafter.

Internal Drainage in a Deeply Wetted Profile

In a deeply wetted profile, the internal drainage rates are normally determined by gravity alone. Therefore, under these conditions, the magnitude of the hydraulic gradient is 1 and we write Darcy's law as

$$q = [-K(\theta_v)d(h + z)]/dz = -K(\theta_v)$$ [41]

Observations have shown that the functional relation between K and θ_v sometimes can be represented by an exponential equation

$$K(\theta_v) = a \exp(b\theta_v)$$ [42]

where a and b are empirical constants. Combining these two equations, the soil water flux becomes

$$q = -K(\theta_v) = -a \exp(b\theta_v)$$

or

$$\ln q = \ln a + b\theta_v$$ [43]

Since q and θ_v are logarithmically related, a small decrease in θ_v can result in a large decrease in q. A plot of $\ln q$ versus θ_v will be linear with a slope of b and an intercept equal to $\ln a$.

For drainage of soil water under the influence of gravity only, we can also write

$$q = -K(\theta_v) = dW/dt \tag{44}$$

where W has units of equivalent depth of soil water. Since $W = \theta_v z_b$, equation [44] can be written as

$$q = -z_b d\theta_v/dt \tag{45}$$

where z_b is the depth of soil at which the flux density of soil water is calculated and θ_v is the average volumetric soil water content over this depth interval.

Internal Drainage in a Shallow Wetted Profile

In some cases, the end of infiltration occurs with a wet upper profile and a dry lower profile. Therefore, the dryer layers "draw" the water from the upper layers because of the high matric potential gradients. The water content profile as a function of time changes during redistribution and will appear as shown in Figure 10.13. Note that the rate of advance and the abruptness of the wetting front decrease with time.

The drainage flux decreases over time because (1) values of the matric potential gradient, dh/dz, decrease as the upper layers lose water and the lower layers gain water, and (2) the K of the upper wet layers decreases as θ_v decreases. The temporal relationship of the drainage flux near the soil surface is shown in Figure 10.14. The rapid decrease in water content of the sandy soil occurs because K decreases rapidly with decreases in water content. The more gradual decrease in water content of the clayey soil occurs because the decrease in θ_v with a corresponding decrease in h is more gradual.

With incomplete wetting of the profile, movement of soil water involves both wetting and drying (or hysteresis). This is caused by water moving out of the region near the soil surface and into the drier region below. Several soil scientists have observed that the decrease in water content in the initially wetted zone obeys the power equation having the form

$$W = a(t + b)^{-c} \tag{46}$$

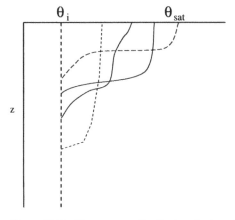

Figure 10.13. Water content distributions in the soil profile after infiltration has ceased.

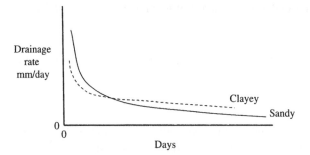

Figure 10.14. Drainage flux at a given depth as a function of time for two soil textures.

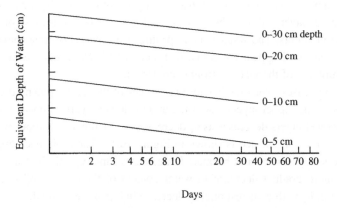

Figure 10.15. Equivalent depth of water in various depth layers within a sandy loam during redistribution following irrigation.

where a, b, and c are empirical constants, and W is the equivalent depth of water. If the parameter b is zero, then equation [46] converts to the linear form

$$\log W = \log a - c \log t \tag{47}$$

which can be used to statistically determine the constants a and c. Plots of log W and log t become straight lines, as shown in Figure 10.15.

The rate of water lost from the wetted profile can be found by differentiation to obtain

$$dW/dt = -ac(t + b)^{-(c+1)} \tag{48}$$

Since the parameter b can be ignored and the constants a and c can be determined from the log-log plot, the drainage rate, dW/dt, can be easily computed.

Preferential Flow

The rapid movement of water in the large pores of undisturbed soil has become known by several terms, including *preferential flow, macropore flow,* and *bypass flow.* This percolating water is a significant component of the flow of water and solute transport in many soils. The rapidly moving water can bypass a large fraction of the soil matrix, thus moving deeper, more rapidly, and with less displacement of the initial soil solution than would be predicted by piston displacement descriptions. Continued developments in our ability to characterize the porous systems of soil pedons permit coupling of the soil structural characteristics to the basic transport equations.

Numerous attempts have been made to make the deterministic soil water flow models more realistic by incorporating important soil characteristics. One example is the mobile-immobile region concept, which attempts to account for known variations that can be observed in soil. Preferential flow under transient, variably saturated conditions results in physical and hydraulic nonequilibrium between the preferential region of flow and the less mobile region. This nonequilibrium results in a discrepancy between resident solute concentration of the soil matrix and the flux concentration of the preferential flow path.

Emphasis in soil structural descriptions is often placed on the solid phase in terms of soil particles and peds; however, in recent years, more attention has been given to characterizing the pore space, where transport processes occur. Pores such as root and worm channels often have a cylindrical shape. Earthworms create channels as they drag food from the surface into their burrows. Decaying roots and cracks due to shrinking also are important avenues for transport of water and solutes. The connectivity of the various pore sizes is strongly impacted by soil water content.

Example

Calculate the maximum depth of penetration into the profile from a 2.5 cm rainfall, assuming that 60% of the water moved only through the macropores and the macroporosity represented only 1% of the soil volume. Compare this depth with the depth of penetration without preferential flow, assuming initial and saturated water contents of 0.25 and 0.50 m^3/m^3, respectively.

With preferential flow,

$$I = (\Delta\theta_v)z_f$$
$$(0.025 \text{ m})(0.6) = (0.01 \text{ m}^3/\text{m}^3)(z_f)$$
$$z_f = 1.5 \text{ m}$$

Without preferential flow,

$$0.025 \text{ m} = (0.50 - 0.25 \text{ m}^3/\text{m}^3)(z_f)$$
$$z_f = 0.1 \text{ m}$$

These calculations show the dramatic impact of preferential flow on the depth of penetration of water.

Field Capacity

Early observations in the field showed that the drainage rate and water content decrease over time. This indicated that the flow rate becomes negligible within a few days or that flow stopped. The water content at which drainage allegedly ceased (i.e., field capacity) was accepted as a physical property of a soil. The common working definition of *field capacity* is the wetness of the initially wetted zone about 2 days after the cessation of infiltration. This definition does not take into account the soil properties and processes such as the antecedent water content of the soil, depth of wetting, presence or absence of a water table, extraction of water by plants, evaporation of water from the surface, and textural changes and the effects on redistribution. In addition, it was commonly assumed that the application of a given quantity of water to the soil would fill the soil up to "field capacity" to a certain depth, and beyond this depth, water would not penetrate.

In recent years with the development of flow theory and more precise experimental techniques, the field capacity concept has been recognized as arbitrary and not an intrinsic physical property of the soil. The redistribution process is continuous; there are no abrupt "breaks" in the q versus t curve (Figure 10.13), and "quasi" equilibrium is approached only after a long time. The redistribution of water in the profile is affected by all those processes that remove water from the profile. Examples include evaporation of water at the surface and extraction by roots from within the soil profile.

In soil science terminology, "field capacity" is considered obsolete in technical work. It is defined as the percentage of water remaining in a soil 2 or 3 days after having been wetted and after free drainage is practically negligible. The percentage of water may be expressed on a weight or volume basis.

A more realistic concept of field capacity incorporates the physical characteristics of the profile. A test plot in the field is irrigated until the soil profile has been wetted for a depth of about 1 m or more. Once the irrigation water has infiltrated into the soil, the plot is covered with a plastic membrane that will prevent evaporation but allow drainage. Then, every 24 h afterward, the soil water content distribution in the profile is measured until there is very little change in water content, thus providing an estimate of the profile water content at field capacity. In humid areas, the soil water content of the profile returns to about the same distribution in early spring every year. This also represents an approximation of the field capacity of the soil profile.

Evaporation

Evaporation is a physical process by which a liquid or solid is transferred to the gaseous state. Evaporation of water can occur from plants, soil surfaces, and free water surfaces such as ponds and lakes. It is one of the critical components in the hydrologic cycle and water balance equations in soil and provides a mechanism for the redistribution of water (and energy) within a region.

Evaporation of water from plants is called transpiration. Transpiration is the principal mechanism of soil water transfer to the atmosphere when the soil surface is covered or partially covered with vegetation. Since evaporation from soil surfaces and transpiration from plant surfaces are similar physical processes and are difficult to separate experimentally, the term *evapotranspiration* (ET) is used to describe the combined loss of water from soil surfaces and from plants.

Physical Conditions Necessary for Evaporation

Three conditions must be met for water to evaporate (Hillel 1982). First, there must be a continual supply of energy to convert the water from the liquid to the vapor state. Evaporation is an endothermic process; that is, it requires heat. About 2.42 MJ/kg of water at 20°C (580 cal/g) are absorbed in evaporation. This energy is the heat of vaporization and is the reason why evaporation is a cooling process. The *latent heat of vaporization* is temperature dependent and can be approximated by the following linear relationship:

$$L = 2490 - 2.134T \qquad [49]$$

where T is the temperature of the evaporating surface (°C) and L is in joules per gram.

Second, the vapor pressure in the atmosphere over the evaporating surface must be lower than the vapor pressure at the surface. This means that there must be a vapor pressure gradient between the surface and the atmosphere. Thus, if the air is completely saturated, no evaporation will occur.

Evaporation is a diffusive process that follows Fick's first law. The evaporation rate can be written as

$$e = K_e v_a (e_s - e_a) \qquad [50]$$

where e is the evaporation rate (m/s), e_s and e_a are the vapor pressures of the evaporating surface and overlying air, respectively (kg/m·s^2), v_a is the wind speed (m/s), and K_e is a coefficient that reflects the efficiency of vertical transport of water vapor by turbulent flow of air (m/s^2·kg). The equation shows that there must be a difference in vapor pressure between the evaporating surface and the atmosphere and that water vapor must be transported by diffusion or convection or both away from the evaporating surface.

The first and second conditions for evaporation are generally considered to be external conditions because they are influenced only by atmospheric factors such as radiation, relative humidity, wind speed, etc. The third condition depends upon the physical characteristics of the medium from which evaporation is occurring.

The third required condition for evaporation is that there must be a continual supply of water from or through the interior of the body to the site of evaporation. This condition depends on the water content and potential energy of water in the body as well as its conductance properties.

Potential Evapotranspiration

Originally, *potential evapotranspiration* (PET) was defined as the rate at which ET would occur from a large area completely and uniformly covered with growing vegetation that has access to an unlimited supply of soil water and without advection or heat storage effects (Dingman 1994). This concept of PET was introduced as part of a scheme for climate classification and to refer to the "drying power" of the atmosphere. Later, PET was defined as the amount of water transpired by a short green crop, completely shading the ground, of uniform height and never short of water.

In practice, PET is defined by the method used to calculate it. According to Jensen et al. (1990), these methods can be grouped on the basis of their data requirements as follows:

Air temperature: Use air temperature and sometimes day length.
Radiation: Use net radiation and air temperature.
Combination: Use net radiation, air temperature, wind speed, and relative humidity.
Pan: Use evaporation from an open pan to estimate free water evaporation.

Note that these methods of estimating PET have differing parameter requirements such as air temperature, net radiation, wind speed, and relative humidity and differing measurement time durations.

Estimating Actual Evapotranspiration

Actual evapotranspiration (AET) is the ET that actually occurs and is usually less than PET. Estimates of AET can be made with lysimeters, water and energy balance equations, and energy transfer methods. In particular, the soil water balance equation can be solved for AET in depth units to give the following:

$$AET = P + I - D + \int_0^{z_r} \theta_1(z)dz - \int_0^{z_r} \theta_2(z)dz \qquad [51]$$

where P is the precipitation, I is the irrigation, D is the internal drainage below z_r, z is the depth, z_r is the depth of the root zone, and $\theta_1(z)$ and $\theta_2(z)$ are the volumetric water content distributions in the profile at the beginning and end of the monitoring period. Use of this equation assumes that there was no runoff, runon, or lateral flow of water from the profile. Values of D can be estimated from Darcy's law.

Evaporation of Water from a Bare Soil Surface

Controlling soil water loss by evaporation is one, if not the greatest, problem facing agricultural scientists and engineers working in dryland regions. Even in the more humid regions, dry periods frequently occur during the crop growing season; therefore, evaporation is a significant component of the water balance. The study of evaporation processes in the soil and atmosphere has over the years been one of the more intensely researched areas in soil physics.

Several initial and boundary conditions influence soil water evaporation. Examples include the following:

water table position: shallow versus deep
profile characteristics: homogeneous versus stratified
flow patterns of water: vertical versus horizontal
temperature of the soil: profile temperature distributions
microclimate
vegetation: canopy closure
time of year

Of the above conditions, the position of the water table and soil profile characteristics are of prime importance.

Evaporation from Homogeneous Soils Having Shallow Water Tables

A soil with a shallow water table from which water can continually be conducted through an unsaturated zone will have an evaporation rate equal to the rate of conduction of water to the surface. The AET rate under steady-state flow conditions depends on

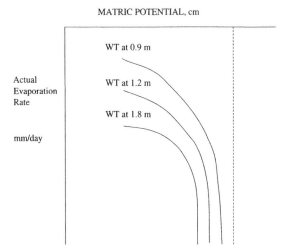

Figure 10.16. Relationships between the matric potential and steady evaporation rate for various positions of the water table (wt).

the ability of the unsaturated soil to conduct water and on the depth of the water table. The steady-state upward flow of water from a shallow water table through the profile toward the surface can be described by the Darcy equation:

$$q = e = -K(h)(dh/dz + 1) \tag{52}$$

where e is the evaporation flux density (m/s), and h is the matric head (m).

The relation between the evaporative flux density e and matric potential at several depths of the water table is shown in Figure 10.16. These curves show that the steady-state rate of capillary rise and evaporation depends on the depth of the water table and on the matric potential, h, at the soil surface. If no evaporation occurs (i.e., e = 0 in equation [52]), then $dh/dz = -1$ and $h = -z$. This means that the magnitude of h is determined by the height or distance above the water table. The water content distribution above the water table is shaped by the water retention curve. However, if e > 0 and $dh/dz < -1$, an increase in e is accompanied by a decrease in dh/dz. Also, $d\theta_v/dz$ increases with an increase in e, especially near the soil surface.

The relationship between h and the depth of the water table at the soil surface for a given evaporation rate can be estimated by rearranging equation [52]:

$$e = -K(h)dh/dz - K(h)$$
$$e + K(h) = -K(h)dh/dz$$
$$[e + K(h)]dz = -K(h)dh$$
$$dz = -K(h)/[e + K(h)]dh$$
$$z = \int dh/\{[e + K(h)]/K(h)\} = \int \{K(h)/[K(h) + e]\}dh \tag{53}$$

To perform the integration of equation [53], the functional relationship between K and h must be known. One empirical relation was given by Gardner (1958) as

$$K(h) = a(h^n + b)^{-1} \tag{54}$$

where the parameters a, b, and n are constants that must be determined for each soil. Substituting equation [54] into the Darcy equation gives

$$e = -[a/(h^n + b)](dh/dz + 1) \qquad [55]$$

which can be used to obtain matric potential distributions with position for different soil water fluxes as well as fluxes for different surface potential values.

Equation [55] predicts that when the water table is near the soil surface, the matric potential at the surface is relatively high, the upward transport of water to the surface is relatively high, and e is determined by the external evaporativity of the atmosphere. When the water table is lowered, the matric potential at the soil surface decreases, upward transport of soil water is reduced, and e approaches a limiting value regardless of the atmospheric evaporativity.

The matric potential at the soil surface is, however, largely determined by external conditions, which are not constant. However, decreasing the matric potential at the soil surface can increase the hydraulic gradient and flux through the soil only up to an asymptotic maximal rate that depends on the position of the water table. Thus, even the driest and most evaporative atmosphere cannot extract water from the surface any faster than the soil profile can transmit water from the water table to the surface. Thus, the physical characteristics of the soil profile can limit the rate of evaporation.

In dry soils at or near the surface, h equilibrates with the water potential of the atmosphere and water moves predominantly as a vapor and transport is by diffusion. This, however, is a much slower transport process than liquid flow even with the same hydraulic gradient.

Transient Evaporation from Homogeneous Soils with Water Table Absent

In many areas, there is no water table or the water table is too deep to affect evaporation of water from the soil surface. In this case, water has infiltrated into soils and is stored for subsequent use by crops. Evaporation of water from soils in which a water table is absent is a transient-state process and proceeds in two distinct stages (Figure 10.17).

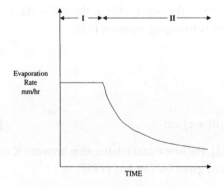

Figure 10.17. Evaporative flux as a function of time in a homogeneous soil.

Stage 1 evaporation (or the constant-rate stage) is characterized by an evaporation rate that depends only on the potential evaporation of the environment and soil surface conditions. Therefore, stage 1 evaporation only depends on the supply of energy reaching the soil surface, that is, the evaporativity of the environment. Under constant external conditions, this evaporation stage continues until the soil can no longer conduct water to the soil surface at the rate dictated by the potential evaporativity.

For the constant-rate stage, the evaporation rate is

$$e = PET \qquad 0 < t < t_1 \qquad\qquad [56]$$

where PET is the potential evaporativity of the atmosphere, and t_1 is the duration of stage 1 evaporation.

The maximum cumulative evaporation from the soil during stage 1 drying is designated as U. Mathematically, we set

$$U = PETt_1 \qquad\qquad [57]$$

where U has SI units of meters (and also may be expressed as centimeters or millimeters) of water evaporated.

The magnitude of PET and the hydraulic properties of the soil appear to cause the main differences in the amount of drying before stage 1 evaporation ends. During stage 1, the volumetric water content decreases at the soil surface; thus, K decreases during this stage. The evaporation rate remains constant and is equal to the potential rate for the soil because the hydraulic gradient increases enough to compensate for the decrease in K. Eventually, dH/dz can no longer increase sufficiently in order to compensate for the decrease in K with the low θ_v at and near the soil surface, and the first stage of evaporation ends.

The duration of stage 1 depends on the intensity of the meteorological factors that determine atmospheric evaporativity, as well as on the hydraulic conductive properties of the soil itself. When the external evaporativity is low, stage 1 evaporation will last longer than when the external evaporativity is high. If PET is low, the profile is dried more uniformly with depth due to the small value of dh/dz required to provide water conduction through the profile to the surface. Under similar external conditions, stage 1 evaporation will last longer in a clayey soil than in a sandy soil and longer in a structureless soil than in a structured soil. This is because a clayey soil retains higher volumes of water and because values of K for unsaturated clayey soils are higher than those for sandy soils, because the water content decreases at the soil surface. At the end of stage 1, the volumetric soil water content can no longer decrease at the soil surface because it has reached its air-dry value.

Stage 2 evaporation is known as the falling-rate stage. The evaporation rate decreases rapidly during stage 2 and is governed by the ability of the soil to transmit water to the soil surface for subsequent evaporation and is less dependent on the available energy. Therefore, this stage of evaporation is characterized by the downward movement of the drying front from the soil surface into the soil profile. During stage 2 drying, water vapor must be transmitted by diffusion through a layer of air-dry soil, which acts as a mulch.

To alter stage 2 evaporation, the hydraulic characteristics of the soil below the surface

Table 10.5. Values of K, U, and a for Several Soils

Soil	K* (cm/d)	U (mm)	a (mm/d$^{1/2}$)
Adelanto clay loam	0.15	12	5.08
Yolo loam	0.10	9	4.04
Houston black clay	0.06	6	3.50
Plainfield sand	0.05	6	3.34

Source: Ritchie 1972.
*The K values were determined at −10 kPa.

have to be altered. At the soil surface, the volumetric soil water content is constant; thus, K must also be constant in this layer. Although K is constant, the evaporation rate decreases with time because near the soil surface the hydraulic gradient decreases with time.

During the falling-rate stage, the cumulative water evaporated, E, can be determined by

$$E = a(t - t_1)^{0.5} \qquad [58]$$

where the parameter a is a constant and is dependent on soil properties, and t_1 is the time since the initiation of stage 2 drying. An example of the relationship between the parameters K, U, and a as given by Ritchie (1972) is presented in Table 10.5.

The evaporation rate during stage 2 also can be calculated from

$$e = D_w \theta_v \pi^2 / 4z_w = D_w D_{eq} \pi^2 / 4z_w^2 \qquad [59]$$

where e is the evaporation rate (mm/d), D_w is the soil water diffusivity (mm^2/d), z_w is the depth to the wetting front (mm), and D_{eq} is the equivalent depth of water stored in the soil profile from z = 0 to x = z_w (mm). This equation gives an estimate of the amount of water that the soil is capable of supplying to the soil surface assuming constant θ_v at the soil surface and negligible gravitational gradients (Gardner and Hillel 1962).

Methods of Reducing Evaporation

Evaporation of water from soil may account for as much as 60% of the precipitation where summer fallow is practiced (Hanks and Ashcroft 1980). In the summer fallow system, only about 30% of the fallow year's precipitation is conserved for use by plants during the next year.

It is relatively easy to reduce evaporation during the constant-rate stage. This can be accomplished by a mulch of straw or gravel, which reduces the reception of energy at the soil surface. It is more difficult to reduce evaporation during the falling-rate stage. The internal conductance properties near the soil surface have to be altered. If the time between rain or irrigation is long, the cumulative evaporation is about the same for mulched soil as for bare soil. If the rains are frequent, mulches may conserve water. Although the amount of soil water that is conserved is greater with more frequent rains, the total amount of water evaporated is greater because the soil surface is wet for longer times.

The attempts of many farmers to conserve soil water against evaporation losses by surface cultivation often fail because by the time the soil dries sufficiently to be cultivated, stage 1 evaporation has already ended.

Water that infiltrates and redistributes to the deeper depths of the soil profile tends not to be lost to evaporation. According to Hanks (1992), in order to minimize evaporation in areas where water is limited, it is advisable to (1) add water in large amounts at infrequent intervals and (2) maintain a thick mulch on the soil surface.

In eastern Arkansas, Vertisols such as the Sharkey have clayey textures at the surface and tend to develop cracks upon drying. These cracks become secondary evaporation planes, penetrating deeply into the soil profile and causing it to desiccate deeply. In general, evaporation from cracks increases as crack depth and width increase, all other conditions remaining the same. Timely cultivations and irrigations can prevent or delay the development of such cracks.

Evaporation from Plant Surfaces

Evaporation from plant surfaces is known as transpiration, T. Transpiration removes water from plant tissues, and unless this water is immediately replaced, the water remaining in the tissue is under increased restraints. The potential energy of the water in the plant tissue is reduced rapidly as the volume concentration of plant water is slowly reduced. When a water source is freely available to the transpiring tissue, water will move into the plant cells to restore the water potential to its original condition. There are, however, resistances to the movement of water through the soil to the plant roots, into the roots, and through the conducting tissue of the plant to the deficient tissue. The rate of replenishment of the water-deficient tissue depends on the hydraulic conductivity (the reciprocal of the resistance) in the plant water system and on the water potential gradients that develop along the supply path.

According to Darcy's law, the difference in total water potential between two places always indicates a tendency for water to flow. Whether or not flow occurs depends on two factors: the opposing potential gradients (such as temperature) and electrical energy barriers (or resistances). The movement of water from a plant to the surrounding atmosphere requires that water change its state, and this change requires a large amount of energy. Thus, the rate of supply of energy for the change rather than the difference in water potential may be the limiting factor that determines the rate of transpiration. The energy comes from the plant's external environment. Consequently, weather conditions usually exert a controlling influence on the rate of evaporation from plants.

Evapotranspiration of Water under Field Conditions

Evaporation and transpiration are physically similar processes. Since it is difficult to measure evaporation and transpiration separately in the field, the two processes are usually combined and called evapotranspiration, ET.

The rate at which ET occurs in the field is strongly dependent on three factors:

1. Micrometeorological factors: the condition of the atmosphere into which water is transpired, including such factors as net radiation, temperature, humidity, and air movement.

2. Plant factors: the kinds of plants and their condition, including the area of plant cover, plant shape, the stage of maturity, the number and arrangement of stomata, and the opening and closing of stomata.

3. Soil factors: the condition of the soil, including soil aeration, the soil water matric potential, and the rate at which water can move to the plant roots.

ET is dictated primarily by micrometeorological factors when soil water in the root zone is nonlimiting and plants provide full cover. As soil water is depleted beyond a threshold value, evaporation rates fall farther below the evaporation potential. When soil water is limited, the relation between soil water content and evaporation rates depends on soil water transmission characteristics and plant properties as well as the evaporative demand.

Loss of water from the soil can be estimated from the equation of continuity, which was derived previously. Now, however, we incorporate a consumption term into the continuity equation to give

$$\partial\theta_v/\partial t = -\partial q/\partial z - r(z, t) \tag{60}$$

where $r(z, t)$ is the soil water extraction rate with SI dimensions of $m^3/m^3{\cdot}s$. The change in θ_v in the profile is now attributed to the change in soil water flux density and to the extraction of water by plant roots. The transpiration, T, is given by

$$T = \int_0^{z_r} \int_0^t r(z, t)dt\, dz \tag{61}$$

where z_r is the depth of the root zone. When equation [61] is combined with the transient-state transport equation for water flow in soils, we obtain

$$\partial\theta_v/\partial t = \partial/\partial z[K(\theta_v)\partial H/\partial z] - r(z, t) \tag{62}$$

When there is incomplete cover and water is freely available to plant roots, plant leaf area must be considered in the evaporation prediction. One model sometimes used is

$$T = PET(-0.21 + 0.70LAI^{1/2}) \qquad 0.1 < LAI < 2.7 \tag{63}$$

where LAI is the leaf area index (m^2 of leaves/m^2 of soil area). The upper limit of 2.7 for LAI is the apparent threshold LAI representing the minimum LAI necessary to constitute an apparent full cover. When the LAI is less than 2.7, potential evaporation rates will not be obtained because of the partial canopy unless soil water is freely available at the surface. Thus, when annual row crop plants are in an early growth stage with little vegetative cover, the evaporation rate from the entire field surface is dominated by the soil evaporation rate.

Evaporation from practically bare, wet surfaces is primarily influenced by the energy available for evaporation. As surface drying proceeds, evaporation becomes more dependent on the hydraulic properties of the soil near the surface. As plant cover increases, the evaporation rate from the field becomes more dependent on the leaf area and the potential evaporation so long as the soil water available to the plant roots is not limited.

Recent studies of ET using precise, well-exposed weighing lysimeters have shown that for cotton, grain sorghum, soybeans, and alfalfa, about 70% of the extractable water from a deep soil profile can be used before stomatal regulation of evaporation begins. One cannot, however, use a single soil water content or soil water potential to describe the entire root environment, because the rhizosphere resistance depends on

the density of the plant roots
the unsaturated hydraulic conductivity of the soil near the absorbing roots
the evaporative demand

The amount of water in the profile that can be used depends on

the atmospheric demand
the plant's ability to regulate the flow of water through the plant system
the exploitation of the subsoil reservoir by the root system
the hydraulic conductivity of the water in the soil

At Stuttgart in eastern Arkansas, Scott et al. (1987) showed that water management affected the evaporation and transpiration losses of water from a soybean field. The soil was a DeWitt silt loam. At the end of the growing season, the 5-year mean ET for the field, transpiration (T) losses, and evaporation (E) losses were 497, 327, and 170 mm of water, respectively. The corresponding values for the nonirrigated soybeans were considerably lower and were 382, 231, and 151 mm of water, respectively. Under irrigated conditions, the lowest seasonal ET (370 mm) was found in a wetter-than-normal season, and the highest ET (539 mm) was found in a drier season. Under nonirrigated conditions, ET during the wetter-than-normal season (370 mm) was similar to the 5-year mean (382 mm) for this same treatment. The results showed that the year-to-year variability of water losses in this humid environment depended on the soil-water management as well as the climatic conditions.

Daniels and Scott (1991) conducted a 3-year study of double-cropped wheat and soybeans at Fayetteville on a Captina silt loam. After March 1, the 3-year mean ET for wheat was 32.8 cm of water, but annual values ranged from 38.7 to 26.7 cm. For soybeans, depletion of soil water by the previous wheat crop and variable precipitation affected the amount of water lost. The annual mean ET was 37.5 cm and 25.5 cm for irrigated and nonirrigated soybeans, respectively. The range in ET under irrigated conditions was from 31.0 to 43.6 cm of water per year. The seasons with greater rainfall had greater values of ET.

Cited References

Celia, M. A., E. T. Bouloutas, and R. L. Zarba. 1990. A general mass-conservative numerical solution for the unsaturated flow equation. *Water Resources Research* 26:1483–1496.

Chow, U. T. (ed.). 1964. *Handbook of Hydrology*. McGraw-Hill. New York.

Daniels, M. B., and H. D. Scott. 1991. Water use efficiency of double-cropped wheat and soybean. *Agronomy Journal* 83:564–570.

Dingman, S. L. 1994. *Physical Hydrology*. Prentice-Hall. Englewood Cliffs, NJ.

Gardner, W. R. 1958. Some steady state solutions of the unsaturated moisture flow equation with application to evaporation from a water table. *Soil Science* 85:228–232.

Gardner, W. R., and D. I. Hillel. 1962. The relation of external evaporative conditions in the drying of soils. *Journal of Geophysical Research* 67:4319–4325.

Green, W. H., and G. A. Ampt. 1911. Studies on soil physics. I. Flow of air and water through soils. *Journal of Agricultural Science* 4:1–24.

Haan, C. T., B. T. Barfield, and J. C. Hayes. 1994. *Design Hydrology and Sedimentology for Small Catchments*. Academic Press. New York.

Hanks, R. J. 1992. *Applied Soil Physics: Soil Water and Temperature Applications*. Springer-Verlag. New York.

Hanks, R. J., and G. L. Ashcroft. 1980. *Applied Soil Physics*. Springer-Verlag. New York.

Hillel, D. 1982. *Introduction to Soil Physics*. Academic Press. New York.

Jensen, M. E., R. D. Burman, and R. G. Allen (eds.). 1990. *Evapotranspiration and Irrigation Requirements*. American Society of Civil Engineering Manuals and Reports of Engineering Practice No. 70. New York.

Kutílek, M., and D. R. Nielsen. 1994. *Soil Hydrology*. Catena Verlag. Cremlingen-Destedt, Germany.

Philip, J. R. 1957. The theory of infiltration. 1. The infiltration equation and its solution. *Soil Science* 83:345–357.

Rawls, W. J., D. L. Brakensiek, and N. Miller. 1983. Green-Ampt infiltration parameters from soils data. *Journal of Hydraulic Engineering, Proceedings of the American Society of Civil Engineering* 109:62–70.

Ritchie, J. T. 1972. Model for predicting evaporation from a row crop at incomplete cover. *Water Resources Research* 8:1204–1213.

Scott, H. D., J. A. Ferguson, and L. S. Wood. 1987. Water use, yield, and dry matter accumulation by determinate soybean grown in a humid region. *Agronomy Journal* 69:870–875.

Soil Conservation Service. 1972. *SCS National Engineering Handbook*, Section 4. Hydrology. USDA.

Viessman, W., G. L. Lewis, and J. W. Knapp. 1989. *Introduction to Hydrology*. 3d ed. Harper Collins. New York.

Additional Reference

Rawls, W. J., L. R. Ahuja, D. L. Brakensiek, and A. Shirmohammadi. 1993. Infiltration and soil water movement. *In* (D. R. Maidment, ed.) *Handbook of Hydrology*. McGraw-Hill. New York.

Problems

1. Assume that the volumetric soil water contents given in the table below were obtained before and after a 7.5 cm rainfall event onto a bare silt loam soil. Fill in the table and calculate the cumulative infiltration and runoff.

Depth interval (cm)	θ_i (m³/m³)	θ_f (m³/m³)	$\Delta\theta$ (m³/m³)	D_{eq} (cm)
0–5	0.08	0.46		
10–15	0.12	0.40		
15–20	0.16	0.24		
20–25	0.19	0.19		

SOIL WATER FLOW PROCESSES IN THE FIELD

2. The data presented in the table below were obtained during an infiltration event. Graph i versus t and numerically integrate the infiltration rates to obtain the cumulative infiltrated water in the profile after 10 minutes.

Time (min)	i (mm/h)
0.1	36.0
1	18.0
2	12.6
3	9.0
4	7.2
6	4.3
8	3.2
10	2.1

3. Fit the Kostiakov model to the infiltration rates in problem 2 using linear regression techniques. Determine the values of the Kostiakov parameters k and n as well as the coefficient of determination. Graph the residual plot and interpret the fit of the model to the data. Use the computed values of k and n to graph the cumulative infiltration versus time.

4. A Captina silt loam is wet to a depth of 10 cm with an average volumetric water content of 0.25 cm^3/cm^3. If the average potential evaporation is 0.6 cm/d and the soil water diffusivity is 30 cm^2/d, calculate the daily soil evaporation rate.

5. For the same Captina soil, if the water content is 0.2 cm^3/cm^3 and the soil water diffusivity is 8 cm^2/d, calculate the daily soil evaporation rate.

6. Assume that a rainfall of 4 cm was recorded on a well-established pasture with a curve number of 60. How much runoff should have been measured? What are the reasons we should expect to find a curve number with this magnitude in this condition in the field?

7. Calculate the amount of water moved into a soil profile if the depth of the wetting front is 0.15 m, the soil bulk density is 1.40 Mg/m^3, and the average initial soil water content is 0.25 m^3/m^3. What is the average infiltration rate if the rainfall event lasted 4 hours?

8. Assume that the initial infiltration rate of water into a soil profile is 15 cm/h and the final rate is 3 cm/h. Calculate the centimeters of water infiltrated into the soil after 3 hours if the exponential rate coefficient of the Horton equation is 2.0 h^{-1}.

9. Determine the influence of antecedent moisture at the soil surface on runoff. Assume that a 4 cm rainfall event occurred on a soil in hydrologic group C with a curve number of 70. Calculate the depth of runoff (mm) for both normal and wet conditions. For the wet condition, assume that the 4 cm rainfall event was preceded by 10 cm of rain within the previous 3 days. For the normal condition, assume that the soil was moist but no significant rainfall occurred during the last 5 days. Based on your calculations, what can be concluded about the effect of antecedent moisture on runoff?

CHAPTER ELEVEN

Solute Transport

Introduction

Transport of solutes in porous materials such as soils is important in understanding problems of (a) chemical residue, (b) salinity, (c) availability of solutes for plant uptake, (d) leaching or redistribution within the vadose zone to groundwater, (e) runoff to surface water, (f) salt water intrusion into coastal aquifers, and (g) seepage from various storage and waste disposal systems. Transport of solutes in soil occurs by two physical processes—diffusion and convective flow—and both transport mechanisms are discussed in detail in this chapter. We begin with a review of solute concentration terms and transport mechanisms, which were previously discussed in Chapter 5 on mass and energy dynamics. As in previous chapters, mathematics will help us to more fully understand the principles of solute transport processes in soil and to quantify the amounts of solutes moved over space and time.

Review of Solute Concentration Terms

All solute concentration terms are not equal in terms of dimensions! Transport of solutes in soil is dependent on the concentration in the three soil phases and on the partitioning among the phases. Therefore, the frame of reference for defining solute concentration is important. The total resident solute concentration in soil is defined as the sum of the volume-weighted concentrations in the three phases and is written as

$$C = \rho_b C_a + \theta_v C_l + f_a C_g \qquad [1]$$

where C is the total resident solute concentration in the soil (kg/m^3 of soil), C_a is the adsorbed concentration (kg adsorbed solute/kg of soil solids), C_l is the solution concentration (kg/m^3 of soil solution), and C_g is the gaseous solute concentration (kg/m^3 of soil gas). The soil physical parameters ρ_b, θ_v, and f_a serve to weight the solute concentrations in the three phases of soil on a volume basis and to convert the different reference dimensions to cubic meters of soil.

Example

Show that the dimension of each component on the right-hand side of equation [1] simplifies to kg solute/m^3 soil.

$\rho_b C_a$ = (kg soil/m^3 of soil)(kg sorbed solute/kg soil solids)
$\theta_v C_l$ = (m^3 water/m^3 soil)(kg solute in solution/m^3 water)
$f_a C_g$ = (m^3 air/m^3 soil)(kg solute in air/m^3 soil air)

Molecular Diffusion

Solute transport by molecular diffusion is due to the random thermal motion of the solute ions and/or molecules. As a result of this random motion, more ions (or molecules) tend to move from areas of higher concentration to areas of lower concentration than from areas of lower to areas of higher concentration. This net movement of ions or molecules from areas of higher to areas of lower concentration is generally proportional to the concentration gradient, the cross-sectional area available for diffusion, and the time during which the transport process occurred.

Fick's First Law

The net amount of solute crossing a plane of unit area in unit time is known as the *flux density* and is given by the empirical relation known as Fick's first law. In one-dimensional form, Fick's first law (1855) for steady-state transport is written as

$$J = -D\partial C/\partial z \qquad\qquad\qquad [2]$$

where J is the solute flux density ($kg/m^2{\cdot}s$), D is the molecular diffusion coefficient (m^2/s), C is the concentration of solute (kg/m^3), and z is the distance (m). Since solute concentration in a porous medium can be expressed in several ways, the dimensions of C should have the same quantity reference as J, and the volume should have the same length reference as z (Kutílek and Nielsen 1994). The flux density of solute is the quantity of solute diffusing across a plane of area A in time t and is equivalent to the ratio Q/At, where Q is the quantity of solute with SI units of either kilograms or moles. In equation [2], the quantity $\partial C/\partial z$ is the solute concentration gradient across the section and represents the driving force. The minus sign in equation [2] arises because solute movement occurs from areas of higher to areas of lower concentration, and it serves to make J positive in the direction of positive z.

Fick's first law is the defining equation for the molecular diffusion coefficient, D. The diffusion coefficient is a measure of the speed that molecules or ions have in diffusing through a porous substance and is a steady-state transport coefficient. It is a quantitative expression of the difficulty that diffusing molecules or ions have in moving through different porous media. It also represents the amount of solute that in unit time and with unit concentration gradient would cross a plane of unit area normal to the plane. Thus, under conditions of unit concentration gradient, higher values of D indicate higher quantities of solute moved. The magnitude of the solute diffusion coefficients in soil varies over a wide range. Experimental studies have shown that D varies with soil texture, soil phase, physical and chemical properties of the ion or molecule, temperature, salt content, concentration, equilibration time, and soil water content.

Diffusion of a solute results in a spreading from the location or depth interval where the solute is introduced. As time proceeds, the spreading results in a lower concentration within the depth interval of application but increasing concentrations outside this interval. The solute concentration follows a normal, or Gaussian, distribution and can be described by two statistical properties, the mean and the variance (Fetter 1993). The depth of penetration of a diffusing ion or molecule in soil, z_σ can be estimated by its root mean-square displacement

$$z_\sigma = (2Dt)^{1/2} \qquad\qquad\qquad [3]$$

The influence of soil phase on the distance moved by diffusion can be seen from the data in Table 11.1. In general, diffusion coefficients depend on the solute species and how they interact with soil surfaces, soil phase, and concentration of the various chemical species; diffusion coefficients tend to increase with increases in temperature and soil water content and decrease with atomic mass.

Table 11.1. Relationship between Soil Phase, Solute Diffusion
Coefficient in the Soil Phase, and Distance Moved
by Diffusion in 1 Day

Soil Phase	Diffusion Coefficient (m²/s)	Distance Moved (m/d)
Gaseous	2×10^{-5}	1.86
Solution	2×10^{-9}	0.0186
Soil (whole)*	2×10^{-11}	0.00186
Solid (adsorbed)	2×10^{-13}	0.000186

*Diffusion in the three phases of soil.

The data in Table 11.1 show that the order of decreasing D is

Gaseous phase > Solution phase > Soil > Solid phase

and that the distance moved in the soil by diffusion in 1 day is low. The conclusion is
that diffusion in soil is a relatively slow transport process and operates over small distances. In situations where soil water movement is slow, diffusion is the dominant solute
transport mechanism.

Transport by diffusion in soil is complicated by the fact that ions must maintain electrical neutrality as they diffuse. For example, if $CaCl_2$ is diffusing in the soil, Ca^{++} cannot diffuse faster than the Cl^- unless there is another anion in the region into which the
Ca^{++} is diffusing.

Example

Assume that a solute with a diffusion coefficient of 2×10^{-10} m²/s in soil is diffusing
under steady-state conditions into a zone originally devoid of the solute. If the concentration of the solute is 25 mg/g of soil and the bulk density is 1.3 Mg/m³, calculate using
Fick's first law the following: (1) the flux density in the z direction at a distance of 0.1 m
and (2) the amount of the solute in 1 hectare that diffused across this boundary in 1 month.

Solute concentration:

$C\rho_b = (25 \text{ mg/g})(1000 \text{ g/kg})(1300 \text{ kg/m}^3) = 3.25 \times 10^7 \text{ mg/m}^3$

Concentration gradient:

$\partial C/\partial z \approx \Delta C/\Delta z = (0 - 3.25 \times 10^7)/(-0.1 - 0) = 3.25 \times 10^8 \text{ mg/m}^4$

The flux density of the solute is obtained by substitution into Fick's law:

$J = -D\partial C/\partial z = -(2 \times 10^{-10} \text{ m}^2/\text{s})(3.25 \times 10^8 \text{ mg/m}^4)$
    $= -0.065 \text{ mg/m}^2 \cdot \text{s}$

Since J is negative, the solute is diffusing in the negative z direction, i.e., downward.

Quantity of solute moved beyond 0.1 m in 1 hectare and 1 month:

$Q = (0.065 \text{ mg/m}^2 \cdot \text{s})(10,000 \text{ m}^2/\text{ha})(86,400 \text{ s/d})(30 \text{ d/mo})$
   $= 1.6848 \times 10^9 \text{ mg/ha} \cdot \text{mo}$

Fick's Second Law

For transient-state conditions, the conservation of mass equation has to be satisfied. In one dimension and without generation and consumption, this equation is

$$\partial C/\partial t = -\partial J/\partial z \tag{4}$$

which implies that the rate of change of solute concentration equals the net rate of flow over the boundary. Combining equations [2] and [4] yields the transient form of the diffusion equation:

$$\partial C/\partial t = (\partial/\partial z)(D\partial C/\partial z) \tag{5}$$

which is known as Fick's second law. If the solute diffusion coefficient is independent of solute concentration and depth, equation [5] simplifies to

$$\partial C/\partial t = D(\partial^2 C/\partial z^2) \tag{6}$$

which has been solved analytically and numerically for several initial and boundary conditions (Carslaw and Jaeger 1960).

Analytical Solutions and Applications

One analytical solution of equation [6] occurs when a solute is uniformly broadcast on the soil surface in quantities high enough to cause saturation in solution. If the initial concentration of solute in the profile is negligible, if only diffusive transport occurs, and if there is no root uptake by plant roots, the solute will diffuse downward into the soil profile. The solute concentration distribution $C(z,t)$ at some depth z and time t after application can be estimated from

$$C(z,t) = C_o \, erfc[z/2(Dt)^{0.5}] \tag{7}$$

where C_o is the initial solute concentration, which remains constant, and erfc is the complementary error function, which is related to the normal, or Gaussian, distribution function. Values of erfc X, where X represents the magnitude of the term in the brackets in equation [7], are presented in Table 11.2.

Solutions of equation [7] for various values of the parameters D and t are shown in Figures 11.1 and 11.2. The plots show the relative concentration, C/C_o, of the solute versus depth in the profile (mm). Diffusion from a constant source with a constant diffusion coefficient ($D = 0.62 \times 10^{-9}$ m^2/s) for five diffusion times is shown in Figure 11.1. These calculations show that as time proceeds, the relative solute concentration increases with distance from the source. The relative concentration distribution with varying values of D for 1 day are shown in Figure 11.2. Higher values of D give higher relative concentrations deeper in the soil profile. This shows that during a fixed time interval, those soil physical, chemical, and microbiological factors that cause D to vary can influence the amount of solute diffusing in the soil profile and the distance moved.

Table 11.2. Values of the Error Function and Its Complement

X	erf X	erfc X
0	0	1.0
0.05	0.056372	0.943628
0.1	0.112463	0.887537
0.15	0.167996	0.832004
0.2	0.222703	0.777297
0.25	0.276326	0.723674
0.3	0.328627	0.671373
0.35	0.378382	0.620618
0.4	0.428392	0.571608
0.45	0.475482	0.524518
0.5	0.520500	0.479500
0.55	0.563323	0.436677
0.6	0.603856	0.396144
0.65	0.642029	0.357971
0.7	0.677801	0.322199
0.75	0.711156	0.288844
0.8	0.742101	0.257899
0.85	0.770668	0.229332
0.9	0.796908	0.203092
1.0	0.842701	0.157299
1.1	0.880205	0.119795
1.2	0.910314	0.089686
1.3	0.934008	0.065992
1.4	0.952285	0.047715
1.5	0.966105	0.033895
1.6	0.976348	0.023652
1.7	0.983790	0.016210
1.8	0.989091	0.010909
1.9	0.992790	0.007210
2.0	0.995322	0.004678
2.2	0.998137	0.001863
2.4	0.999311	0.000689
2.6	0.999866	0.000134
2.8	0.999925	0.000075
3.0	0.999978	0.000022

Source: Carslaw and Jaeger 1960.
Additional characteristics of the error function are

$$\text{erf } X = 2(\pi)^{-0.5} \int_0^X e^{-n^2} d\eta$$

$$\text{erf}(\infty) = 1$$

$$\text{erf}(-X) = -\text{erf } X$$

$$\text{erfc } X = 1 - \text{erf } X = 2\pi^{-0.5} \int_x^\infty e^{-n^2} d\eta$$

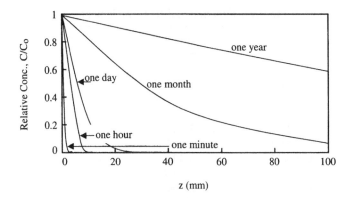

Figure 11.1. Diffusion of a solute from a constant source. The value of $D = 0.62 \times 10^{-9}$ m^2/s and $t = 1$ min, 1 h, 1 d, 1 mo, and 1 y.

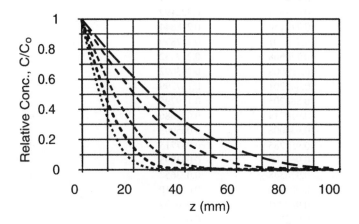

Figure 11.2. Diffusion from a constant source at 1 d, for several values of the diffusion coefficient (curves left to right): $D/(10^{-9}) = 0.62$ (P$_2$O$_7^{4-}$), 1.07 (SO$_4^{2-}$), 2.08 (br$^-$), 5.26 (OH$^-$), and 9.3 (H$^+$).

Example

Assume that the diffusion coefficient of a given solute in soil was 1×10^{-9} m^2/s. Find the value of the relative concentration, C/C_o, at a depth of 5 m after 100 years of diffusion only.

$$C/C_o = \text{erfc}[5/2(1 \times 10^{-9} \times 86{,}400 \times 365 \times 100)^{0.5}] = \text{erfc } 1.408$$
$$= 0.0477$$

Thus, in 100 years, diffusion over a distance of 5 m would result in a solute concentration that is 4.77% of the original concentration at the surface.

Equation [7] can also be used to predict the concentration distribution curve for a solute moving by diffusion only in the soil as a function of time and depth. This is demonstrated in the exercise below.

Example
If we assume that the source stays constant, that there is no convective flow, and that the diffusion time is 100 years, calculate the concentration distribution curve for the same solute as in the previous example. The approximations are as follows:

z (m)	0.1	0.2	0.4	0.6	0.8	1.0	1.5
	2.0	2.5	3.0	3.5	4.0	4.5	5.0
X	0.0282	0.0563	0.1126	0.2689	0.2252	0.2816	0.4223
	0.5631	0.7039	0.8447	0.9855	1.1262	1.267	1.408
erfc X	0.97	0.94	0.87	0.81	0.75	0.69	0.55
	0.43	0.32	0.23	0.16	0.11	0.07	0.05

This calculation shows that the relative concentration distribution curve is smooth with depth in the profile, and the relative concentration decreases with depth in the profile. This also shows that diffusion will cause a solute to spread away from the location where it is introduced, thereby resulting in a lower concentration away from the source.

Convection

If water in the soil is also moving, the solute is swept along by the physical transport process known as *mass flow advection* or *convective flow*. The simplest way to visualize convection is called *piston flow* or *simple convection*.

Steady-State Convection
Consider a column of soil with pure water uniformly moving through it (Figure 11.3). Assume that we quickly replace the water at the intake end of the column with a solution containing a solute at some known concentration, C_0. In piston flow, the solute will travel through the column, at the same rate as the flow of water and will not spread. At the outflow end of the column, the concentration of the solute is zero until the front appears, at which time the concentraton of the solute becomes the same as that at the inflow end of the column.

The solute flow rate in the column can be described by the product of the macroscopic flow rate of water (i.e., the Darcy flux density) and the concentration of the solute in that water. In the z dimension, this solute flow rate is determined by

$$J_z = q_z C_1 \tag{8}$$

where J_z is the solute flux density (kg/m^2·s of soil) in the z direction (m), q_z is the macroscopic water flow rate per unit area perpendicular to the z direction (m/s), and C_1 is the concentration of the solute in that water (kg/m^3). The flux density of soil water is calculated by the Darcy equation for steady-state flow of water.

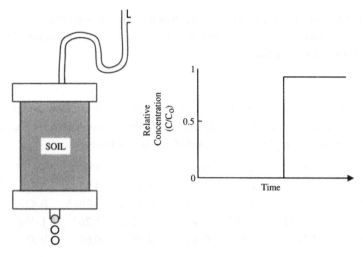

Figure 11.3. A column of soil with water moving through as piston flow.

Since in soil $q = \theta_v v$, we can also write equation [8] as

$$J_z = v\theta_v C_1 \hspace{6cm} [9]$$

where v is the average water velocity in the soil pores in the z direction (m/s), and θ_v is the average volumetric soil water content (m^3 water/m^3 soil).

Example

If the soil water flux density is 0.045 cm/d downward, the soil water content is 0.35 m^3/m^3, and the soil solution concentration of nitrate-N is 3 mg/L, calculate the pore water velocity and the amount of nitrate-N moved by convective flow per unit area below the plant root zone in 1 day.

$q = 0.045$ cm/d $= 0.00045$ m/d
$v = q/\theta = 0.045/0.35 = 0.129$ cm/d $= 0.00129$ m/d

$J_c = qC = (0.00045$ m/d$)(3$ mg/L$)(1000$ L/m^3)
    $= 1.36$ mg/$m^2 \cdot$d

$Q = J_c At = (1.36$ mg/$m^2 \cdot$d$)(1$ m^2)(1 d) $= 1.36$ mg

Transient-State Convection

Combining equations [4] and [9], we obtain the one-dimensional conservation equation for convective flow:

$$\partial C/\partial t = -\partial(v\theta_v C_1)/\partial z \hspace{4cm} [10]$$

Solution of equation [10] gives a sharp concentration front in which, on the advancing side of the front, the solute concentration is equal to the inflow C_o and, on the other side, it is equal to the initial concentration C_i. This is known as piston (or plug) flow, where all of the resident soil water is replaced by the invading solute front. The sharp interface

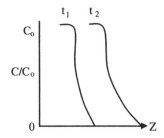

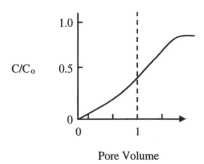

Figure 11.4. Relationship between relative concentration and position in the profile and between relative concentration and pore volume.

that results from piston flow is shown in Figure 11.3. It should be emphasized that piston flow very seldom, if ever, occurs in field soils, but it is a useful first approximation of the potential movement of solutes. Nonuniform flow mechanisms such as hydrodynamic dispersion, preferential flow, and fingering lead to solute concentration distributions that differ significantly from those predicted by piston flow. Fingering is the preferential movement of water in soil along a vertically elongated path. These heterogeneous concentration distributions are usually attributed to the influences of nonhomogeneity of soil properties, such as soil structure and biological channels, resulting in water and solute flowing more rapidly in large pores than in small pores.

Mechanical Dispersion

As a solute moves through the soil profile, it mixes with the soil solution, which has a different chemical composition. Initially, there is a sharp boundary or interface between the solute concentrations of the new and old solutions. Eventually, a transition zone develops between the two solutions where the solute concentration varies as shown in Figure 11.4. Assuming no interaction between the solute and soil surfaces (i.e., no sorption), the solute concentration curve varies with time and has the shape of an "S curve." The width of the transition zone increases with time as the solute moves through the profile.

The tendency for solute molecules to become more diffuse with time when water is moving is known as *mechanical dispersion*. Dispersion occurs when a transition zone

a. Influence of velocity distribution within a pore

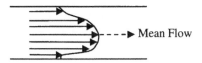

b. Influence of pore size

c. Influence of microscopic flow direction

Figure 11.5. Examples of the three physical mechanisms that contribute to hydrodynamic dispersion of solutes.

develops between two solutions of differing chemical compositions. The change in width of the front is due to both diffusion and mechanical dispersion. However, it is not generally possible to separate the effects of the two processes.

Hydrodynamic dispersion can be visualized as occurring in soil when the water is moving. Its velocity distribution is not uniform because of three boundary effects. One cause of dispersion occurs when the velocity is zero at the soil particle surfaces. This creates a velocity gradient in the solution phase, with the maximum velocity at the center of the pore or at the air-water interface under unsaturated conditions (Figure 11.5a). Another boundary effect is due to the variation of pore size. This causes differences in the average flow velocity between the pores of the soil, with the velocity in the large pores greater than that in the small pores (Figure 11.5b). The third boundary effect occurs when the actual flow path of the element of water fluctuates with respect to the mean direction of flow (Figure 11.5c).

The mixing or dispersion that occurs along the direction of the flow path is called *longitudinal* dispersion. The dispersion in directions normal to the flow path is called *transverse* dispersion. Both of these mechanical dispersion coefficients are the result of the product of the linear velocity of the fluid in the principal direction of flow and a property of the soil called the *dispersivity*.

These three mechanisms, shown in Figure 11.5, combine to yield the contribution of hydrodynamic dispersion to solute transport. The mathematical equation for the flux density of solutes by hydrodynamic dispersion is

$$J_h = -D_h \partial C_l / \partial z \qquad\qquad [11]$$

where D_h is the hydrodynamic dispersion coefficient (m^2/s). Note that equation [11] has the same form as Fick's first law. The magnitude of the dispersion coefficient depends on the pore water velocity and the solute diffusion coefficient. Since the effects of diffusion and dispersion on the concentration distribution are similar, they are often combined even though the mechanisms of transport are different.

Miscible Displacement and Breakthrough Curves

The displacement of one fluid by another miscible fluid in soil has been studied extensively over the last several decades in disturbed and undisturbed soil columns. Usually, one-dimensional soil columns have been constructed to provide a simple means of quantifying the mixing, dispersion, and attenuation of a solute as it moves through soil. The column is designed to maintain steady-state flow conditions when the initial soil solution, with a solute concentration of C_i, is invaded and eventually displaced by a second miscible solution with solute concentration C_o. No mixing of the two solutions should occur at the ends of the column and sidewalls. Samples of the effluent are collected and analyzed for the solute without disturbing the flow.

We let the volume of the soil column occupied by resident solution be V_o (m^3) and the steady-state rate of inflow and outflow of the solution be Q/t (m^3/s). The initial solute concentration, C_i, is suddenly displaced by an incoming solute concentration solution, C_o. The fraction of this incoming solute in the effluent at time t(s) will be $(C - C_i)/(C_o - C_i)$ or, for an initial concentration (C_i) of zero, simply C/C_o.

Plots of the relative concentration, C/C_o, at a fixed location (usually the outlet of the column) versus pore volume of effluent, Q/V_o, are called *breakthrough curves* (BTCs). Pore volume, a unitless number, is defined as the ratio of the column effluent volume to the total pore water volume in the column. The parameter t is the time since the dissolved solute was introduced to the effluent. The BTCs describe the transport behavior of the solute in the soil and provide information on interactions between the soil matrix and solute. Examples of BTCs are shown in Figures 11.6 and 11.7.

If there is no dispersion and the dissolved solute is conservative (i.e., no sorption or microbial or chemical transformation), at 1 pore volume the effluent breakthrough relative concentration will instantaneously reach 1.0, being zero until 1 pore volume is displaced from the column (Figure 11.3). This ideal shape of the BTC is known as piston, or plug, flow. If the dissolved solute is conservative but dispersion occurs in the soil column, some solute will appear in the column effluent prior to 1 pore volume, and the second BTC shown in Figure 11.6 will be obtained. Notice that the curve passes through a one-half relative concentration. If sorption occurs, the relative effluent concentration C/C_o will not equal 1.0 until more than 1 pore volume is displaced and all sorption sites on the soil are occupied, as is shown with BTC 3 in Figure 11.6.

Breakthrough curves of the simultaneous transport of a pulse of two herbicides, metolachlor and metribuzin, in the Ap horizon of a Memphis silt loam are shown in Figure 11.7. The column parameters were $\rho_b = 1.38$ Mg/m^3; $v = 0.0163$ m/h; $\theta_v = 0.479$ m^3/m^3; $L = 0.15$ m; and pore volume = 5.64×10^{-7} m^3. The shapes of these BTCs show that there were significant differences in the transport of the herbicides in the Memphis soil.

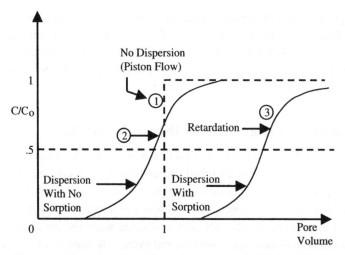

Figure 11.6. Examples of breakthrough curves for miscible displacement. C/C_o is the relative concentration of the solute measured in the effluent, and pore volume is the ratio of the volume of effluent to the volume of fluid in the soil column.

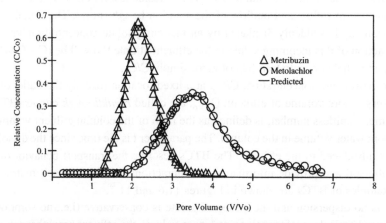

Figure 11.7. Breakthrough curves for metolachlor and metribuzin in Memphis silt loam. (The experimental data were kindly supplied by Dr. Alton Johnson.)

Both BTCs were delayed by the soil primarily through sorption by soil surfaces. The greater delay and slightly greater tailing by the metolachlor compared with metribuzin are consistent with its greater sorption by the Memphis soil. The higher peak concentration for metribuzin can be attributed to it lower sorption compared with metolachlor.

Derivation of the Solute Transport Equation

The equations used for the prediction of solute transport in soils will be derived first by summing the steady-state flow processes in the three soil phases. Then, with the aid of the law of conservation of mass, the transient-state processes will be accounted for. We will concentrate primarily on solute transport in the liquid and gas phases. Working assumptions are that the soil is homogeneous and isotropic and that the flow conditions are such that Darcy's law is valid and that the solute is conservative.

Steady-State Solute Transport

Under steady-state conditions, the flux density of solutes in soil is the sum of the flux densities in the three phases of soil. Mathematically, this can be written as

$$J_s = J_l + J_g + J_a \qquad [12]$$

where J_s is the total solute flux density in soil, J_l is the flux density in the liquid phase, J_g is the flux density in the gas phase, and J_a is the flux density in the solid or adsorbed phase. Values of J_a are usually considered negligible compared to the overall macroscopic transport process of solutes and often ignored when long-distance transport is considered.

For solute transport in the soil gas phase, diffusion is the primary transport mechanism. We use Fick's first law to write

$$J_g = -f_a D_g \partial C_g / \partial z \qquad [13]$$

where f_a is the aeration porosity (m^3/m^3), D_g is the molecular diffusion coefficient in the gas phase (m/s), and C_g is the concentration of the solute in the gas phase (kg/m^3 of soil air).

In the liquid phase, both diffusion and convective flow transport processes operate. The convective flux density, J_c, can be represented by

$$J_c = J_{lc} + J_{ld} = v\theta_v C_l - \theta_v D_h v \partial C_l / \partial z \qquad [14]$$

where J_{lc} is the convective flux density, J_{ld} is the diffusion-dispersion flux density, v is the pore water velocity (m/d), C_l is the concentration of solute in the soil solution (kg/m^3), and D_h is the hydrodynamic dispersion coefficient. Often, it is difficult to separate diffusion from dispersion, and these terms are combined and collectively called diffusion-dispersion.

The total solute flux density in the liquid phase, J_l, is the sum of the diffusion and convective-dispersion flux densities:

$$J_l = -\theta_v D \partial C_l / \partial z + v\theta_v C_l \qquad [15]$$

where D is the diffusion-dispersion coefficient of the solute. The first term on the right-hand side is the contribution of dispersion (molecular diffusion + mechanical dispersion), and the second term is the contribution of convective flow.

Neglecting J_a, the steady-state solute flux density in soil, J_s, is written as

$$J_s = -\theta_v D \partial C_l / \partial z + v\theta_v C_l - f_a D_g \partial C_g / \partial z \qquad [16]$$

which combines the solute flux densities in both liquid and gaseous phases of soil.

Transient-State Solute Transport

The general form of the mass balance equation for one-dimensional solute flow in the same direction as water and without generation and consumption can be represented as

$$\partial(\theta_v C_l)\partial t = -\partial J_s / \partial z \qquad [17]$$

where C_l is the solute concentration in solution (kg of solute per m^3 of soil water), θ_v is the volumetric soil water content (m^3/m^3), t is the time (s), z is the distance (m), and J_s is the solute flux density in soil (kg/m^2·s). Note that $\theta_v C_l = C$ and that J_s is expressed on a soil basis. For a conservative solute with negligible diffusion in the gaseous phase, combining equations [15] and [17] gives the convective-dispersion equation (CDE):

$$\partial C/\partial t = \partial/\partial z(\theta_v D \partial C_l/\partial z - v\theta_v C_l) \qquad [18]$$

For steady-state transport of water in a homogeneous soil with uniform soil water content, equation [18] can be simplified to

$$\partial C_l/\partial t = D\partial^2 C_l/\partial z^2 - v\partial C_l/\partial z \qquad [19]$$

which is the general transport equation for transient, isothermal flow of solutes in soil.

For solutes such as pesticides and fertilizers, consumption and generation are significant terms in the conservation equation. Under these conditions, equation [17] can be written as

$$\partial(\theta_v C_l)/\partial t = -\partial J_s/\partial z \pm r \qquad [20]$$

The quantity r is composed of both source and sink terms and has the same units as $\partial C/\partial t$. If the solute is generated within the unit volume of soil, r is called a source and has a positive sign; if the solute is consumed, then r is called a sink and has a negative sign. Examples of sources and sinks include

- release of ions from nonexchangeable to exchangeable form
- release or fixation (sorption) of solutes by soil surfaces
- degradation of organic solutes by chemical or microbial action
- uptake by plants
- generation by microbial and enzymatic processes

The parameter r is a collective term that combines one or more generation and consumption rate processes. Assuming loss of the solutes by sorption by soil surfaces only, r can be represented by

$$r = -\rho_b \partial S/\partial t \qquad [21]$$

where ρ_b is the soil bulk density, and S is the amount of solute in the solid or adsorbed phase (kg solute/kg of soil). Normally, the symbol S is used in soil physics literature instead of C_a.

For a given unit volume of soil, the total concentration of a nonvolatile solute (kg/m^3) is represented by the sum of the amounts retained by soil surfaces and that present in the solution phase. According to the revised form of equation [1],

$$C = \rho_b S + \theta_v C_l \qquad [22]$$

Differentiating equation [22] with respect to t, assuming that ρ_b is constant with t, and substituting the result into equation [19], gives the following form of the CDE:

$$\partial(\theta_v C_l)/\partial t + \rho_b \partial S/\partial t = \theta_v D\partial^2 C_l/\partial z^2 - v\theta_v \partial C_l/\partial z \qquad [23]$$

Using the chain rule, collecting $\partial C/\partial t$ terms, and assuming that θ_v does not change with z or t, we define

$$R = 1 + (\rho_b/\theta_v)(dS/dC_1) \tag{24}$$

where R is called the *retardation coefficient*. For organic and inorganic cations and neutral molecules, R will be greater than 1. However, if there is no sorption of the solute by soil surfaces, then the term dS/dC_1 is 0 and R becomes 1. This assumption is often made of anionic and neutral tracers such as chloride, bromide, and tritium. However, if there is anion exclusion, R may be less than 1 (Gilmour and Scott 1973).

Finally, the CDE that accounts for sorption of a conservative solute by soil surfaces is written as

$$R\partial C_1/\partial t = D\partial^2 C_1/\partial z^2 - v\partial C_1/\partial z \tag{25}$$

Many organic and inorganic solutes also undergo decomposition by biological and chemical processes. The loss of the solute by decomposition is frequently represented by

$$r = -kC \tag{26}$$

were k is the first-order decay constant (s^{-1}). Substitution of equation [26] into equation [25] results in the CDE for a solute that both is sorbed by soil surfaces and undergoes first-order decomposition:

$$R\partial C_1/\partial t = D\partial^2 C_1/\partial z^2 - v\partial C_1/\partial z - k\theta_v C_1 \tag{27}$$

Furthermore, we define the apparent diffusion-dispersion coefficient as

$$D_e = D/R \tag{28}$$

and the apparent pore water velocity as

$$v_e = v/R \tag{29}$$

and the apparent degradation rate coefficient as

$$B = k/R \tag{30}$$

Finally, the solute CDE becomes

$$\partial C_1/\partial t = D_e \partial C_1^2/\partial z^2 - v_e \partial C_1/\partial z - B\theta_v C_1 \tag{31}$$

This transient-state equation connects the one-dimensional flow of soil water in homogeneous soil with the spatial and temporal transport of solutes as modified by sorption and degradation.

Prediction of Solute Transport in Soil

We consider the transport of solutes in soil under two steady-state conditions of water flow. Steady-state flow conditions occur when the flux density is constant, which implies that $\partial\theta_v/\partial t$ is zero. In contrast, transient-state conditions occur when the flux density is not constant.

Convection under Steady-State Conditions

When soil water is moving under steady-state flow conditions and when diffusion and dispersion are zero, all of the solute moves at the same velocity, v, and the front arrives as one discontinuous jump to the final concentration C_o at $t = t^*$. This ideal condition is known as piston flow, in which the residual water in the column is replaced by the incoming water similarly to a piston pushing the residual water through the column (Figure 11.3).

In piston flow, all of the center of the solute front arrives at the end of the column at the same time. We will designate this time with the symbol t^* and as the "breakthrough time," "residence time," or "travel time." It is calculated from

$$t^* = L/v \qquad [32]$$

where L is the length of the column of soil (m). The distance of travel of a solute at any time is calculated from

$$L = vt^* \qquad [33]$$

which gives the straight-line length of the path traversed within the soil. In piston flow, 1 pore volume is required to transport a noninteracting conservative tracer through the soil column. A conservative tracer does not undergo generation or consumption processes.

Example

Calculate the amount of time required to transport the anion NO_3 from the bottom of the root zone of a pasture to groundwater that is 60 m below if the average soil water content is $0.25 \ m^3/m^3$ and the average drainage rate is 0.30 m/y and where no diffusion and dispersion occur. Use the piston flow model for the calculations.

The total depth of water in the vadose zone is

$$\theta_v z = 0.25 \times 60 = 15 \ m$$

It would require 15m/(0.3 m/y) = 50 y to replace this amount of water. The NO_3 velocity is

$$v = q/\theta_v = 0.3/0.25 = 1.2 \ m/y$$

The breakthrough time is

$$t^* = L/v = 60/1.2 = 50 \ y$$

Example

Calculate the amount of time required to transport a pesticide from the bottom of the root zone to the groundwater 60 m below if the average soil water content is $0.25 \ m^3/m^3$ and the average drainage rate is 0.3 m/y. Assume that dS/dC_l is $2 \ cm^3/g$ and the bulk density is $1.5 \ Mg/m^3$. Use the piston flow model with retardation for the calculations.

The retardation factor is

$$R = 1 + (\rho_b/\theta_v)dS/dC_l$$
$$= 1 + (1.5 \times 2/0.25) = 13$$

The average pesticide velocity is

$v = q/\theta_v R = 0.3/(0.25 \times 13) = 0.092$ m/y

The breakthrough time is

$t^* = L/v = 60/0.092 = 650$ y

These calculations show the influence of sorption on retarding the transport of solutes in soil. The adsorbing pesticide had a breakthrough time that was R times longer than that of the nonadsorbing anion NO_3.

Example
Show the influence of depth to the groundwater in the previous example for L = 10, 20, 30, and 60 m. For NO_3 and the pesticide, the predicted breakthrough times are

	Breakthrough Time (y)	
L(m)	Nitrate	Pesticide
10	8.33	108.3
20	16.67	216.7
30	25.00	325.0
60	50.00	650.0

This shows that the breakthrough time increases directly with distance to the water table.

In reality, piston flow seldom, if ever, occurs in the field. Usually, the boundary front between the resident and invading solutions is gradually mixed. This mixing occurs because of the combined effects of diffusion and hydrodynamic dispersion. The boundary front becomes increasingly diffuse about the mean position with time and pore water velocity (Figure 11.4). Ideally, BTCs should be symmetrical about the front of the advancing solution, with the inflection representing 50% displacement of the cumulative flow at 1 pore volume. In most cases, however, a symmetrical distribution is not found because of retardation of the solute and dispersion into pores of restricted flow.

Convection-Diffusion under Steady-State Flow
The CDE can be solved by either analytical or numerical methods. Analytical methods involve the solution of the partial differential equation using advanced calculus techniques and require information on the initial and boundary conditions. There are only a few analytical solutions of the CDE, and these are for the most simple cases and require assumptions that may be inappropriate for many field conditions.

Numerical methods involve the solution of the partial differential equation by numerical techniques. They are more powerful than analytical solutions in the sense that the geometry of the heterogeneous soil profile can be analyzed. However, they may have problems with numerical errors, which can cause solutions to show excess spreading of solute from plumes that are not related to the dispersion of the tracer solute. A brief review of numerical modeling is given in Appendix B.

In some cases, the solute is applied as a broadcast application to the soil surface or, in column experiments, as a pulse. The mathematical solution of equation [19] and the appropriate boundary conditions were given by Jury et al. (1992). The working equation is

$$C(z, t) = [C_o z/2(\pi D t^3)^{0.5}] \exp[-(z - vt)^2/4Dt] \qquad [34]$$

This equation predicts that as D gets larger, the solute concentration becomes more dispersed. For a given value of D and t, as v increases, the pulse moves deeper in the soil profile.

Van Genuchten and Wierenga (1986) listed four analytical solutions: two for semi-infinite and two for finite systems. The analytical expressions were for the relative concentration c, which was defined as

$$c(z, t) = [C(z, t) - C_i]/(C_o - C_i) \qquad [35]$$

where C_o is the solute concentration of the applied solution and C_i is the initial solute concentration. Both C_o and C_i are assumed to be constant.

For semi-infinite systems in the field, the initial and boundary conditions are

$$C(z, 0) = 0 \qquad [36a]$$
$$(-D\partial C/\partial z + vC)|_{z=0} = vC_o \qquad [36b]$$
$$\partial C/\partial z\ (\infty,t) = 0 \qquad [36c]$$

Equation [36b], which is the third-type boundary condition at (or close to) the soil surface, was assumed to apply (i.e., when $z \to \infty$). The lower boundary condition is assumed to be described by

$$\partial C/\partial z(\infty, t) = 0 \qquad [37]$$

The analytical solution for these boundary conditions, given as A-2 in Van Genuchten and Wierenga (1986), is

$$c(z, t) = (0.5)\text{erfc}[(Rz - vt)/2(DRt)^{0.5}]$$
$$+ (v^2t/\pi DR)^{0.5} \exp[-(Rz - vt)^2/4DRt]$$
$$- 0.5(1 + vz/D + v^2t/DR)\exp(vz/D)\text{erfc}[(Rz + vt)/2(DRt)^{0.5}] \qquad [38]$$

This analytical solution of the CDE correctly evaluates volume-averaged, in situ (resident) concentrations in semi-infinite field profiles.

Variations in the parameters in equation [38] affect the relative concentration distributions. These are shown in Figures 11.8 through 11.11.

Prediction of Solute Transport Coefficients

Solute transport coefficients in soil are complex parameters that depend on physical and chemical properties of both the soil and the solute molecule. Soil properties known to affect the magnitudes of the transport coefficients include water content, water flux, tem-

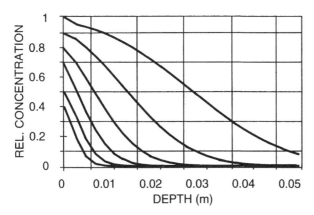

Figure 11.8. Effects of variations of the retardation factor, R, on the relative concentration distribution in soil. These predictions were made assuming that $D = 5 \times 10^{-11}$ m^2/s, v = 1 $\times$ 10^{-8} m/s, t = 8 d, and R = 0.25, 0.5, 1, 2, 4, and 8, for the curves from the upper right to lower left, respectively.

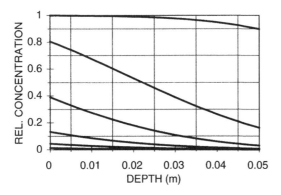

Figure 11.9. Effects of variations of velocity on the relative concentration distribution in soil. These predictions were made assuming that $D = 5 \times 10^{-11}$ m^2/s, R = 1, t = 100 d, and $v = 3 \times 10^{-11}$, 1×10^{-10}, 3×10^{-10}, 1×10^{-9}, 3×10^{-9}, and 3×10^{-8} m/s for the curves from bottom to top, respectively.

perature, and those surface properties influencing the adsorption and desorption of the solute, such as organic-matter content and pH. Molecular properties implicated include vapor pressure, solubility in water, and those structural properties influencing sorption.

Molecular Diffusion Coefficients

Diffusion coefficients of solutes in soil are influenced by soil physical properties such as the water content, temperature, and bulk density, and by variables affecting sorption such as organic-matter content, pH, salinity, and clay content of the soil. Diffusion coefficients obtained from transient-state experiments and mathematical solutions of Fick's second law usually are labeled as apparent diffusion coefficients, D_e. Since most solutes react with the soil, the apparent diffusion coefficient can be derived from the product of the molecular diffusion coefficient in a pure phase of the soil and the soil physical and

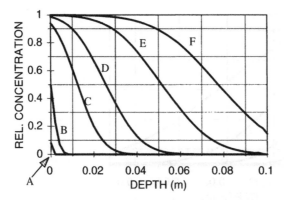

Figure 11.10. Effects of variations of time on the relative concentration distribution in soil. These predictions were made assuming that $D = 1 \times 10^{-9}$ m²/s, $R = 1$, $v = 3 \times 10^{-7}$ m/s, and $t = 1$ min, 1 h, 12 h, 1 d, 2 d, and 3 d for the curves A, B, C, D, E, and F, respectively.

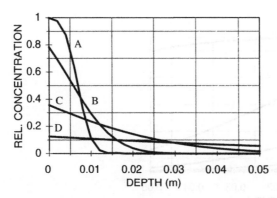

Figure 11.11. Effects of variations in D on the relative concentration distribution in soil. These predictions were made assuming that $R = 1$, $v = 1 \times 10^{-9}$ m/s, $t = 8$ d, and $D = 5 \times 10^{-12}$, 5×10^{-11}, 5×10^{-10}, and 5×10^{-9} m²/s for the curves A, B, C, and D, respectively.

chemical properties. For nonvolatile, nonsorbed solutes, the apparent diffusion coefficient can be written as

$$D_e = wD_o \hspace{6cm} [39]$$

where D_o is the solute molecular diffusion coefficient (m²/s) in water, and w is an empirical coefficient that takes into account the tortuous pathlength and other effects of the solid phase on diffusion. Values of w range from 0.5 to 0.01. Values of D_o for cations and anions are presented in Table 11.3 (Kemper 1986). Values of the molecular diffusion coefficients calculated for several organic chemicals in water and in air are tabulated in Table 11.4 (Thibodeaux and Scott 1985). With the exception of H^+ and OH^-, the major ions in soil have diffusivities in the range 0.5×10^{-9} to 2×10^{-9} m²/s. In general, D_o decreases as the valence increases and is about 10,000 times greater in air than in water (Table 11.4). For the larger organic molecules, D_o in water tends to be lower than for the cations and anions.

Table 11.3. Diffusion Coefficients of Ions, D_0, Calculated from Their Equivalent Conductivities in Water at 25°C

Cation	D_0 ($m^2/s \times 10^{-9}$)	Anion	D_0 ($m^2/s \times 10^{-9}$)
H^+	9.30	OH^-	5.26
Li^+	1.03	F^-	1.48
Na^+	1.33	Cl^-	2.03
K^+	1.96	Br^-	2.08
Rb^+	2.07	I^-	2.05
Cs^+	2.06	NO_3^-	1.90
Ag^+	1.65	ClO_3	1.72
NH_4^+	1.96	BrO_3^-	1.49
Mg^{2+}	0.71	IO_4^-	1.45
Ca^{2+}	0.79	HCO_3	1.19
Sr^{2+}	0.79	formate	1.45
Ba^{2+}	0.85	acetate	1.09
Cu^{2+}	0.72	SO_4^{2-}	1.07
Zn^{2+}	0.70		
Co^{2+}	0.73	CO_3^{2-}	0.92
Pb^{2+}	0.93	$P_3O_9^{3-}$	0.74
La^{3+}	0.62	$P_2O_7^{4-}$	0.64

Source: Adapted from Kemper 1986.

Table 11.4. Molecular Diffusivities of Selected Chemicals in Water and Air at 25°C

Chemical	Molecular Diffusivity	
	Water ($m^2/s \times 10^{-9}$)	Air ($m^2/s \times 10^{-5}$)
Methylchloroform	0.92	0.83
Trichlorofluoromethane	1.02	0.90
Monochlorobenzene	0.91	0.81
1,2,4-triclorobenzene	0.76	0.68
2-chlorobiphenyl	0.65	0.59
2, 4, 2′, 4′-Tetrachlorobiphenyl	0.55	0.52
Quinoline	0.81	0.72
p-Cresol	0.87	0.78
2, 4-Cresol	0.65	0.60
Chlorpyrifos	0.47	0.46
DDT	0.49	0.47

Source: Adapted from Thibodeaux and Scott 1985.

For solutes sorbed by soil surfaces, the apparent diffusion coefficient is further retarded and can be calculated from

$$D_e = (w/R)D_o \hspace{5cm} [40]$$

where R is the retardation coefficient, which was discussed earlier. Equation [40] states that the molecular diffusion coefficient of a solute in soil is equivalent to the molecular diffusion coefficient of the solute in free water times the ratio of the effects of the solid phase on the transport path to the effects of retardation due to sorption.

Hydrodynamic Dispersion Coefficients

All of the spreading or mixing of the solute due to mechanical dispersion associated with varying pore water velocity is included in the dispersion coefficient, D. In soil, the velocity distributions depend on the pore geometry, water content, and solute concentration, and therefore, empirical relationships between D and other macroscopically observable parameters have been established. It is commonly assumed that D can be expressed in terms of two components:

$$D = D_e + D_h \hspace{5cm} [41]$$

where D_e represents the apparent molecular diffusion and D_h represents mechanical (or hydrodynamic) dispersion. This assumption results from statistical considerations of solute transport as a consequence of a random distribution of pore water velocities; application of the central-limit theorem results in a Gaussian distribution of solute concentration. Thus, although the parameters D_e and D_h account for different physical processes, they are additive.

Several empirical models have been formulated to describe the relation between D_h and v. For example, if we assume that the effects of molecular diffusion are negligible and that the mechanical dispersion coefficient varies linearly with pore water velocity, we may write

$$D_h = \epsilon v \hspace{5cm} [42]$$

where ϵ is a proportionality factor frequently referred to as the *dispersivity* (m). The magnitude of ϵ depends on the flow system. For laboratory columns, ϵ is often on the order of 0.01 m; for field experiments, ϵ often ranges from 0.1 to 1 m. In unsaturated soil, however, and at low values of v, molecular diffusion is not negligible, and therefore, we may write

$$D = wD_o + \epsilon v \hspace{5cm} [43]$$

Often at high pore water velocities, the effects of molecular diffusion can be neglected, and the dispersion coefficient is proportional to the power of the pore water velocity:

$$D_h = \epsilon v^n \hspace{5cm} [44]$$

where n is an empirically determined constant characterizing the porous medium and with a magnitude slightly higher than 1.0 in soil systems. Most often, the model chosen to predict D in soil incorporates the influences of both molecular diffusion and pore water velocity and is written as

$$D = wD_o + \epsilon v^n \tag{45}$$

For unsaturated soils, the values of w, n, and D_e all depend on the water content and are not universally applicable to all soils.

Sorption by Soil Surfaces

Sorption processes are important in modifying the movement of many solutes through soil. The mobility in soil of any solute is a function of the fraction of the substance in the solution phase at any given time. Sorption is the major mechanism for retention of many organic and inorganic compounds. As a result, sorption and its associated desorption phenomena play a major role in regulating the availability of plant nutrients, the rate of leaching (redistribution), volatilization, and decomposition of organic solutes.

To quantify the effects of sorption on solute transport, the definitions of several terms should be understood.

Terminology

Adsorption occurs when the concentration of a liquid or gas is greater at a solution-soil or solution-air interface than in the bulk or gaseous phase. Adsorption is the process through which a net accumulation of a substance occurs at the common boundary of two contiguous phases. In other words, it is the adhesion of ions or molecules in an extremely thin layer to the surfaces of soil solids with which they are in contact. In practical terms, adsorption refers to the existence of a higher concentration of the solute at the surface of a solid phase than is present in the bulk solution.

The *adsorbent* is the surface or material receiving the substance.

The *adsorbate* is the substance or adsorbed molecule.

An *adsorption isotherm* is a plot of the amount adsorbed versus the amount in solution when determined at the same temperature.

There are two general types of adsorption. *Positive adsorption* occurs when the concentration of the adsorbate is higher at the interface than in the adjoining phases. *Negative adsorption* occurs when the adsorbate concentration is greater in the bulk solution than at the interface.

The concentration of solute as a function of distance from soil particle surfaces for two types of sorption is illustrated in Figure 11.12.

Types of Forces Active at Interfaces

The forces that are active at or near soil-water interfaces are basically electrical and are the same as the forces active at the molecular level in all kinds of matter. Characteristics of these forces are given in Table 11.5. The forces active at the surface are short ranged and vary with the reciprocal of the distance of separation raised to a power.

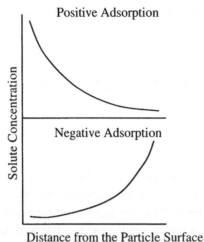

Figure 11.12. Relationship between solute concentration and distance from the particle surface for positively adsorbed and negatively adsorbed solutes.

Table 11.5. Types, Energies, and Mathematical Expressions of the Forces Operating at the Particle Surface of Soil

Origin of Force	Expression of Force as a Function of Distance	Example of Force	Internal Energy (J/mol)
Ion–ion	$F = f(1/X^2)$	Ionic crystal-salt	27,170
Ion–dipole	$F = f(1/X^3)$	Water hull around ion	6,186
Dipole–dipole	$F = f(1/X^4)$	Water-to-water attraction	1,881
Ion–induced dipole	$F = f(1/X^5)$	Attraction of ion for nonpolar gas or liquid	334
Dipole–induced dipole	$F = f(1/X^7)$	Adsorption of polar liquid or gas on surfaces	167

Source: Adapted from Taylor and Ashcroft 1972.
Note: The symbol X represents the distance from the particle surface.

Sorption Models

Several models have been proposed in an attempt to quantify the sorption of solutes by soil surfaces. Some of these models were initially derived for the adsorption of gases by solids, others are empirical models, and the rest are kinetic or rate models. Selim and Amacher (1997) provide a listing of commonly used equilibrium and kinetic sorption models, and these are given in Table 11.6.

Based on the multireaction approach, Selim (1992) proposed that sorption of solutes by soil may be regarded as involving three types of sites:

$$S = S_e + S_k + S_{ir} \qquad [46]$$

Table 11.6. Selected Equilibrium and Kinetic Models for Heavy-Metal Retention in Soils

Model	Formulation
Equilibrium Type	
Linear	$S_e = K_d C_1$
Freundlich	$S_e = K_d C_1{}^n$
General Freundlich	$S/S_{max} = [\omega C/(1 + \omega C)]^\beta$
Rothmund-Kornfeld ion exchange	$S_i/S_T = K_{RK}(C_i/C_T)^n$
Langmuir	$S/S_{max} = \omega C/(1 = \omega C)$
General Langmuir-Freundlich	$S/S_{max} = (\omega C)^\beta/[1 + (\omega C)^\beta]$
Langmuir with sigmoidicity	$S/S_{max} = \omega C/[1 + \omega C = (\sigma/C)]$
Kinetic Type	
First-order	$\partial S_k/\partial t = k_f(\theta_v/\rho_b)(C_1 - k_b S)$
n^{th} order	$\partial S/\partial t = k_f(\theta_v/\rho_b)(C_1{}^n - k_b S)$
Irreversible (sink/source)	$\partial S/\partial t = k_s(\theta_v/\rho_b)(C - C_p)$
Second-order irreversible	$\partial S/\partial t = k_s(\theta_v/\rho_b)C(S_{max} - S)$
Langmuir kinetic	$\partial S/\partial t = k_f(\theta_v/\rho_b)C(S_{max} - S) - k_b S$
Elovich	$\partial S/\partial t = A \exp(-BS)$
Power	$\partial S/\partial t = K(\theta_v/\rho_b)C^n S^m$
Mass transfer	$\partial S/\partial t = K(\theta_v/\rho_b)(C - C^*)$

Source: Selim and Amacher 1997.

Note: A, B, b, C^*, C_p, K, K_d, K_{RK}, k_b, k_f, k_s, n, m, S, S_{max}, ω, ß, and σ are adjustable model parameters, ρ_b is soil bulk density, θ_v is volumetric soil water content, C_T is total solute concentration, and S_T is total amount sorbed of all competing species.

Empirical and mechanistic models have been developed to describe sorption for each type of site. The first type of site, S_e, assumes that equilibrium is rapidly established between the amount of solute in the solution and sorbed phases. The relationship between the concentration in the sorbed phase and time is linear with a slope of zero (Figure 11.13a). The second type of site is kinetic, S_k. Kinetic sorption models (Figure 11.13b) also treat adsorption as a time-dependent phenomenon (Skopp 1986; Travis and Etnier 1981). The third type, S_{ir}, refers to sorption sites that are subjected to irreversible retention.

We now consider an example of each type of sorption model and discuss some of the mathematics and assumptions associated with each. Incorporation of any of these models into the CDE requires that S be differentiated with respect to C_1.

Equilibrium Sorption Models

The simplest equilibrium model is the linear equation, which can be represented by

$$S_e = mC_1 + b \qquad [47]$$

where S_e is the amount of solute adsorbed by the soil (kg/kg), C_1 is the concentration of solute in the solution phase (kg/m^3), and m and b are empirical constants given by the slope and intercept, respectively, of a plot of S_e versus C_1. Graphically, this model is shown in Figure 11.14.

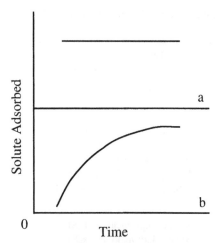

Figure 11.13. Relationship between the solute adsorbed and equilibration time for (a) equilibrium and (b) kinetic conditions.

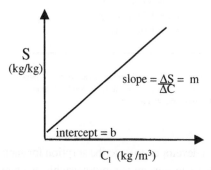

Figure 11.14. Adsorption isotherm for the equilibrium model.

Often, the constant b is assumed or forced to be zero. Under these conditions, we have

$$S_e = mC_1$$

In the soil science literature, we often see this model presented as

$$S_e = K_dC_1 \qquad\qquad [48]$$

where K_d is the equilibrium distribution (or partition) coefficient. Graphically, a plot of S_e versus C_1 results in a line whose slope is equal to the value of K_d. This is shown in Figure 11.15. Note that as K_d increases, the amount of solute sorbed also increases. Therefore, the sorption of a given solute can be compared on a number of soils, or a number of different solutes can be compared on a given soil. The units of K_d are (kg of solute/kg of soil)/(kg of solute/m^3 of soil solution), or m^3 of water/kg of soil.

Many studies have shown that the equilibrium sorption of organic molecules is dominated by the organic fraction of the soil. These potential pollutants tend to be hydrophobic with limited solubility in water but are easily dissolved into oils, fats, nonpolar organic

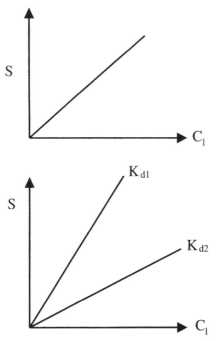

Figure 11.15. Sorption isotherm under equilibrium conditions showing the influence of K_d, where $K_{d1} > K_{d2}$.

solvents, and organic carbon in soil. To account for this, values of K_d may be divided by the organic carbon content, expressed as a fraction, to give values designated as K_{oc}:

$$K_{oc} = K_d/f_{oc} \tag{49}$$

where f_{oc} is the organic-carbon fraction of the soil. This equation shows that K_{oc} is the ratio of the solute concentration sorbed by the organic carbon to the solute concentration in water. The units of K_{oc} are m^3 of water/kg of organic carbon.

Another type of empirical equilibrium adsorption model is the Freundlich equation. The model can be expressed mathematically as

$$S_e = K_f C_1^N \tag{50}$$

where K_f is the Freundlich equilibrium constant $(m^3/kg)^{-N}$, and N is an exponent that often has a value less than 1.0. For most organic pesticides, N is approximately 0.9. The Freundlich adsorption isotherm is shown in Figure 11.16a. The Freundlich model assumes that there is no limiting concentration of the adsorbate as the solution concentration is increased without limit. This is unreasonable since any adsorbent should have a limited amount of surface and, hence, a maximum amount of adsorption. The logarithmic form of equation [50] is linear:

$$\log S_e = \log K_f + N \log C_1 \tag{51}$$

which is shown graphically in Figure 11.16b. To find K_f, note where $C_1 = 1$ (i.e., $\log C_1 = 0$), then $\log S_e = \log K_f$.

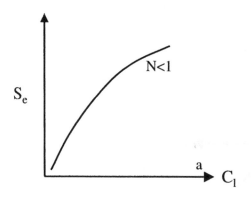

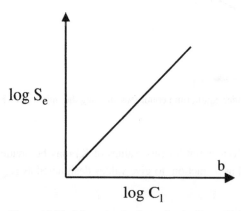

Figure 11.16. Adsorption isotherm for the Freundlich model.

The last equilibrium sorption model that we consider is the Langmuir equation, which has a sound conceptual basis. It was developed from the kinetics of adsorption of gases on solid surfaces. Three assumptions are made in the derivation:

1. The energy of adsorption is constant and is independent of surface coverage.
2. Adsorption occurs on localized sites and there is no interaction between adsorbate molecules.
3. The maximum adsorption possible is that of a complete monolayer.

The Langmuir sorption equation is written as

$$S_e = a\,Q\,C_l/(1 + aC_l) \hspace{3cm} [52]$$

where S_e is the number of adsorbed molecules per unit mass of soil, C_l is the solute concentration in solution, Q is the total number of adsorption sites available (i.e., the maximum amount of the ion or molecule that will be sorbed), and the parameter a is the ratio of the adsorption rate constants (a constant related to the bonding energy). Note that the

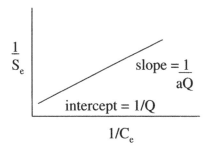

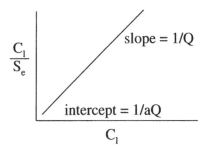

Figure 11.17. Adsorption isotherm for the Langmuir model.

parameter a represents the value of S_e that is approached asymptotically as C_1 gets large. The product aQ determines the magnitude of the initial slope of the isotherm. Usually the following transformation is made

$$1/S_e = (1 + aC_1)/aQC_1 = (1/aQC_1) + (1/Q)$$

or

$$C_1/S_e = (1/aQ) + (C_1/Q) \qquad [53]$$

The data are then plotted as shown in Figure 11.17. If the data approximate a straight line, then one can say that the Langmuir model can be used to describe the sorption data.

The disadvantages of the Freundlich and Langmuir equilibrium sorption models are as follows:

1. They provide little information about mechanisms of sorption.
2. They assume uniformity of sorption sites.
3. They assume noninteraction laterally of sorbed molecules.
4. They do not work well at very low and at very high concentrations, probably because different types of sorption sites and mechanisms operate in these regions.
5. Many adsorption processes are time dependent.

Kinetic Sorption Models

Reactive solute transport must also consider chemical nonequilibrium, which occurs when the sorption process is rate dependent (i.e., not instantaneous). This may include both

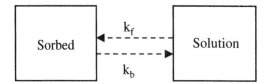

Figure 11.18. Diagram of the relationship between the solute partitioned in the sorbed and solution phases.

reversible and irreversible sorption. Kinetic sorption models listed by Selim and Amacher (1997) are given in Table 11.6. We will discuss only one of these kinetic models.

One of the early kinetic sorption models used in the CDE was the first-order kinetic reaction. This is given as

$$\rho_b \partial S_k / \partial t = k_f \theta_v C_l - k_b \rho_b S_e \qquad [54]$$

where S_k and S_e are the kinetic and equilibrium sorbed solute, and k_f and k_b are the forward and backward sorption rate coefficients. This model indicates that sorption is completely reversible and that the magnitudes of the rate coefficients dictate the extent of the kinetic behavior of the reaction (Figure 11.18). For small values of k_f and k_b, the rate of sorption is slow, and strong kinetic dependence is anticipated. In contrast, for large values of k_f and k_b, sorption is rapid and approaches a quasi equilibrium in a relatively short time. At long times, when the rate of sorption approaches zero, equation [54] reduces to

$$S_k = (\theta_v / \rho_b)(k_f / k_b)C_l = K_d C_l \qquad [55]$$

which results in $K_d = (\theta_v k_f)/(\rho_b k_b)$, with units of m^3/kg. Under these conditions, this kinetic equation reduces to a form that is similar to that of the linear sorption isotherms under equilibrium conditions.

The first-order sorption model, equation [54], has been modified slightly by assuming that the C_l term in equation [55] is raised to a power. This results in a kinetic model that is similar to the Freundlich equilibrium model. At long times when the rate of sorption approaches zero, the modified first-order kinetic equation reduces to the Freundlich model for equilibrium sorption (Table 11.6).

Degradation

Introduction

The rate of decomposition of solutes is of great importance because the rate strongly influences the chances for harmful accumulations of solutes. The decomposition process may be photolytic, chemical, or biological.

Photodecomposition is a common occurrence in air and water and may be important in some pesticides that remain on the soil surface. Little photodecomposition would be expected within the soil profile since radiant energy is strongly attenuated at the soil surface.

Chemical transformations have widespread occurrence, particularly in acid soils. All of the reactions are mediated through water functioning as a reaction medium, as reactant,

Table 11.7. Pesticides and Their Half-Lives in Soil

Pesticide	Half-Life (years^{-1})
Malathion	0.6
Aldicarb	16.0
Linuron	24.0
DDT (anaerobic)	35.0
Simazine	100.0
Terbacil	180.0
Lindane	570.0
Heptachlor	2000.0
DDT (aerobic)	3800.0

Source: De Haan and Bolt 1978.

or both. Examples include hydrolysis and oxidation. The chemical reactions that occur in soils are catalyzed by clay surfaces, metal oxides, metal ions, and organic surfaces.

Microbial decomposition is by far the most important process for the degradation of organic solutes such as pesticides. The decomposition rate depends on conditions in the soil and the bonding of the pesticide to soil surfaces. For most pesticides, aerobic decomposition proceeds much faster than anaerobic decomposition; however, there are classic exceptions to this. An example is DDT, whose decomposition proceeds 10 times faster under anaerobic conditions than under aerobic conditions. The rate coefficient of decomposition is related to the "half-life" or as a "persistency value." Both parameters represent the number of days required for decomposition of 50% of the solute under the specified conditions. Examples of half-lives of some pesticides as presented in Table 11.7.

Rate Models

The quantitative aspects of degradation of solutes in soil can be characterized over a certain concentration range by kinetic equations. Although there are many possible types of equations, two are used most often in modeling solute transport in soil.

The first of the degradation rate models is the zero-order model. Mathematically, this model is expressed as

$$dC_l/dt = -k_0 \qquad [56]$$

where k_0 is the zero-order rate constant (mg/m^{3}d). Integrating both sides gives the working equation

$$C_l = C_{l0} - k_0 t \qquad [57]$$

where C_l is the solute concentration in solution, C_{l0} is the solute concentration in solution at $t = 0$, and t is the time. Note that this equation assumes that the degradation process proceeds independently of solute concentration and at a constant rate. Thus, a plot of C_l versus t should give a straight line with an intercept of C_{l0} and a slope of $-k_0$.

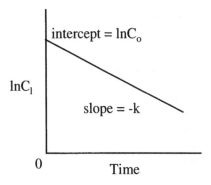

Figure 11.19. The relationship between the natural logarithm of solute concentration and time.

The second type of degradation rate model is the first-order kinetic model. Mathematically, this becomes

$$dC_l/dt = -k_1C_l \qquad [58]$$

where k_1 is the first-order degradation rate coefficient (d^{-1}). This equation can be integrated to give the following working equations:

$$C_l = C_o \exp(-k_1t) \qquad [59a]$$

$$\ln C_l = (\ln C_o) - k_1t \qquad [59b]$$

This equation assumes that the decomposition rate proceeds at a rate proportional to the first power of the solute concentration. These results show that the rate constant can be calculated for the slope of a semilog plot of solute concentration versus time. Graphically, this relationship is shown in Figure 11.19.

The half-life of the solute in the soil system is computed using equation [59]:

$$t_{1/2} = (\ln 2)/k_1 = 0.693/k_1 \qquad [60]$$

No one degradation rate law or type of rate law is likely to be completely adequate over the entire solute concentration versus time curve. Nevertheless, the use of the first-order coefficients to describe pesticide degradation in soil has reasonable validity. Other types of rate equations have been proposed, but it seems doubtful that a single rate equation that is universally and precisely applicable to all or even most pesticides will be found.

The processes by which organic pesticides are degraded in soils also involve microbial utilization of the pesticide as an energy source. Multiplication of the number of microorganisms involved in the process occurs, and the rate of degradation often increases in proportion both to their numbers and to any increase in efficiency due to adaptation by the microorganisms. Eventually, multiplication of the microorganisms slows and finally ceases due to the disappearance of the food and energy supply. The rate of degradation of the solute first ceases to increase and then decreases, approaching proportionality with time and the remaining pesticide concentration (i.e., first order). During the transition, the apparent order will be somewhere between zero order and first order depending on the concentration of the solute.

First Approximation of the Long-Term Leaching of Solutes

The assessment of human health risks from toxic solutes in terrestrial environments invariably includes exposure pathways that involve transport through contaminated surface soils. In many instances, the relative contribution of these pathways to the total exposure is a function of the residence time of the contaminants in soils. Quick removal of toxic solutes from the root zone by plant uptake suggests that exposure pathways from ingestion of food produced on contaminated soils may be relatively important. However, pathways involving drinking water drawn from springs or wells near contaminated soils may be significant if leaching is involved. Thus, removal processes such as radioactive decay, microbial and chemical degradation, or leaching from the root zone are important to assessment modelers and to regulators who determine compliance with standards or propose remedial actions to mitigate exposures. A simple model that predicts the net effect is often preferable to a more complex formulation that simulates the processes but requires a heavier investment of input parameters, time, and expense to use. We examine the approximation method proposed by Baes and Sharp (1983), which adopts several of the soil physical concepts discussed in previous chapters. The relatively simple model relates the annual average infiltration to bulk density, water content, and the distribution coefficient of the solute.

The traditional assessment modeling approach has been to approximate the solute removal processes in soil by first-order rate constants. In such an approach, the concentration of a pollutant in soil at a given time is given as

$$C = C_o \exp(-r_r t) \tag{61}$$

where C_o is the initial solute concentration in soil, and r_r is the sum of all first-order removal processes (y^{-1}). For example, such removal processes may include microbial decay (r_m), plant uptake (r_u), and leaching (r_l). Assuming independence between these processes, we then write r_r as

$$r_r = r_m + r_u + r_l \tag{62}$$

Using relationships defined previously, the removal rate constants can be made by approximating observed data with exponential regression equations or by calculating mean residence times of pollutants in soil. The latter approach gives removal constants via the relationship

$$r_r = 1/t_r^* \tag{63}$$

where t_r^* is the mean residence time of the pollutant in soil (y). The mean residence time of a solute leaching from a soil layer, t_l^*, may be given by

$$t_l^* = L/v_e \tag{64}$$

where L is the depth from which the solute is removed via leaching (cm), and v_e is the apparent solute migration velocity (cm/y) defined by equation [29]. The rate of leaching of the solute may be related to the infiltration rate of water by the retardation factor (R), such that

$$v_e = i/\theta_v R \tag{65}$$

where i is the infiltration rate (m/y), and θ_v is the volumetric soil water content (m^3/m^3). Assuming a linear sorption isotherm, the retardation factor as defined previously is

$$R = 1 + \rho_b K_d/\theta_v \qquad [66]$$

where K_d is the distribution coefficient, which is an equilibrium partitioning coefficient between the concentration of a solute in the solid (S_e) and solution (C_l) phases. That is,

$$S_e = K_d C_l \qquad [67]$$

Solute removal from the soil solution may also occur by uptake by plant and microorganisms, precipitation, etc.

By combining equations [64] to [67], the first-order leaching constant is given by

$$r_1 = (i/\theta_v)/LR \qquad [68]$$

It should be recognized that the leaching model represented by equation [68] is a relatively simple approximation of extremely complex physical and chemical processes. However, it serves as a first approximation of contaminant removal from soil via leaching. Measurement of the influence of the input parameters requires knowledge of their statistical central and dispersion tendencies.

Water infiltration rate, i, is dependent on climatic, soil, and management factors. It is site specific and time dependent. The total annual average rainfall plus irrigation provides an estimate of the amount of water applied to the soil in a given year. Mathematically, this is approximated from a reduced form of the water balance equation:

$$i = P + I - ET - R - \Delta W \qquad [69]$$

where P is the precipitation rate (cm/y), i is the net irrigation rate (cm/y), ET is the evapotranspiration rate (cm/y), R is the surface runoff (cm/y), and ΔW is the annual average soil storage of water. Surface runoff is dependent on climate, soil texture, vegetative cover, initial soil water content, slope, and anthropogenic interactions. These influences are difficult to predict in large landscapes. As a first approximation on an annual basis, we assume that R and ΔW are negligible compared to the water applied. In essence, i is the net infiltration rate of water or internal drainage through the soil profile.

Values of soil bulk density were summarized as a function of textural classes by Baes and Sharp (1983) and are presented in Table 11.8. This summary shows that slight, but distinct, differences in bulk density exist among the soil textural classes. The order of magnitude is sandy loams > loams > silt loams > clays and clay loams.

The actual quantity of water in soil at any time is dynamic and a function of soil, environmental, vegetative, and management effects. In humid regions, it is probable that an average annual value of θ_v is near "field capacity" or "drained upper limit," whereas in arid or semiarid regions, an average annual value of θ_v is likely near or below the wilting point. Usually θ_v near field capacity is used. Baes and Sharp (1983) summarized data for statistical values of θ_v at -30 and -1500 kPa by textural class. These data are presented in Table 11.9. Water retention is roughly in the order of clay loams > silt loams and loams > sandy loams.

Table 11.8. Summary of the Bulk Density of Soils of Varying Textural Classes

Textural Class	N	m (Mg/m^3)	m + s (Mg/m^3)	m + 2s (Mg/m^3)	Range (Mg/m^3)
Silt loams	99	1.33	1.49	1.67	0.86–1.67
Clays and clay loams	49	1.30	1.45	1.62	0.94–1.54
Sandy loams	37	1.50	1.64	1.81	1.25–1.76
Loams	22	1.42	1.53	1.66	1.16–1.58

Source: Baes and Sharp 1983.
Note: The parameters N, m, and s are the number of samples, mean, and standard deviation, respectively.

Table 11.9. Summary of the Volumetric Soil Water Contents at Two Pressures as a Function of Textural Class

Textural Class	N	m	m + s	m + 2s	Range
–30 kPa					
Silt loams	76	0.345	0.396	0.455	0.243–0.454
Clays and clay loams	33	0.360	0.423	0.423	0.255–0.448
Sandy loams	24	0.217	0.284	0.372	0.124–0.329
Loams	17	0.319	0.371	0.431	0.226-0.394
–1500 kPa					
Silt loams	76	0.127	0.176	0.244	0.060–0.297
Clays and clay loams	33	0.218	0.269	0.331	0.145–0.325
Sandy loams	24	0.077	0.121	0.188	0.029–0.158
Loams	17	0.131	0.170	0.220	0.082–0.167

Source: Baes and Sharp 1981.
Note: The parameters N, m, and s are the number of samples, mean, and standard deviation, respectively.

Values of K_d are relatively easy to calculate from equilibrium sorption isotherms but are quite variable among soils and solutes. Sources of variability of K_d include the choice of method used to determine the value and the time until equilibrium is attained. Also, since the calculations are based on the ratio of two concentrations, a small error in the measurement of either one may produce a large error in the ratio. Values of K_d vary with soil texture, pH, clay content, organic-matter content, and free iron and manganous oxide contents. Values of K_d are also related to the variability of these soil properties in the field (Wood et al. 1987) and with the physical and chemical properties of the solute molecule or ion. Values of K_d can range from less than 1.0 to much greater than 10.

The variability of model parameters is ordered approximately as $\rho_b \approx \theta_v < I \ll K_d$. This shows that the most variable and unpredictable parameter in the model is K_d.

The goal of assessment modeling is to provide order-of-magnitude, average-condition predictions rather than process-level simulations. In real systems, processes that render solutes unavailable for transport through exposure pathways involving soil are neither continuous nor necessarily of a first-order nature. However, this assessment model was shown by Baes and Sharp (1983) to reasonably approximate the net effect of leaching on solute concentration in the root zone over a period of several years with a minimum of user-input requirements.

Cited References

Baes, C. F., and R. D. Sharp. 1983. A proposal for estimation of soil leaching and leaching constants for use in assessment models. *Journal of Environmental Quality* 12:17–28.

Carslaw, H. S., and J. C. Jaeger. 1960. *Conduction of Heat in Solids.* 2d ed. Clarendon Press. Oxford.

De Haan, F. A. M., and G. H. Bolt. 1978. Pollution. In R. W. Fairbridge and C. W. Finkl Jr. (eds.), *The Encyclopedia of Soil Science,* part 1. Dowden, Hutchinson, and Ross. Stroudsburg, PA.

Fetter, C. W. 1993. *Contaminant Hydrogeology.* Macmillan. New York.

Gilmour, J. T., and H. D. Scott. 1973. The effect of exclusion volume on potentiometric nitrate measurement. *Soil Science Society of America Proceedings* 37:959–960.

Jury, W. A., W. R. Gardner, and W. H. Gardner. 1992. *Soil Physics.* 5th ed. John Wiley and Sons. New York.

Kemper, D. 1986. Solute diffusivity. In A. Klute et al. (eds.), *Methods of Soil Analysis,* part 1, *Physical and Mineralogical Methods,* p. 1009. 2d ed. Agronomy Monograph 9. Soil Science Society of America. Madison.

Kutílek, M., and D. R. Nielsen. 1994. *Soil Hydrology.* Catena Verlag. Cremlingen-Destedt, Germany.

Scott, H. D., D. C. Wolf, and T. L. Lavy. 1982. Apparent adsorption and microbial degradation of phenol by soil. *Journal of Environmental Quality* 11:107–112.

Selim, H. M. 1992. Modeling the transport and retention of inorganics in soils. *Advances in Agronomy* 47:331–384.

———. 1998. *Reactivity and Transport of Heavy Metals in Soils.* Lewis Publishers, Boca Raton, FL.

Selim, H. M., and M. C. Amacher. 1997. *Reactivity and Transport of Heavy Metals in Soils.* CRC Press. Boca Raton, FL.

Skopp, J. 1986. Analysis of time-dependent chemical processes in soil. *Journal of Environmental Quality* 15:205–213.

Taylor, S. A., and G. L. Ashcroft. 1972. *Physical Edaphology.* W. H. Freeman and Co. San Francisco.

Thibodeaux, L. J., and H. D. Scott. 1985. Air/soil exchange coefficients. In W. B. Neely and G. E. Blau (eds.), *Environmental Exposure from Chemicals,* vol. 1, pp. 65–89. CRC Press. Boca Raton, FL.

Travis, C. C., and E. L. Etnier. 1981. A survey of sorption relationships for reactive solutes in soil. *Journal of Environmental Quality* 10:8–17.

Van Genuchten, M. Th., and P. J. Wierenga. 1986. Solute dispersion coefficients and retardation factors. In A. Klute et al. (eds.), *Methods of Soil Analysis,* part 1, *Physical and Mineralogical Methods,* pp. 1025–1054. 2d ed. Agronomy Monograph 9. Soil Science Society of America. Madison.

Wood, L. S., H. D. Scott, D. B. Marx, and T. L. Lavy. 1987. Variability in sorption coefficients of metolachlor on a Captina silt loam. *Journal of Environmental Quality* 16:251–256.

Problems

1. A nondegradable, nonvolatile, nonreactive contaminant was applied to the soil surface in an area where the annual rainfall is 113 cm, the annual evaporation is 100 cm, the water table is 60 m below the surface, and the water content of the vadose zone is 0.25 cm^3/cm^3. Estimate the travel time for the contaminant to reach the groundwater table. Assume that one-dimensional piston flow occurs toward the water table and that steady-state conditions exist.

2. Calculate the travel time for piston flow of solute moving through a vertical, 100 cm saturated soil column. Assume that the Ksat is 5 cm/d and θ_{sat} is 0.50 cm^3/cm^3 and that 10 cm of water is ponded on the soil surface.

3. Calculate the annual travel distance of nitrate if the soil water flux density is 2 cm/d and the soil water content is 0.25 cm^3/cm^3.

4. A scientist found the following relationships for the sorption of a solute by Captina soil. The oven-dry soil to solution ratio by weight was 1:5.

	Solute Treatment				
	1	2	3	4	5
Amount of solute added (μg/g oven-dry soil)	4,700	470	47	4.7	0.47
Amount of solute in solution (μg/ml)	897	88.7	9.03	0.9	0.9

Quantify the influence of sorption of the solute by the Captina soil by calculating the Freundlich constants K and N.

5. A scientist from Kentucky found the following relationships between the herbicides 2,4-D acid and Atrazine sorbed by a Lanton silty clay loam:

2,4-D acid:

Amount added (μg/g oven-dry soil)	0.342	0.684	2.051	3.419	6.833
Amount in solution (μg/mL)	0.202	0.413	1.217	2.217	4.387

Atrazine:

Amount added (μg/g)	0.130	0.260	0.780	1.300	2.600
Amount in solution (μg/mL)	0.031	0.063	0.201	0.362	0.793

Show the influence of sorption of these herbicides by the Lanton soil by doing the following:

  a. Make a plot of the amount sorbed as a function of the amount added.
  b. Make a Freundlich plot of the amount sorbed as a function of the equilibrium solution herbicide concentration.
  c. Calculate the Freundlich constants K and N.

6. Assume that a herbicide was broadcast on the surface of a bare soil at a rate of 5 kg/ha, that the initial profile concentration of this herbicide was zero, and that there was no influence of water movement on the transport process. Show the influence of the magnitude of the

diffusion coefficient and diffusion time on the concentration distribution curve in a soil profile using the equation

$$C(z, t) = M/[2(\pi Dt)^{0.5}]\exp(-z^2/4Dt)$$

where $C(z, t)$ is the herbicide concentration (kg/m^3), M is the amount of herbicide applied to the soil surface (kg/m^2), D is the molecular diffusion coefficient (m^2/s), and t is the time after application (s).

7. Show the influence of the dispersion coefficient on the relative solute concentration as predicted by equation [33]. Use values of $D = 1 \times 10^{-4}$, 1×10^{-6}, and 1×10^{-8} m^2/s, respectively, along with $t = 10$ d and $v = 1 \times 10^{-1}$ m/d.

8. A solute with a molecular diffusion coefficient of 2×10^{-9} m^2/s is diffusing at steady state into a zone devoid of the solute. If the initial solute concentration is 25 mg/kg of oven-dried soil and the bulk density is 1200 kg/m^3, calculate the flux density at a distance of 0.05 m and the amount of solute diffused across this boundary in 1 mo.

9. In a soil profile, the matric potential is –200 cm at both the 15 cm depth and the 25 cm depth, and the average hydraulic conductivity in this depth interval is 0.10 cm/d. If the soil water content is 0.25 m^3/m^3 and the average soil solution concentration of solute is 5 mg/L, calculate the magnitude of the pore water velocity and the flux density of the solute moved across the depth interval in 1 week. In which direction in the profile is the solute moving?

10. Calculate the amount of time required to transport chloride from the bottom of the root zone to the alluvial aquifer 50 m below the root zone if the average soil water content of the vadose zone is 0.40 m^3/m^3 and the annual drainage rate is 0.2 m/y.

11. An equilibrium sorption experiment with metribuzin yields the following table of C_l and S. Determine the type of isotherm (linear, Freundlich, or Langmuir) and the appropriate equation parameters (to two significant digits) and their units.

C_l (µg solute/cm³ solution)	S (µg solute/µg soil)
0.23	0.063
0.45	0.16
0.69	0.29
1.54	0.90
2.63	1.90
4.14	3.58
6.20	6.30
11.52	15.00

12. An equilibrium sorption experiment with metachlor produces the following table of values for solution concentration, C_l, and sorbed concentration, S. Determine whether the sorption isotherm is linear, Freundlich, or Langmuir. What are the appropriate equation parameters (to three significant digits) and their units?

C_l (µg solute/cm³ solution)	S (µg solute/µg soil)
0.53	0.93
1.21	2.12
3.27	5.72
5.34	9.34
7.89	13.81
9.50	16.62

13. A sorption experiment with metribuzin produces the following table of C_l and S. Is the appropriate isotherm linear, Freundlich, or Langmuir? What are the appropriate equation parameters (to three significant digits) and their units?

C_l (µg solute/cm³ solution)	S (µg solute/µg soil)
0.64	0.23
1.13	0.40
4.80	1.62
9.75	3.11
15.84	4.74
30.26	7.89
75.64	13.99
105.9	16.41
237.5	21.57

14. If the results of problem 13 had been limited to $C_l < 16$ µg solute/cm³ solution, what would the likely answer have been?

CHAPTER TWELVE

Soil-Plant-Water Relations

Introduction

Interest in soil-plant-water relations can be attributed to the serious dwindling of water supplies in arid regions such as the western United States and the Great Plains and in humid regions where short-term droughts occur during the growing season. The impetus for this work is the need to continually increase crop production in the face of growing human populations around the world. Wherever plants grow, their development is limited to some extent by either too little (water deficit or drought stress) or too much water (water excess or oxygen stress). For example, many deserts and grasslands could support luxuriant plant growth if only an adequate supply of water were available. Wetlands, where growth and development of most plants are limited by poor soil aeration, can sometimes be made more productive by drainage. Although plants grow best in regions of high and recurrent rainfall, all land plants periodically undergo internal water stress even if the soil is well watered.

In most regions of the United States, soil water supplies are at optimal levels only rarely during the crop growing season. Therefore, large losses in plant growth and yield occur annually because of recurrent or sustained internal water deficits in plants. The extent of such losses in plant productivity is often not realized because data for many areas are not available to indicate how much more growth would occur if plants had favorable water supplies throughout the growing season.

The scientific literature indicates that water stress affects the following plant processes: water uptake from the soil, root water potential and leaf water potential, leaf temperature, cell expansion, seed germination, stomatal aperture, transpiration, photosynthesis, enzymatic activity, nutrient relations, nitrogen metabolism, and flow of oleoresin and latex. Within the plant, water enters into a variety of mixtures and solutions:

1. Water is the principal medium for the chemical and biochemical processes that support plant metabolism.

2. Water acts as the *solvent* and the *medium* by which the basic building blocks of growth—dissolved sugars and minerals—are transported throughout the plant.

3. Evaporation of water from plant cell walls absorbs much of the excess radiant energy unused in photosynthesis and, thus, cools the plant. This process aids in maintaining the plant temperature at values below those that might inhibit the plant's metabolic processes.

4. The presence of additional water vapor decreases air density and promotes circulation of air within the plant canopy. This probably aids in replenishing the supply of CO_2 in the near vicinity of the plant.

5. Water acts as a reagent or reactant in photosynthesis and in hydrolytic processes such as the hydrolysis of starch to sugar.

6. Water acts to maintain turgidity, which is essential for cell enlargement and growth, for maintenance of physical morphology, for opening of stomata, and for movement of leaves, flowers, and various specialized plant structures.

Supplies of water and essential mineral nutrients are controlled largely by belowground characteristics of the soil and plant. The demand for water is controlled by aboveground plant and environmental characteristics. Soils and plant species differ in their combination of supply and demand relationships. Differences in the ability to integrate the supply and demand of water also exist within plant species.

In this chapter, the components of the soil-plant-atmosphere system and the interactions among these components are discussed. Mechanisms governing the movement of water from the soil to plant roots, through the plant, and finally from the plant to the atmosphere are examined in detail. Parameters to evaluate the productivity and efficiency of water use by plants are presented. Many of the physical and mathematical concepts developed previously for the transport of water, heat, and solutes in soil also apply to transport in the plant and atmosphere.

Measurement of Plant Water Status

There are numerous techniques for the estimation of plant water status (Barrs 1968). We will concentrate only on the measurement of the relative water content and water potential.

Relative Water Content

Previously, we showed that soil water content could be expressed on a dry-weight and a wet-weight basis. For plants, the use of these two bases is unsatisfactory because the dry and wet weights of plants change with plant age. This has led to the development of a plant water stress method based on the water content of a turgid plant. The *relative water content, RWC,* is defined as

$$RWC = [(FW - ODW)/(TW - ODW)] \times 100 \tag{1}$$

where FW is the field weight (kg), ODW is the oven-dried weight (kg), and TW is the turgid weight (kg). The turgid weight is determined by equilibrating the plant tissue with water for several hours at constant temperature. Calculation of RWC represents the relative saturation of the plant tissue and is expressed as a percentage of full-turgor water content.

Water Potential

Water potential is the most frequently used measure of plant water status and is particularly relevant in studies on water movement. Two instruments have been used for its determination: the thermocouple psychrometer and the pressure chamber.

The principle of the psychrometer is the same as given in Chapter 8. For plants, a sample of tissue is allowed to come to water vapor equilibrium with a small volume of air in a closed chamber, and the humidity of this air is measured using thermocouples positioned to measure the wet-bulb depression or the dew point of the air. The water potential is given by

$$\psi_w = (RT/V_w)\ln(e/e_s) \tag{2}$$

where R is the universal gas constant (8.314 J/mol·K), T is the absolute temperature (K), V_w is the volume of a mole of water (0.000018 m³/mol), and e and e_s are the water vapor partial pressure and saturation partial pressure of water vapor, respectively. The SI units of e and e_s and ψ_w are Pascals. The water vapor pressure increases curvilinearly with temperature. The ratio of e/e_s is commonly referred to as the *relative humidity*.

The pressure chamber is another widespread method used to measure the water potential of leaves. It is a rapid, accurate, but destructive method. If a leaf is cut from a plant, the tension in the xylem causes the xylem sap to be withdrawn from the cut surface. This leaf may then be sealed into a gas-tight pressure chamber with the cut end exposed. The chamber is pressurized until the sap just wets the cut surface, thus restoring the sap to its original position. The negative of the applied pressure is equal to the original water potential of the leaf. The method is rapid and largely insensitive to temperature.

Components of Plant Water Potential

Water in the soil and plant is in equilibrium when it is at the same chemical potential. A general statement of the components of the chemical potential is

$$d\psi = V_{T,\theta,n}dP - S_{P,\theta,n}dT + (\partial\psi/\partial\theta_v)_{T,P,n}d\theta_v + (\partial\psi/\partial n)_{T,P,n}dn + mgh \qquad [3]$$

where ψ is the chemical potential, V is the molar volume, P is the external pressure, S is the molar entropy, T is the temperature, n is the mole fraction of solute, θ_v is the volumetric water content, m is the molar mass of water, g is the gravitational acceleration, and h is the height above a reference level (Rendig and Taylor 1989). The chemical potential is the amount by which the Gibbs free energy in the system changes as water is added or removed while the temperature, pressure, and other constituents remain constant.

The first term on the right-hand side of equation [3] is the pressure component, $V_{T,\theta,n}dP$, which may be the hydrostatic pressure of a column of water, an increase in pressure above atmospheric pressure within plant cells, or the overburden pressure where part of the load is carried by soil water. The entropy term, $S_{P,\theta,n}dT$, is not precise because it usually is derived at constant pressure. The third term is the matric component, $(\partial\psi/\partial\theta_v)_{T,P,n}\,d\theta_v$, which includes all interactions of water with the solid matrix, either surface tension or surface adsorption effects. The fourth term is the osmotic or solute component, $(\partial\psi/\partial n)_{T,P,n}dn$. Considerable care should be exercised in the use of this term because liquid flow carries both dissolved ions and molecules, but the phase change from liquid to vapor leaves the solutes behind. The last component, mgh, accounts for the effects of gravity, which may be significant when the soil has a high water content (Rendig and Taylor 1989). The chemical potential can be expressed on a mass (J/kg), volume (Pa), or weight basis (m).

Water Potentials within the Plant

Many of the water potential concepts that were originally developed for the soil also apply to the plant. We have already seen that the movement of water occurs from regions of higher total potential energy, ψ_t, to regions of lower total energy potential, and that unsaturated soil has a negative soil water pressure. Since plants obtain significant amounts of water from the soil, ψ_t must be lower within the plant than in the soil surrounding the roots.

Total Water Potential

The total water potential may be partitioned into several component potentials relevant in the soil-plant-atmosphere system:

$$\psi_t = \psi_p + \psi_o + \psi_m + \psi_g \tag{4}$$

where ψ_p, ψ_o, ψ_m, and ψ_g are components of the total water potential due to pressure, osmotic, matric, and gravitational forces, respectively. The pressure component potential, ψ_p, represents the difference in hydrostatic pressure from the reference and can be positive or negative. It arises from the forces exerted on the walls from the water attracted to the cell by the solutes and solids in the protoplast. The osmotic component potential, ψ_o, results from dissolved solutes lowering the free energy and is always negative. The matric potential, ψ_m, is similar to the osmotic potential except that the reduction of the activity of water results from capillary and adsorption forces at cell walls and the surfaces of solids. In plant systems, the distinction between ψ_o and ψ_m is to some extent arbitrary, since it is often difficult to decide whether the particles are solutes or solids, so ψ_m is often included in ψ_o (Jones 1983). The gravitational component, ψ_g, results from differences in potential energy due to a difference in height from the reference level. Values of ψ_g are positive if above the reference level and negative below. The contributions of ψ_g are often neglected in agronomic crop systems and are not significant at the cellular level (Hopkins 1995).

Plant Water Potentials

In plant cells, the principal components of water potential are the osmotic and pressure components. The plant water potential (ψ_w) is defined as

$$\psi_w = \psi_p + \psi_o \tag{5}$$

where ψ_p is the pressure difference between that inside and that outside the cell wall. It is usually positive and commonly called the *turgor pressure*. For a given plant cell solute content, ψ_p decreases as ψ_w decreases. Visible wilting of leaves is usually observed when ψ_p is zero. In the xylem conduits, ψ_p is the dominant component of ψ_w, which can reach very large negative values, less than –4 MPa in some severely stressed plants. The xylem vessels are rigid enough to withstand these tensions without undergoing serious deformation (Jones 1983).

The ψ_o is indirectly related to the concentration of solutes, which is related to the activity of water. The governing equation is

$$\psi_o = (RT/V_w)\ln a_w \tag{6}$$

where a_w is the activity of the water, which is related to the relative concentration of solutes. As the solute concentration increases, ψ_o decreases. Another useful approximation of ψ_o is the van't Hoff relation

$$\psi_o = -RTC_s \tag{7}$$

where C_s is the concentration of solute (mol/m^3 solvent or mol/1000 kg of solvent), R is the universal gas constant (8.314 J/mol·K), and T is the absolute temperature (K). The

ψ_o of typical cell sap for many plants is about -1 MPa. Solute concentrations in the cell walls and in the xylem are usually low, giving rise to values of ψ_w less than 0.1 MPa (Jones 1983).

Water Potential of Plant Component Parts

For plant water potential, the reference state is arbitrarily taken as pure water at atmospheric pressure. Under *equilibrium* conditions, ψ_w within the root cell must be the same as ψ_w of the water in the cell walls or in the soil with which the root cell is in contact. Within the cell, there are no curved air-water interfaces, so except possibly for some water tightly bound to the colloidal particles within the cell, ψ_m is small. The liquid within the cell, is an aqueous solution of sugars and minerals. Thus, within the cell C_s may be quite large, and therefore, the effects of ψ_o may be appreciable.

The ψ_o plays an important role in the maintenance of turgor in some plant tissues. Plant cells are invariably pressure vessels, since it is the internal pressure of the collection of cells composing the plant that gives the plant form and shape. Since the cell is an enclosed vessel, ψ_p must exist and is positive in turgid plant tissues.

The value of ψ_g, except in special cases, such as for certain very tall trees, is negligible compared to the other component potentials. Therefore, for most crops, ψ_g is small compared with the other components of ψ_t and is often ignored.

As a result, we can state the following general conditions. In a *turgid plant,*

$$\psi_w \approx 0$$

In a *partly turgid* plant, typical values are

$$\psi_w \approx -1000 \text{ kPa} = -1 \text{ MPa}$$

Example

Suppose that the component plant water potentials were determined to be $\psi_o = -2000$ kPa and $\psi_p = 1000$ kPa. What is the plant water potential?

$$\psi_w = \psi_o + \psi_p = -2000 \text{ kPa} + 1000 \text{ kPa} = -1000 \text{ kPa} = -1 \text{ MPa}$$

In a *flaccid* or wilted plant,

$$\psi_w = -2000 \text{ kPa} + 0 \text{ kPa} = -2 \text{ MPa}$$

Example

If $\psi_o = -2000$ kPa and $\psi_p = 0$ kPa, calculate the plant water potential.

$$\psi_w = -2000 \text{ kPa} + 0 \text{ kPa} = -2000 \text{ kPa} = -2 \text{MPa}$$

Note that $\psi_p = 0$ when the tissues are flaccid or wilted. By considering the components of ψ_w within a plant, it is possible to obtain a good idea of the plant's water status at any given time.

Example

Suppose that a xylem vessel is located next to three cells attached in series. The ψ_w is 0, –0.1, –0.2, and –0.3 MPa for the xylem, cell A, cell B, and cell C, respectively. In which direction is water moving?

The direction of water movement is determined solely by the value of ψ_w in adjacent cells. Water moves from locations of higher total potential energy to locations of lower total potential energy. Thus, in this case, water moves from the xylem toward cell C.

Example

A root cell is in contact with a soil that has a soil ψ_w equal to $\psi_o + \psi_m$, where $\psi_o = -180$ kPa and $\psi_m = -1120$ kPa. If the concentration of solutes within the cell is 8.65×10^{-4} mol/cm^3 and the temperature is 23°C, at equilibrium what is the water potential of the plant root cell? What is ψ_p?

At equilibrium, we can write

$$\psi_{soil} = \psi_{cell}$$
$$= \psi_o + \psi_m = -180 \text{ kPa} + (-1120 \text{ kPa})$$
$$= -1300 \text{ kPa} = -1.3 \text{ MPa}$$

Since

$$\psi_{cell} = \psi_p + \psi_o = -1.3 \text{ MPa}$$
$$\psi_o = -RTC_s = (-8.314 \text{ J/mol·K})(296 \text{ K})(8.65 \times 10^{-9} \text{ mol})$$
$$= -2.13 \text{ MPa}$$

Therefore,

$$\psi_p = -1.3 \text{ MPa} - (-2.13 \text{ MPa}) = 0.83 \text{ MPa}$$

Water within the xylem vessels of plants usually possesses a low concentration of solutes. Since ψ_w must be as low (or lower) in the plant as that of the soil surrounding the roots, this xylem water is under tensile stresses. The tension in water within stems is supported by either (a) attraction of water molecules for each other within the xylem tubes or (b) the highly hydrated cell walls through which a continuous film of water penetrates from cells near the root tip to those in the palisade layer and spongy mesophyll of the leaf. At the sites where transpiration occurs, ψ_w on the exterior of the cell walls exposed to the intercellular spaces must be less than that within the xylem system.

Example

The ψ_w of a soil is found to be –0.8 MPa. The leaves of a plant growing in the soil are in an environment with a temperature of 30°C and a relative humidity of 70%. Within the intercellular spaces of the leaves, the relative humidity is 98.5% and the leaf temperature is 27°C. If the solute concentration within a leaf cell is 8.6×10^{-4} mol/cm^3, find the following:

a. ψ_w in the atmospheric environment surrounding the leaf. At 30°C (303 K) the value of (RT/V) is 139.95 MPa. Therefore, for this problem,

$\psi_{atm} = 139.95 \times \ln(0.7) = -49.9$ MPa

b. ψ_w in the intercellular spaces of the leaf.

$\psi_{spaces} = [(8.314)(300)\ln(0.985)]/0.000018 = -2.09$ MPa

c. The pressure component of ψ of a leaf cell, assuming that the interior of the leaf is in equilibrium with the ψ of the cell walls, which, in turn, is 0.2 MPa greater than the water potential within the intercellular spaces.

$\psi_{cell\ wall} = -2.09$ MPa $+ 0.2$ MPa $= -1.89$ MPa

d. The difference in ψ between the leaf xylem water and the xylem water near the root tips.

$\psi_{leaf\ cell\ wall} - \psi_{soil} = \Delta\psi = -1.89$ MPa $- (-0.8)$ MPa $= -1.09$ MPa

Soil-Plant-Atmosphere Continuum

Definition

The approach usually taken to study soil-plant-water relations is based on recognition that all components of the field environment (the soil, the plant, and the atmosphere), when taken collectively, form a physically unified and dynamic system. In the system, the various water flow processes occur interdependently. This unified system has been dubbed the

Soil:Plant:Atmosphere:Continuum

or SPAC by J. R. Philip (1966). Flow of water in the SPAC occurs from regions of higher total potential energy to regions of lower total potential energy (Figure 12.1).

The concept of "water potential" is equally valid and applicable in the soil, plant, and atmosphere alike. The important principle to remember is that the various terms used to characterize the state of water in different parts of the SPAC are alternative expressions of the potential energy of the water. The difference in potential energy from one location to another is responsible for the tendency of water to flow. In order to characterize the SPAC physically, it is necessary to evaluate the potential energy of water and changes in potential energy with distance and time along the entire path of water movement.

Flow of Water and Distribution of Water Potentials in the SPAC

The flow rates of water in the SPAC are essentially steady state and are inversely proportional to the resistance R_p (i.e., Δz/hydraulic conductivity). Mathematically, this flow rate of water is approximated by an analogy with Ohm's law (a Darcy-like equation) and can be represented by

$$q = -\Delta\psi/\Sigma R_p$$
$$= -\Delta\psi_1/R_1 = -\Delta\psi_2/R_2 = -\Delta\psi_3/R_3 = -\Delta\psi_4/R_4 \qquad [8]$$

where q is the volume flux density of water ($m^3/m^2 \cdot s$ or m/s) and the subscripts represent different pathways of the system. In equation [8], the subscript 1 represents the soil

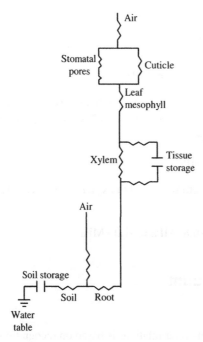

Figure 12.1. Representation of pathways of water transport in the soil, plant, and atmosphere. (Adapted from Cowan 1965.)

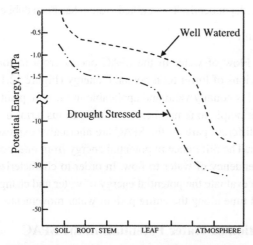

Figure 12.2. Distribution of water potentials in the SPAC for well-watered and drought-stressed plants.

to plant root pathway, the subscript 2 represents the root to xylem pathway, the subscript 3 represents the plant to leaves pathway, and the subscript 4 represents the leaves to atmosphere pathway. Equation [8] assumes that there is liquid continuity between the water in the soil and the water evaporating from the stomatal chambers of the leaf and that water flow occurs from locations of higher potential energy to locations of lower potential energy with the hydraulic gradient as the sole driving force.

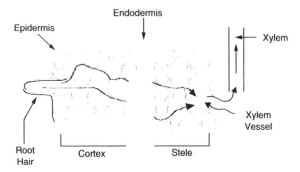

Figure 12.3. Cross-sectional view of a root with the water flow pathway. (Adapted from Papendick and Campbell 1975.)

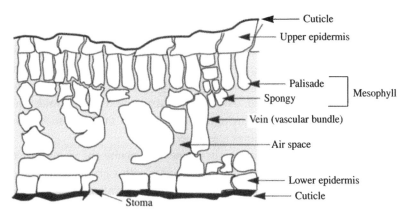

Figure 12.4. Cross-sectional diagram of a typical leaf.

Movement of water through the SPAC can be thought of as occurring along a gradient of decreasing water potential, ψ_w (Figure 12.2). Gradients of ψ_w between parts of the system constitute the driving force for flow within the system. However, the resistances to the flow of water are not equal throughout the SPAC.

The root presents a major resistance to water flow, particularly where water passes through the endodermis. Figure 12.3 shows a diagrammatic cross section of the major path of water flow across the root into the xylem. The shaded band across the endodermis represents the Casparian strip. Major flow of water occurs along and between the cell walls except at the endodermis, where water must pass through the cells.

Resistance to water flow within the main xylem is negligible compared with the other resistances in the SPAC. Transport of water through xylem vessels of the plant has generally been regarded as conforming to Poiseuille's law for capillary flow.

The greatest resistance to flow of water, however, occurs in the leaf (Figure 12.4). This is because in the leaf, water changes from a liquid to a vapor state, a process governed by the latent heat of vaporization. At 20°C, this change requires approximately 244 MJ/kg. Therefore, R_4, which is the sum of the stomatal resistance and the aerial resistance, may be as much as 50 times greater than R_1. Leaf resistance (R_L) is low when the stomata are open and the leaf water potential (ψ_L) responds directly to transpirational flux. However, when plants are stressed for water, the stomata close or partially close, and R_L increases significantly so that ψ_L reflects R_L more closely than it reflects the demand placed on the leaves by the atmosphere. At night, the atmospheric demand is low, the stomata are closed, and R_L tends to be high so that the transpirational flux is generally negligible. Therefore, ψ_L tends to reflect supply factors at night.

Water Potentials during the Plant Life Cycle

The ψ_w varies during the life cycle of the plant. For example, ψ_w of dry seed may be –20 MPa or more and is probably near equilibrium with the relative humidity of the surrounding atmosphere.

After the seeds are planted in soil, they imbibe water primarily due to large gradients in ψ_m. This results in a rehydration of the seed tissues, which causes swelling of the tissues and rupturing of the seed coat. Imbibition continues until ψ_w approaches that of the surrounding soil (e.g., –10 kPa). Germination is completed when the radicle emerges from the seed coat. As the seedling develops, ψ_w of the conducting tissue may become positive, up to 100–200 kPa, through active adsorption, particularly with roots in a high soil water environment. In this case, the shoot may exude nearly pure water while still in the soil, causing the soil microenvironment adjacent to the plant to become wet. Guttation is also common on seedlings after emergence.

The ψ_w values of most plants are highest when the plants are young and tend to decrease with age. Early crop growth usually coincides with high soil water content and lower evaporative demand, and both conditions favor high plant water potentials. In addition, young plants tend to have fewer transpiring surfaces and have higher tissue water contents. As the growing season progresses, water stored in the soil profile is lowered, and water may be available only from deeper depths in the soil, where the rooting density is less. Often coupled with this are increases in evaporative demand and the transpirational capacity, all of which lower the ψ_w. As the plant cells mature, their walls thicken and the amount of dry matter increases so that the percentage of water in the plant decreases. With the beginning of senescence, the ψ_w of the plant tissue increases again and continues to do so until reversed by death and desiccation (i.e., complete equilibration with the atmosphere). Presumably, the *increase* in ψ_w at senescence results from a *decrease* in solute content of the cells due to translocation to other plant parts.

The response of plants during their life cycle to variations in water is reflected in the temporal variability of ψ_w, which in turn responds to the temporal supplies of and demands for water. Plant response to the supply of and demand for water is related to the availability of soil water to plant roots, the energy load and water vapor diffusion gradients between the leaf and the surrounding microclimate, and the resistance at the leaf surface, which is the primary control over water loss.

The Dynamics of the Plant Water Balance

Since nearly all land plants undergo temporal changes in water status during their life cycle, the manner in which they endure these changes in water determines their produc-

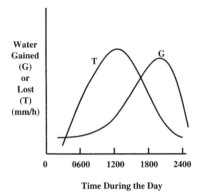

Figure 12.5. General relationship between the water gained or lost by crop plants and time during the day.

tivity and ability to survive. Measurement techniques to quantify plant water status and response to variations in soil water are described below.

Plant Water Balance

A qualitative model of plant water status may be expressed as a simple mass balance equation:

$$W \approx (G - T) + M \tag{9}$$

where W is the water status of the plant; G is the water gained by absorption, principally through the roots; T is the water lost, primarily by transpiration through the leaves; and M is the water stored within the plant. Since the amount of water stored is small in most herbaceous plants, the plant water status is generally proportional to the difference between water gained and water lost (Figure 12.5).

During the early morning hours, the radiant energy load onto the leaves increases and water vapor diffuses outward, mostly from the stomata. The leaf water potential, ψ_L, decreases in response to the internal deficits of water. Since uptake of soil water by roots, G, lags behind T during the morning, T normally exceeds G, resulting in a progressive decrease in W. This difference is termed *water deficit* or *drought stress*.

During the latter afternoon hours, there is less transpiration from the leaves due to the reduction in radiant energy load and, perhaps, to partial closure of the stomata. During this time, uptake of soil water by the roots is normally greater than transpiration. Therefore, the partial stomatal closure and reduced atmospheric demand for water retards T and thereby allows residual hydraulic gradients within the plant to equilibrate and G exceeds T.

During the night, the rates of absorption are greater than rates of transpiration, which results in a reduction or elimination of the water deficit. The plant tissues rehydrate during the night if sufficient water is supplied by the roots. Thus, if soil water is not limiting on a daily basis, $G \approx T$.

Transpiration may be considered a two-stage process: (1) evaporation of water from the moist cell walls into the stomatal air space and (2) the diffusion of water vapor from the stomatal space into the atmosphere (Hopkins 1995). Once the water vapor has left the cell surfaces, it diffuses through the stomatal space and exits the leaf through the stomatal pore (Figure 12.4). The driving force for diffusive transport is the vapor pressure gradient: the substomatal air space of the leaf is very nearly saturated with water vapor, and the atmosphere surrounding the leaf is usually unsaturated. Resistance to transport is encountered by the vapor molecules as they diffuse through the intercellular spaces. Resistance is also caused by a layer of undisturbed air on the surface of the leaf, which is called the boundary layer and has a thickness that is a function of wind speed. Therefore, transpiration is largely controlled by the aerial environment and by the plant. Parameters that affect transpiration include solar radiation, temperature, humidity, wind speed, leaf structure, and stomatal opening.

Absorption of water by the roots is controlled by (1) plant factors such as the rate of transpiration, (2) the size, distribution, and activity of the root system, and (3) soil factors such as soil temperature, soil water matric potential, and soil aeration.

Recurrent temporary wilting of leaves in the afternoon because of excessive transpiration is not serious in well-watered soils because the leaves usually recover turgidity at night. Wilting becomes serious when the soil profile begins to dry, because the leaves are increasingly less likely to recover turgidity at night, and permanent wilting often results.

As a result of the changes in plant water content, turgidity, and water potential, both diurnal and seasonal shrinkage and expansion of plants occur. Most plants show marked shrinkages in stem diameter in the afternoon and expansions of stem diameter at night. Similar results, but to a lesser extent, have been shown with plant reproductive tissues.

The transpirational flux density can be defined in terms of a resistance and a driving force, that is, a form of Darcy's law. Mathematically, this can be presented as

$$T = -\frac{\psi_L - \psi_s}{R_s + R_r + R_x + R_L} = -\frac{\psi_L - \psi_s}{R_p} \tag{10}$$

where T is the transpirational flux density (i.e., water loss from unit soil area per unit time; cm/d), ψ_L is the leaf water potential, ψ_s is the soil water potential (kPa), and R_p is the resistance to flow (kPa·d/cm) along the path from the soil (s), root (r), xylem (x), and finally to the leaf (L). As equation [10] indicates, the water flow rate from soil through the roots to the leaves is roughly proportional to the difference between the water potential of the plant leaves and that of the soil.

The water potential difference between any two points in the SPAC is simply the total water potential difference multiplied by the fraction of the total resistance to flow between the two points. Thus, determining ψ_w at any point along the transpiration stream involves measuring soil and leaf water potentials and resistances to water flow in each pathway of the system.

Assuming that the plants are well watered and the stomata are open, we can approximate the transpirational flux by

$$T \approx \frac{\text{atmospheric demand}}{\text{leaf resistance}} \approx \psi_L$$

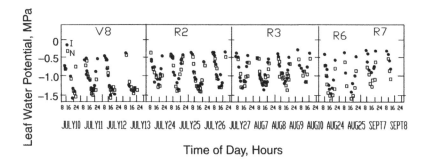

Figure 12.6. Diurnal leaf water potential of Forrest soybeans grown under irrigated (●) and nonirrigated (□) conditions on a Crowley silt loam. (After Jung and Scott 1980.)

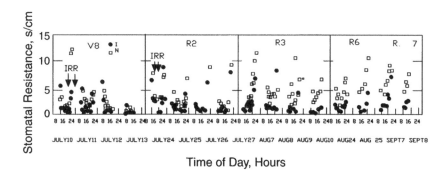

Figure 12.7. Diurnal stomatal resistance of Forrest soybeans grown under irrigated (●) and nonirrigated (□) conditions on a Crowley silt loam. (After Jung and Scott 1980.)

This shows that when soil water availability is adequate, diurnal fluctuations in ψ_L should reflect the diurnal pattern of the atmospheric demand, which is a function of radiation, vapor pressure, and humidity. Therefore, under well-watered conditions, ψ_L serves as a sensitive integrator of simultaneous soil, atmospheric, and physiological effects on water flow.

An example of the seasonal changes of ψ_L and R_L for irrigated (I) and nonirrigated (N) soybeans grown in the field on a DeWitt silt loam at three growth stages is shown in Figures 12.6–12.9 (Jung and Scott 1980). These results indicate the expected diurnal shifts in these plant parameters and also the seasonal trends. The maximum ψ_L occurred at predawn and for the season averaged –420 and –520 kPa for the I and N soybeans, respectively. The minimum ψ_L occurred at midday and averaged –1160 and –1290 kPa for the I and N soybeans, respectively. As soil water became limiting during the growing season, the diurnal curves for I and N soybeans assumed different characteristics. As

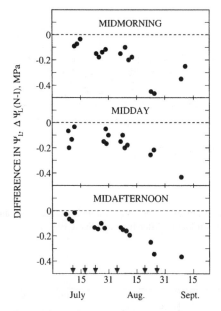

Figure 12.8. Seasonal changes in the differences in leaf water potential between irrigated (I) and nonirrigated (N) soybeans at three times during the day grown on a DeWitt silt loam. (After Jung and Scott 1980.)

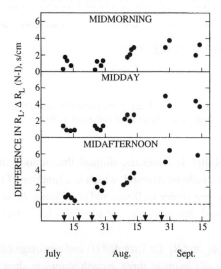

Figure 12.9. Seasonal changes in the differences in stomatal resistance between the nonirrigated (N) and irrigated (I) soybeans at three times during the day grown on a DeWitt silt loam. (After Jung and Scott 1980.)

the drought intensified, the curves of ψ_L for the N soybeans decreased earlier in the day and increased later in the afternoon than for the I soybeans (Figure 12.6). The stomata of N soybeans were partially closed during the daylight hours and tended to close earlier in the afternoon than those of the I soybeans (Figure 12.7).

Also, as the season progressed and drought intensified, differences in ψ_L and R_L between the N and I soybeans increased (Figures 12.8 and 12.9). Maximum seasonal differences during the day were approximately $\Delta\psi_L = -400$ kPa and $\Delta R_L = 6.0$ s/cm and were usually found at midday and/or midafternoon late in the growing season. The cumulative effects of the differences in these plant water parameters over the season were reflected in lower total dry matter and seed yield of the N soybeans.

Daily Cycle of Plant Water Components

Quantifying soil-plant-water relationships requires simultaneous measurements of atmospheric, plant, and soil factors because these dynamic factors work together to influence plant growth and water use. Ritchie (1974) summarized the diurnal cycle of various factors influencing the plant water balance under conditions of variations in evaporative demand and soil water potential. The diurnal relationships of several plant water parameters are shown in Figures 12.10 and 12.11.

The curves in Figure 12.10 represent two successive days when ψ_m of soil in the root zone is approximately -50 kPa. This is a relatively high soil water matric potential and should cause few significant drought stresses. Ritchie (1974) summarized the scenarios as follows:

On the first day, atmospheric demand was assumed to be 7 mm of water/day. Assuming soil evaporation of water (E_s) is small and a well-developed canopy exists, T begins soon after sunup and generally follows the trend of the solar radiation flux density throughout the day. Values of ψ_L decline rapidly as transpiration (T) increases and the resulting differences in ψ between the leaf and the soil are proportional to the transpiration rate. At the beginning of the day, leaf water potential (ψ_L) and soil water matric potential (ψ_m) are approximately equal. The possible diurnal variations in ψ_w at the root-soil interface are also indicated (not verified). The larger water potential differences between the leaves and the roots than between the soil and the roots indicate that the resistance to water flow is greater through the plant organs than through the soil to the root surfaces. Leaf resistance values, R_L, represent the harmonic mean of the stomatal resistance of both sides of a leaf and are typical for many plants not experiencing severe water deficits. They are lowered rapidly after light intensity exceeds a certain critical value and remain low throughout the day until light levels again become low and the stomata close. The relative water content is almost 100% before transpiration begins and decreases gradually in response to the lag between transpiration and adsorption of water.

On the second day, the atmospheric demand of 3 mm was lower than the previous day. Values of ψ_s changed little from the previous day because of the large storage volume of water in the soil profile. The lower evaporative demand during the day results in significantly different ψ_L and ψ_r (root water potential) patterns from the previous day. Because T is low, the minimum ψ_L is only about half the minimum value for the previous day. The leaf resistance values are about the same, indicating open stomata on this day too.

Figure 12.11 shows two successive days of high and low atmospheric demand similar to the days in Figure 12.10 except that ψ_m in the effective root zone has now been lowered to about -800 kPa. When T starts on the day with high evaporation potential (Figure 12.11a), ψ_L decreases rapidly and causes a severe water deficit in the plant leaves. The

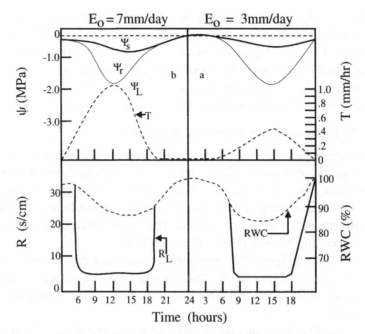

Figure 12.10. Daily cycle of various plant water status factors affecting the plant water balance when the root zone water potential is –50 kPa. (Adapted from Ritchie 1974.)

leaf stomata respond to this severe deficit by partially closing, as indicated by the relatively high leaf resistance values of about 10 cm/s. Because of the partially closed stomata, T is considerably less than when the soil water matric potential was higher. The relative water content of the plant leaves decreases more than for plants growing in wet soil, and the plant never regains full turgidity during the following night.

On the next day, which has a relatively low potential evaporation and a dry soil, another pattern develops (Figure 12.11b). Because of low evaporative demand, the potential gradient between the leaf and the soil is less, but ψ_L still falls below a critical level and the stomata partially close. Values of T for this day are expected to be about the same as for the day with a low evaporative demand when soil water is plentiful (Figure 12.10b). The leaf resistance, R_L, for the day is considerably greater but the atmospheric demand is low enough to become the primary factor limiting T, even though the stomata are partially closed.

Illustration of the soil-water potential distribution patterns for relatively wet and dry soil and the simultaneous daily root-extraction rate patterns expected for the same four days examined previously was given by Ritchie (1974) and are presented in Figure 12.12.

In Figure 12.12 (top) values of ψ_m are typical of those existing in the profile a few days after a sizable rain or irrigation. The ψ_m near the soil surface is low because of evaporative drying and a relatively high root density near the soil surface. The ψ_m is fairly uniform throughout the remainder of the profile.

An attempt to assign an average or "integrated" matric potential for such a ψ_w distribution in the soil profile is difficult. The low ψ_m near the soil surface does not greatly

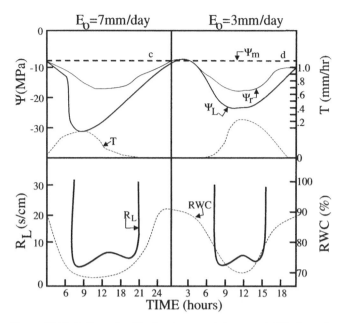

Figure 12.11. Daily cycle of leaf water potential, stomatal diffusive resistance, transpiration, and relative water content for two consecutive days when the root zone water potential is − 800 kPa. (Adapted from Ritchie 1974.)

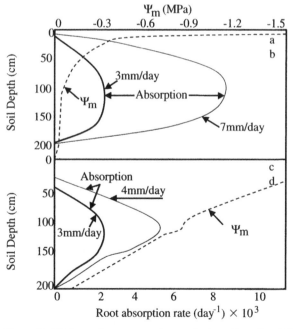

Figure 12.12. Extraction patterns of water in the soil profile. (Adapted after Ritchie 1974.)

influence the root extraction of water. The high ψ_m deep in the profile has little influence on the root extraction of water because of low root densities. If root extraction patterns are known, the determination of the approximate soil ψ_m influencing the plant water status becomes a simpler problem. The best single ψ_m value should be weighted in proportion to the amount of root absorption at each depth.

In Figure 12.12, the maximum extraction rate for the day with a high evaporative demand (i.e., 7 mm/d) was in the 50–125 cm soil depth interval. The water extraction rate pattern was about the same on the next day, which had a lower transpiration rate of 3 mm/d, except that the actual extraction rate was less than half that of the day before. The ψ_s weighted for root extraction distribution with depth for the wet soil was approximately –50 kPa.

In Figure 12.12 (bottom) the soil water matric potential, ψ_m, distribution shown indicates that the soil has dried to below –1500 kPa from the surface to the 25 cm depth. The ψ_m then decreases rather uniformly with depth down to the lower part of the root zone. In this scenario, the ψ_m weighted with depth is approximately –800 kPa. The maximum extraction rate is deeper in the profile, centered at the 100 cm depth. Water could not be supplied to the roots rapidly enough on the high potential evaporation day because of the limited root density below 100 cm. This reduced rate of extraction of soil water caused the extremely low plant ψ_L noted previously and resulted in stomatal regulation of T on that day.

Except when soil water deficits are severe, ψ_w in the early morning hours is approximately equal to the ψ_m influencing root water extraction. Therefore, measurements of ψ_L are important indicators of plant water stress.

Movement of Soil Water to Roots and Extraction by Roots

Water movement from soil to plant roots is not well understood because of the complexity of the plant root system and the difficulty of measuring changes in the parameters of the SPAC over small distances. However, it is well known that plant species differ greatly in the spatial distribution and size of their root systems and in their water extraction patterns (Taylor and Klepper 1978). The rooting pattern of each species is genetically controlled and is modified by the soil physical and chemical characteristics in the profile. Two approaches to measuring water flow to plant roots have been considered: water uptake by a root system and water uptake by individual roots.

Macroscopic Root System Models

Root system models are classified as macroscopic models in which the uptake of soil water is represented by a distributed source function. The flow of water to individual roots is ignored, and the overall root system is assumed to extract water from each differential volume of the root zone at a rate r(z, t, θ_v). Thus, geometrical complications involved in analyzing the distributions of water fluxes and potential energy gradients on a microscopic scale are avoided. The extraction rate of water by the roots depends on position in the soil profile, z; the volumetric soil water content, θ_v; and time during the growing season, t. Boundary conditions of the composite soil-plant system are specified at the soil surface and at the bottom of the root zone or the water table, if present.

The extraction of water by plant roots can be determined by adding to the continuity equation the volumetric sink term. In one dimension, this leads to

$$\frac{\partial \theta_v}{\partial t} = -\frac{\partial q}{\partial z} - r(z, t, \theta_v)$$

[11]

where θ_v is the volumetric water content of the soil (m^3 water/m^3 soil), t is the time (d), z is the vertical distance from the reference level (m), q is the soil water flux density (m^3 water/m^2 soil·d), and r is the volumetric rate of water extraction by the roots per unit volume of soil (m^3 water/m^3 soil·d). The root extraction term has a negative sign because it is a sink term. In essence, it is a function of the root activity. Equation [11] shows that the root extraction term can be quantified if both $\partial \theta_v/\partial t$ and $\partial q/\partial z$ are known. An example of the distribution of the extraction rate of water by plant roots in a soil profile was given by Ritchie (1974) and is presented in Figure 12.12.

In the vertical dimension, substitution of the general differential equation governing water flow in the soil profile into equation [11] gives

$$\frac{\partial \theta_v}{\partial t} = \left(\frac{\partial}{\partial z} K(\theta_v) \frac{\partial H}{\partial z} \right) - r(z, t, \theta_v)$$

[12]

where $K(\theta_v)$ is the hydraulic conductivity (m/s), which is a function of the volumetric soil water content.

The transpiration, T, can be calculated by integrating the above equation over z and taking the difference between measured total flow and calculated flow of soil water:

$$\int_0^{z_r} r(z, t, \theta_v) dz \approx T$$

[13]

where z_r is the depth to the bottom of the root zone. This equation shows that T is approximately equal to the cumulative root extraction of water in the soil profile. Under conditions of low evaporation from the soil, this quantity also is approximately the actual evapotranspiration (AET) rate since most of the water extracted is transpired and very little is retained by the plant (Rice 1975).

Frequently, information on the root extraction characteristics in a soil profile can be gained by assuming that in equation [11] the flux term in the root zone, $\partial q/\partial z$, is zero. Under these conditions, the temporal change in water content with time, $\partial \theta_v/\partial t$, is equal to the root extraction term $-r(z, t, \theta_v)$. As a result, soil water depletion patterns are created in the profile, with shapes that depend on the climate, soil physical and chemical properties, and water extraction rates by the roots. An example is given in Figure 12.13, which shows changes in volumetric soil water content for soybeans grown in the field on a Sharkey clay loam (Scott et al. 1998). Measurement of the volumetric water contents in the Sharkey profile over time results in the determination of the depletion rates and serves as an indication of root extraction activity.

In general, plant growth habit also influences the rooting pattern. In perennial plants, the root system may be extensive before drought begins, so a large soil water deficit may be developed before T is appreciably affected. In contrast, the root system of annual crops often ceases to develop appreciably after the initiation of reproductive growth, so the crop

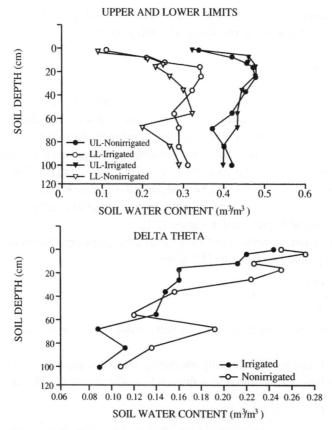

Figure 12.13. Changes in volumetric soil water content beneath a crop of soybeans growing in a Sharkey clay loam. (Adapted from Scott et al. 1998.)

does not have much time to develop a deep root system. This often results in considerable amounts of water remaining in the soil profile after the crop has been harvested.

Use of simulation models of water movement in the field often requires a mathematical function for the root extraction term, r. The problem comes down to postulating the form of the root extraction model for $r(z, t, \theta_v)$. Several have been proposed and three models are briefly discussed below. These macroscopic models were developed to describe the extraction of soil water from the root zone as a whole without considering explicitly the effects of individual roots.

Model 1: Molz (1971) proposed the root extraction term

$$r(z, t, \theta_v) = T \frac{D(\theta_v) R(z, t)}{\int D(\theta_v)R(z, t)dz} \qquad [14]$$

where T is the volumetric transpiration rate per unit soil surface area, D is the soil water diffusivity, and R(z, t) is the effective root distribution (proportional to the length of roots per unit volume of soil that are "actively" absorbing water).

Model 2: Nimah and Hanks (1973) defined the root extraction term as

$$r(z, t) = [H_{root} + RRES(z) - \psi_m(z, t) - \psi_o(z, t)]\, RDF(z) \left(\frac{K(\theta_v)}{\Delta x \Delta z} \right) \qquad [15]$$

where H_{root} is an effective water potential in the root at the soil surface, where z is considered zero; and RRES is a root resistance term equal to $1 + Rc$, where Rc is a flow coefficient in the plant root system to account for longitudinal resistance in the xylem and is assumed to be 0.05. When RRES is multiplied by z, the product accounts for the gravity term and friction loss in the root water potential. The quantity $\psi_m(z, t)$ is the soil water matric head, $\psi_o(z, t)$ is the osmotic potential or head, RDF(z) is the proportion of total active roots in depth increment Δz, $K(\theta_v)$ is the hydraulic conductivity at depth z, Δx is the distance between the plant roots at the point in the soil where $\psi_m(z, t)$ and $\psi_o(z, t)$ are measured (usually assumed to be 1.0), and Δz is the depth increment of soil.

Model 3: Feddes et al. (1974) proposed the following model:

$$r = \frac{-K(h_r - h)}{b} \qquad [16]$$

where r is the volumetric rate of water uptake per unit volume of soil (day^{-1}), h_r is the pressure head (m) of the root-soil interface, h is the pressure head (cm) in the soil, and b is the empirical coefficient of proportionality (m^2) that represents the geometry of the flow and is thought to be proportional to the specific area of the soil-root interface (total surface area of roots per unit volume of soil) and inversely proportional to the impedance (thickness divided by the hydraulic conductivity) of the root soil-interface. It is similar to RDF(z)/($\Delta x \Delta z$) of Nimah and Hanks (1973).

The primary disadvantage of these macroscopic models for root extraction of soil water is that they utilize a gross spatial average of ψ_m and ψ_o and do not account for the decrease in water potential and the changes in concentration of salts at the soil-root interface (Rendig and Taylor 1989).

Single-Root Models

Single-root models are microscopic models in which the root system is conceptually represented as an array of parallel cylindrical sinks withdrawing water from soil cylinders each with an outer radius equal to one-half the mean distance between roots. The plant root is often viewed as a hollow, elongated cylinder of uniform radius and infinite length with uniform water-absorbing properties. Flow of water is two-dimensional and largely radial. In this case, the root extraction term does not appear and the governing Darcy flow equation, in cylindrical coordinates, can be written as

$$\frac{\partial \theta_v}{\partial t} = \frac{1}{r} \frac{\partial}{\partial r} \left(\frac{r D(\theta_v) \partial \theta_v}{\partial r} \right) \qquad [17]$$

where θ_v is the volumetric soil water content (m^3/m^3), t is the time(s), r is the radial distance from the center of the root (m), and $D(\theta_v)$ is the soil water diffusivity (m^2/s). Equation [17] is a nonlinear, partial differential equation.

For steady-state conditions ($\partial\theta_v/\partial t = 0$) with soil water flowing from a distance r_{soil} (m) to a root with radius r_{root} (m), the analytical solution with the assumption of constant K is

$$q_r = -2\pi K \frac{h_r - h_s}{\ln(r_{soil} / r_{root})} \qquad [18]$$

where q_r is the rate of water uptake per length of root (m^3 of water/m of root/d), K is the unsaturated hydraulic conductivity (m/d), h_r is the matric potential of the water at the root surface (m), h_s is the matric potential of the soil water at a distance surrounding the root, r_{root} is the radius of the root, and r_{soil} is the radius of the cylinder of soil surrounding the root through which water is moving (m) and is a function of rooting density, L_v, according to

$$r_{soil} = \frac{1}{\sqrt{(\pi L_v)}} \qquad [19]$$

L_v is the root length (m) divided by the volume of soil (m^3) and has units of meters per cubic meter. Equation [18] shows that water uptake per unit length of root is assumed to be proportional to the hydraulic conductivity of the soil and to the difference in hydraulic head in the plant and that in the soil. Note that the ratio of the radius of influence, r_{soil}, to the radius of the root, r_{root}, enters the equation only as a logarithmic term, and therefore, root radius has less effect on the potential difference than does the flux density of water or the initial soil water matric potential, ψ_m.

Gardner (1960) showed that with these assumptions the single-root model predicts that there is little water potential gradient in the vicinity of the roots until the soil becomes dry. In a dry soil, the water potential at the plant root surface will have to be much lower to maintain a given q than in an initially moist soil, which has a water potential that is much higher. It has been recognized that values of K less than 10^{-8} m/d seriously impede water movement from soil to roots.

In a numerical simulation study, McCoy et al. (1984) found that the radial ψ_m distribution around individual roots depends on $D(\theta)$, which decreases as the soil dries, creating steep ψ_m gradients adjacent to the roots. Recall that $D(\theta)$ is the ratio of $K(\theta)$ to the specific water capacity $C(\theta)$. Therefore, $D(\theta)$ combines the resistance to water flow in the soil and the ability of the soil to release water. The relative rate of decrease of ψ_m around the root varies with time as the soil dries. The drawdown around the root is reduced by increasing root density. At low rooting density, the root surface potential depends more closely on $D(\theta)$, whereas at high rooting densities, it depends more closely on $C(\theta)$. In addition, roots of smaller radii exhibit greater drawdown as the total root surface area decreases and the flux of water at the root surface increases.

The water extracted by the plant roots in a specific soil volume is the product of q_r and L_v. The transpiration rate, T, assumed to be equal to the extraction rate of water by roots in the profile, is calculated by summing the extraction of water in all soil horizons from the deepest, i, to the soil surface, n, by using the following equation:

$$T = \sum_i^n q_{ri} L_{vi} \qquad [20]$$

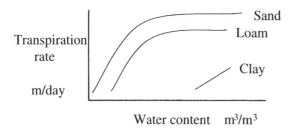

Figure 12.14. The relationship between transpiration rate and the percent soil water content. (Adapted from Hsiao and Acevedo 1974.)

One major disadvantage of the single-root model is the insufficient quantitative knowledge of the appropriate value for the hydraulic conductivity, K. The general consensus seems to be that the best K to use is a "rhizosphere" K and not strictly a soil K.

Drought-Stressed Plants

In dry conditions, the soil can limit the rate of water movement to plant roots just as it does to an evaporating surface. Transpiration from wilted plants appears to be limited by soil hydraulic properties and by root distribution. The stomata open just wide enough and often enough to permit water loss only as fast as it is obtained. Transpiration rate decreases with decreasing soil water content and/or soil water potential below a threshold water content (Figure 12.14). The nature of the transpiration rate over much of the range in soil water content is linear; however, there are differences among soil textural classes.

As the initially wet soil profile dries, there is a tendency for water to first be extracted near the soil surface, where the greatest root density occurs. However, as the plant ages, the zone of maximum extraction rate tends to occur deeper in the soil profile. For many deep well-drained soils, soil water may be extracted below the measured root zone, which can be attributed to the difficulty in measuring low root densities.

For conditions where the initially dry root zone has been partially wetted by rainfall and/or irrigation, T may not be limited for a few days. During this time, extraction of soil water occurs primarily near the surface due to higher root densities in this region of the soil profile and to the higher soil water content, which results in lower resistance to flow of water.

Effects of Drought Stress on Plant Productivity

Because water is lost through the stomata at the same time as CO_2 is taken in for use in photosynthesis, water loss through T is an inevitable cost of dry-matter production. Under field conditions, drought stress usually does not develop suddenly but increases gradually (changes requiring several hours are gradual relative to molecular events in the cell). Therefore, the physiological process most sensitive to drought stress is normally altered first, and such alterations, in turn, may lead to many secondary and tertiary changes. The order of sensitivity to drought stress is given below (in decreasing order):

- cell growth
- wall synthesis and protein synthesis
- stomatal opening and CO_2 assimilation
- respiration
- proline and sugar accumulation

One of the early macroscopic effects of drought is a reduction in vegetative growth, a consequence of a reduction in photosynthesis from the closure of the stomata. Cell growth (leaf extension) is defined as the irreversible enlargement or expansion of cells. It is the plant process most sensitive to water stress. After cell division and when the cell is metabolically prepared to expand, an internal hydrostatic pressure, turgor pressure, is necessary for the final expansion process, acting as a push from inside.

Shoot growth, especially the growth of leaves, is generally more sensitive to drought than root growth (Hopkins 1995). Many mature plants subjected to prolonged drought stress respond by rolling or curling the leaves and by accelerated senescence and abscission of the older leaves. These responses reduce exposure of leaf area and transpiration during times of limited water availability.

Definition of Water Use Efficiency

One of the most frequently used methods to determine the macroscopic effects of drought on plant growth and productivity is water use efficiency (WUE). Crops with high productivity and low water use are particularly advantageous in areas where water is scarce. WUE has been defined in several ways, but probably the most frequently used definitions involve the ratio of some aspect of the production of the crop to the water used. WUE may be defined as follows:

WUE = dry matter produced (DM)/actual evapotranspiration (AET)   [21]

WUE = seed yield (SY)/AET   [22]

The SI units of WUE are kilograms per hectare of dry matter (or marketable yield) produced per equivalent depth of water used (m, cm, mm). Other definitions of WUE include the following:

- tons of dry weight produced per acre-inch of water evapotranspired
- pounds of marketable crop per acre-inch of consumptive use (or AET)
- kilograms of dry weight or yield per kilogram of AET
- kilograms of dry weight or yield per centimeter of AET

The numerator of equations [21] and [22] represents the utilization of radiation in the absorption spectra of chlorophyll and other pigments of photosynthesis minus respiration and losses to pests. To some extent, the plant or crop productivity can be changed by management practices. The many years of effort devoted to increasing crop yields—for example, by selection and breeding of higher-yielding cultivars and hybrids, row spacing and plant density studies of crop spacing in the row and between rows, correction of nutrient deficiencies, irrigation, and protection of the crop against diseases and pests—have increased the numerator of equations [21] and [22] without changing or only slightly increasing the denominator, or AET portion, of the equation (Viets 1966).

For example, one of the primary management techniques of obtaining more efficient water use is the elimination of weeds in crops. Weeds compete with crops for soil water, light, nutrients, oxygen, and space. Competition for water begins when the root systems of the weed and crop overlap (Geddes et al. 1979). Competition for light is primarily important when the weed species are as tall as or taller than the crop, the weed density is high, and shading of the crop canopy occurs.

The denominator of equations [21] and [22] depends on the same three factors governing evaporation. These include energy available for ET (most of which comes from solar and longwave sky radiation), a vapor pressure deficit, and a supply of water to the evaporating surface. In general, AET increases when the air is dry, warm, and moving and when the surface is wet; AET decreases when the opposite conditions occur; that is, air is motionless with a high relative humidity, temperatures are cool, and the soil is dry. When plants are grown in the field, evaporation of water from soil surfaces usually occurs simultaneously with transpiration from plant surfaces.

Management practices to increase storage of water in the soil profile have tended to increase the yield of crops, particularly the marketable portion, more than they have AET. To a considerable extent, this dependence of the numerator of equations [21] and [22] on factors under the grower's control, and the dependence of the denominator on the physical environment, have been fortunate in that most of the agricultural management information and practices developed through the years for obtaining high crop yields have led automatically to improved WUE. This demonstrates that both soil and crop management practices affect the efficiency of energy interception in photosynthesis, the distribution of heat between the soil and the crop, and the distribution of water loss between evaporation from the soil and transpiration from plants. Therefore, all three components of the SPAC must be considered.

To illustrate that WUE varies with environmental conditions and can be changed, Viets (1966) related AET and WUE under six possible conditions of plant growth, assuming there is "readily available" water (i.e., soil water is nonlimiting). These models are shown in Figure 12.15, where WUE is the ratio of Y to AET.

The seasonal AET and the yield of the total or marketable crop are an integration of the many factors in the SPAC, such as (1) the amount of vegetative cover, (2) the interception of energy by the foliage so it does not reach the soil surface to evaporate water, (3) the wetness of the soil surface, and (4) the availability of water at critical stages of growth.

The actual WUE varies with climate, soil, and crop factors, but between 200 and 800 kg of water is used to produce 1 kg of dry matter. This indicates that the mass (and volume) of water required to produce a crop is several hundred times its dry weight. For the most part, this water has to be extracted from the soil profile.

Example

Calculate the WUE and the amount of water used in a soybean field if the seasonal AET was 497 mm and the dry matter produced was 9492 kg/ha. Convert this WUE to kilograms of water used per kilogram of dry matter produced.

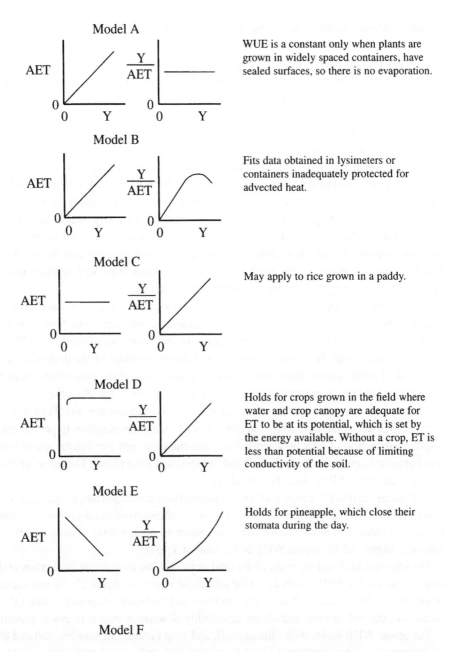

Model A

WUE is a constant only when plants are grown in widely spaced containers, have sealed surfaces, so there is no evaporation.

Model B

Fits data obtained in lysimeters or containers inadequately protected for advected heat.

Model C

May apply to rice grown in a paddy.

Model D

Holds for crops grown in the field where water and crop canopy are adequate for ET to be at its potential, which is set by the energy available. Without a crop, ET is less than potential because of limiting conductivity of the soil.

Model E

Holds for pineapple, which close their stomata during the day.

Model F

Applies where increasing the availability of water increases ET, but has little effect or depresses yield.

Figure 12.15. Relationships between actual evapotranspiration (AET) and crop yield (Y) under various scenarios (Adapted from Viets 1966).

WUE = (9492 kg/ha)/49.7 cm = 191 kg of water/ha cm
volume = (0.497 m of water)(10,000 m^2/ha)
$\quad\quad$ = 4970 m^3 of water/ha

Conversion to mass basis:

(4970 m^3)(1000 L/m^3)(1.0 kg/L) = 4.97 × 10^6 kg of water/ha
(4.97 × 10^6 kg of water/ha)/(9492 kg of dry matter/ha)
$\quad\quad$ = 523.6 kg of water lost/kg of dry matter produced

Therefore, almost 5000 m^3 of water was used by 1 ha of soybeans during the growing season.

Example

Calculate the equivalent depth of water needed to grow the soybeans given in the example above.

(4970 m^3/ha)(1 ha/2.47 acre)(0.009729 acre-inch/m^3)
$\quad\quad$ = 19.6 in = 49.7 cm

This depth is the water that must come from soil storage, rainfall, and irrigation during the growing season.

Crop Yield and Transpiration

For many climates, the relationship between biomass production (i.e., crop yield) and transpiration can be empirically related by the equations

$$Y = nT \quad\quad\quad\quad [23]$$

$$Y = m(T/Tmax) \quad\quad\quad\quad [24]$$

where Y is the crop yield (such as the total dry matter per area; kg/ha), T is the cumulative transpiration (cm), and Tmax is the mean daily pan evaporation for the same period (mm/d). The parameter n is the transpiration ratio and has SI units of kg/ha·cm. The parameter m is a crop growth rate coefficient (kg/ha·d) that depends on plant species and the soil-plant system. The close relationship between transpiration and biomass produced occurs because the stomata simultaneously control the inward flux of CO_2 and the outward flux of water vapor from the leaf surface. In addition, the same radiation that drives photosynthesis also supplies energy to evaporate the water during transpiration. Plants that have low transpiration ratios are more efficient users of water; that is, they require lower amounts of water to produce the same amount of dry matter. Thus, managing water and soil to conserve water produces the highest economic yield and leads to the optimum WUE.

Example

Scott et al. (1987) reported that over a 5-year period, the seasonal average transpiration was 327 and 231 mm for irrigated and nonirrigated soybeans grown in the field, respectively. The average seed yields were 2707 and 1884 kg/ha for the irrigated and nonirrigated soybeans, respectively. Calculate the transpiration ratio for each water management treatment.

n = seed yield/cumulative transpiration

For the irrigated soybeans,

n = (2707 kg/ha)/(32.7 cm of water) = 82.8 kg/ha·cm

For the nonirrigated soybeans,

n = (1884 kg/ha)/(23.1 cm of water) = 81.6 kg/ha·cm

Since the transpiration ratio of both water management treatments was similar, this indicates that about the same amount of transpiration was required to produce the soybean seed. The irrigated soybeans, however, transpired more water.

The Plant-Available Water Capacity of a Soil

For any given well-drained soil there appears to be an upper limit (UL) and a lower limit (LL) of water in the soil profile that is available for plant root extraction. Originally, these limits were translated roughly into the concepts of "field capacity" and "permanent wilting point," and the difference between these limits was designated the "available water content."

The *plant-available water capacity (PAWC)* is a soil characteristic related to both soil and plant response. It is the maximum quantity of water that can be extracted from a given soil profile by a given plant (or crop) during the growing season. PAWC has SI units of meters. The concept of PAWC, however, has changed over the years. Initially, Veihmeyer (1927) defined it as the amount of water held by a soil between field capacity and its permanent wilting point. He found that the yields of many irrigated fruit orchards in central California were little affected by drought if the top 1.8 m of soil was allowed to dry to its permanent wilting point before the next irrigation was applied. This concept of available water capacity has proved useful in practice only under limited conditions, and then only for soils that hold most of their available water at matric potentials close to field capacity. These are typically coarse-textured soils. However, since fine sandy or silt loams, which commonly occur on alluvial river terraces, constitute a large proportion of all irrigated lands of the world, this definition has never been useful for irrigated fine-textured soils.

Problems occur in calculating the magnitude of the PAWC in a particular soil. The main difficulty is in identifying the upper limit of PAWC, sometimes called the field capacity, which indicates the amount of water held in the soil after the excess water has drained away and after the rate of downward movement of water has materially decreased. The difficulty posed by this definition is the interpretation of the time when the downward movement has "materially decreased." Initially, a time of 2 or 3 days after an irrigation or heavy rain was chosen. However, many fine-textured soils continue to drain for weeks rather than days.

As a practical means of overcoming this difficulty, the soil water content at field capacity was defined in terms of matric potential. Here again, there is considerable variability in the matric potential to be used among soils and within soil profiles. Colman (1947) found a matric potential of −33 kPa to be an acceptable value for his soils, while Haise (1955) found that −10 kPa was more appropriate. In coarse-textured soils, the matric potential in the surface horizons of well-drained soils tends to be somewhat higher, typically about −5 kPa.

The LL of available water, the permanent wilting point, was first defined by Veihmeyer and Hendrickson (1928) as the water content of the soil when the leaves of plants growing in the soil first reached a stage of wilting from which they did not recover when placed in a saturated atmosphere without the addition of water to the soil. This definition incorrectly assumed that all plants behave in the same way in all soils. Usually −1.5 MPa is the standard matric potential chosen for the permanent wilting point. However, Savage et al. (1996) showed that −1.5 MPa was probably somewhat high; they measured values with carefully calibrated thermocouple hygrometers in the range of −2.5 to −3.0 MPa. However, because of the low specific water capacity of the soil water characteristic curve in the −1.5 to −3.0 MPa region, the exact value for the permanent wilting point matters less than that for the field capacity.

Cassel et al. (1983) used regression techniques to model the UL and LL of soil water availability of 61 soil profiles in 15 states. The soils were moderately to well drained without root-limiting restrictions such as indurated layers or toxic elements. Parameters in their statistical model included a complete textural analysis, bulk density, −33 and −1500 kPa water retention, and soil chemical properties such as pH, cation exchange capacity, and percentage of $CaCO_3$. The primary purpose for developing the regression equations was to estimate UL and LL with minimum input data and to use the equations in estimating PAWC.

To avoid the problem of defining the UL and LL of available water in relation to a specific value of matric potential, Ritchie (1981) proposed the use of the concept of *plant-extractable water*. Plant-extractable water was defined as the difference between the highest volumetric water content in the field (after drainage) and the lowest measured volumetric water content when plants are very dry and leaves are either dead or dormant. Extractable water has the advantage of being a field-based measurement, which takes into account the distribution of roots in the soil profile and physical and chemical characteristics of the profile. It is soil (site), crop, and season specific. Therefore, there is a large variability among soils and crops in plant-extractable water. The soils with the greatest volume of PAWC usually have loamy textures containing appreciable quantities of silt.

In the short term, the availability of soil water to a plant is determined by the rate at which water can be conducted through the soil to the root. Hydraulic conductivity decreases dramatically as the matric suction increases, with the result that water held at high matric potentials is more readily available than water held at low matric potentials. Typical water retention curves for three contrasting textures show that a much greater proportion of the available water is held at high matric potentials in loamy sands than in silt loams or clays, and this greater proportional availability at low suctions in the loamy sands counterbalances, to some extent, the advantage of the higher total available water content of the finer-textured loams.

The PAWC of a soil can be increased marginally by increasing the soil's organic-matter content or by increasing the proportion of finer-textured particles in a coarse-textured soil. Reviews have shown that although additions of organic materials to soil may promote aggregation, resulting in increased pore space, the increase in PAWC is small. However, in dry years, the effects may be important in delaying the need for irrigation.

Clearly, the depth of rooting of a crop will influence how much of the water in a soil profile is available to the crop. Crops with deeper root systems tend to extract water from deeper depths in the soil profile. Thus, more of the water in the profile is available for plant uptake.

Concluding Remarks

Water in plants is part of a dynamic hydrologic system that includes the soil and the atmosphere. Plants that are actively photosynthesizing use substantial amounts of water, largely through transpiration from leaf surfaces. Very little of this water is stored within the plant. Since water moves from regions of higher total potential to regions of lower total potential energies, large quantities of water must be extracted from the soil and moved through the plant to satisfy deficiencies that develop in the leaves. For the most part, the flow rates of water from the soil to the leaves of plants can be described by a Darcy-like equation. This flow of water through plants also plays a significant role in the growth and survival of the plant by contributing to temperature regulation; it helps ensure that plants do not cool or warm too rapidly.

The concept of plant-available water and its associated terms remain an inexact part of soil science. Plant-available water is an equilibrium concept and, therefore, is inadequate to describe the rate at which water becomes available over short periods. However, for a rough estimate of the soil water balance for irrigation purposes over long periods, it has distinct and useful advantages.

Finally, it should be obvious that an understanding of the physics of the soil is important for understanding the physics of the plant and atmosphere. The interesting conclusion is that the language used to describe water and heat relations in all three components of the SPAC (i.e., the mathematical equations) is analogous.

Cited References

Barrs, H. D. 1968. Determination of water deficits in plant tissues. In T. T. Kozlowski (ed.), *Water Deficits and Plant Growth,* vol. 1:235–368. Academic Press. New York.

Cassel, D. K., L. F. Ratliff, and J. T. Ritchie. 1983. Models for estimating in-situ potential extractable water using soil physical and chemical properties. *Soil Science Society of America Journal* 47:764–769.

Colman, E. A. 1947. A laboratory procedure for determining the field capacity of soils. *Soil Science* 63:277–283.

Cowan, I. R. 1965. Transport of water in the soil-plant-atmosphere system. *Journal of Applied Ecology* 2:221–239.

Feddes, R. A., E. Bresler, and S. P. Neuman. 1974. Field test of a modified numerical model for water uptake by root systems. *Water Resources Research* 10:1199–1206.

Gardner, W. R. 1960. Dynamic aspects of water availability to plants. *Soil Science* 89:63–73.

Geddes, R. D., H. D. Scott, and L. R. Oliver. 1979. Growth and water use by common cocklebur (*Xanthium pennsylvanicum*) and soybeans (*Glycine max.*) under field conditions. *Weed Science* 27:206–212.

Hopkins, W. G. 1995. *Introduction to Plant Physiology.* 2d ed. John Wiley and Sons. New York.

Hsiao, T. C., and E. Acevedo. 1974. Plant responses to water deficits, water-use efficiency, and drought resistance. *Agricultural Meteorology* 14:59–84.

Jones, H. G. 1983. *Plants and Microclimate.* Cambridge University Press. New York.

Jung, P. K., and H. D. Scott. 1980. Leaf water potential, stomatal resistance and temperature relations in field grown soybeans. *Agronomy Journal* 72:986–990.

_____. 1974. Fifty years of progress in water relations research. *Plant Physiology* 54:463–471.

McCoy, E. L., L. Boersma, M. L. Ungs, and S. Akratanakul. 1984. Toward understanding soil water uptake by plant roots. *Soil Science* 137:69–77.

Molz, F. J. 1971. Interaction of water uptake and root distribution. *Agronomy Journal* 63:608–610.

Nimah, M. N., and R. J. Hanks. 1973. Model for estimating soil water, plant, and atmospheric interrelations: I. Description and sensitivity. *Soil Science Society of America Journal* 37:522–527.

Papendick, R. I., and G. S. Campbell. 1975. Water potential in the rhizosphere and plant and methods of measurement and experimental control. In *Biology and Control of Soil-borne Plant Pathogens.* Third International Symposium on the Factors Determining the Behavior of Plant Pathogens in Soil, pp 139–149. University of Minnesota Press. Minneapolis.

Philip, J. R. 1966. Plant water relations: some physical aspects. *Annual Review of Plant Physiology* 17:245–268.

Rendig, V. V., and H. M. Taylor. 1989. *Principles of Soil-Plant Interrelationships.* McGraw Hill. New York.

Rice, R. C. 1975. Diurnal and seasonal soil water uptake and flux within a bermudagrass root zone. *Soil Science Society of America Proceedings* 39:394–398.

Ritchie, J. T. 1974. Atmospheric and soil water influences on the plant water balance. *Agricultural Meteorology* 14:183–198.

_____. 1981. Soil water availability. *Plant and Soil* 58:327–338.

Savage, M. J., J. T. Ritchie, W. L. Bland, and W. A. Dugas. 1996. Lower limit of soil water availability. *Agronomy Journal* 88:644–651.

Scott, H. D., J. A. Ferguson, and L. S. Wood. 1987. Water use, yield, and dry matter accumulation by determinate soybean grown in a humid region. *Agronomy Journal* 79:870–875.

Scott, H. D., J. A. Ferguson, T. Fugitt, L. Hanson, and E. Smith. 1998. Agricultural water management in the Mississippi Delta region of Arkansas. Arkansas Agricultural Experiment Station Bulletin 959.

Taylor, H. M., and B. Klepper. 1978. The role of rooting characteristics in the supply of water to plants. *Advances in Agronomy* 30:99–128.

Veihmeyer, F. J. 1927. Some factors affecting the irrigation requirements of deciduous orchards. *Hilgardia* 2:125–290.

Veihmeyer, F. J., and A. H. Hendrickson. 1928. The relation of soil moisture to cultivation and plant growth. *Proceedings and Papers of the First International Congress of Soil Science* 3:498–513.

Viets, F. G., Jr. 1966. Increasing water use efficiency by soil management. In W. H. Pierre, D. Kirkham, J. Pesek, and R. Shaw (eds.), *Plant Environment and Efficient Water Use.* American Society of Agronomy and Soil Science Society of America.

Additional References

Blad, B. L. 1983. Atmospheric demand for water. In I. D. Teare and M. M. Peet (eds.), *Crop–Water Relations*. John Wiley and Sons, New York.

Kramer, P. J. 1969. *Plant and Soil Water Relationships: A Modern Synthesis*. McGraw-Hill. New York.

Pendleton, J. M. 1965. Increasing water use efficiency by crop management. In W. H. Pierre, D. Kirkham, J. Pesek, and R. Shaw (eds.), *Plant Environment and Efficient Water Use*. American Society of Agronomy and Soil Science Society of America.

Slatyer, R. O. 1967. *Plant Water Relationships*. Academic Press. New York.

Turner, N. C., and G. J. Burch. 1983. The role of water in plants. In I. D. Teare and M. M. Peet (eds.), *Crop–Water Relations*. John Wiley and Sons, New York.

Problems

1. Assume that the sap from a plant cell has an osmotic potential of -1 MPa. Use the van't Hoff equation to calculate the total cell sap solute concentration at 20°C.

2. The leaf water potential of a soybean canopy varied from -0.47 to -1.73 MPa during one day in August. If the maximum and minimum air temperatures for this day were 30° and 21°C, respectively, calculate the maximum and minimum relative humidities of the leaf air spaces. If the atmospheric relative humidities varies from 95 to 48%, at which measurement time is the potential gradient the greatest?

3. If the soil water potential is -0.125 MPa and the leaf water potential is -1.242 MPa, calculate the resistance to flow if the transpirational flux is 0.0125 cm/h. What is the hydraulic conductivity?

4. Describe the relative contributions of the component water potentials in the cell, cell walls, and xylem.

5. Rice (1975) determined that the water uptake rates were distributed in the soil profile for two days as given below. For each day, plot the water uptake rates versus soil depth, estimate the daily AET, and compute the fraction of the total extraction of water by depth. Comment on the similarities and differences in the extraction rates for the two days.

Soil Depth (cm)	Water Uptake Rate (d^{-1})	
	June 28	Sept. 24
5	0.026	0.14
10	0.048	0.12
15	0.005	0.011
20	0.004	0.008
40	0.008	0.006
60	0.004	0.004
80	0.003	0.003
120	0.001	0.002

6. Scott et al. (1987) found that water management affected the AET and dry matter produced by soybeans grown on a DeWitt silt loam. The 5-year average ET was 497 and 382 mm for irrigated and nonirrigated soybeans, respectively. If the average dry-matter production was 9492 and 6650 kg/ha for the irrigated and nonirrigated soybeans, respectively, determine the most efficient water use treatment.

7. In the same experiment as given in problem 6, the seed yields averaged 2707 and 1884 kg/ha in the irrigated and nonirrigated plots, respectively. Calculate the WUE for the water management treatments. How many grams of water were used to produce 1 g of soybean seed?

8. Assume that a 1 m² crop canopy transpires at a rate of 5 mm/d at 25°C. Calculate the heat energy lost by transpiration.

9. Comment on the following statements:
   a. Errors associated with the water balance approach are likely to invalidate the use of this approach for quantification of normal daily evapotranspiration in the field.
   b. The major problem associated with the water balance method used to calculate evapotranspiration, especially for large areas, is not in the method itself but in the lack of adequate spatial averaging of inputs and outputs due to the variation in rainfall over extended areas and to a lack of homogeneity in land use and the soils that underlie them.

10. Assume that at midday a crop is transpiring according to the following equation:

$$\text{Transpiration flux density} = \frac{\psi_L - \psi_a}{r_s + r_a}$$

where ψ_L and ψ_a are the water potentials of the leaves and atmosphere, respectively, and r_s and r_a are the resistances due to the stomata and aerial boundary layer, respectively. Assume that ψ_L is −1.5 MPa, the relative humidity of the air above the canopy is 0.6 at 30°C, and that r_s and r_a are 4 s/cm and 0.6 s/cm, respectively. Calculate the transpiration flux density in kg/m²·s. What is the transpiration flux density in mm/s?

Appendix A
Review of Geostatistics and the Variability of Soil Properties

Soil physicists are concerned with phenomena that are the results of interactions between the solid, liquid, and gaseous phases of soil. The relationships between these phases often are unknown, are complex, and may exhibit continuity within the landscape. The variability of soil properties results from the factors of soil formation and changes induced by humans. Soil properties vary in four dimensions: vertically within the profile (z) and along the horizontal plane (x, y) and with time (t). Since soil physical properties play a central role in the soil's capacity to transmit, store, and react to water, solutes, and gases, knowledge of the variability of these properties in the landscape is important. For further study, the student is encouraged to read the latest texts in statistics as well as Kempthorne and Allmaras (1986), Petersen and Calvin (1986), Warrick et al. (1986), and Dixon (1986). Here we will focus on the statistical methods most frequently used to characterize the variability of soil physical properties.

Summary Statistics and Distribution of Continuous Random Variables

A *random variable* is a variable whose value is a numerical outcome of a random phenomenon. A *continuous random variable* takes all values within an interval. Examples include bulk density, hydraulic conductivity, soil water content, and pH. A *discrete random variable* has a finite number of values. Examples of discrete variables include soil color, structural class, number of nematodes per unit volume of soil, and whether a drinking-water well is contaminated.

The population of soil properties to be sampled should be subdivided both horizontally and vertically into sampling strata that are as homogeneous as possible. All sources of variation within the population should be sampled if valid inferences are to be made about the population from the sample. The population values of a soil physical population parameter can be characterized by calculating measures of central tendency and measures of dispersion.

Measures of Central Tendency

Several statistics can be used to quantify the tendency of a soil parameter to be centrally located. The most common of these and the one most frequently used is known as the *arithmetic mean*, arithmetic average, or expected value. The arithmetic mean, which serves as an estimate of the population mean μ, can be calculated from

$$\mu = \sum_{i=1}^{N} y_i / N \qquad\qquad [1a]$$

where N is the total number of observations. We estimate the population μ with the sample mean $\bar{y}$.

$$\bar{y} = \sum_{i=1}^{n} y_i / n \qquad\qquad [1b]$$

where $\bar{y}$ is the expected value (i.e., the mean), n is the number of samples, and y_i is the measured value of sample $\sum$. The summation, $\sum$, is over all values of n.

The *geometric mean,* G, may be useful for soil properties with skewed frequency distributions or when working with rates and ratios. It is calculated from

$$G = (y_1 y_2 \cdots y_n)^{1/n} \qquad\qquad [2]$$

From a practical standpoint, G is determined by averaging the logarithms of both sides of equation [2], then taking the antilog of that numerical value or by taking the nth root of the product of the measured values. Logarithms either to the base 10 or to base e can be used. Both arithmetic and geometric sample means give estimates of the "average value" of the parameter in the population. Higher sample means are indicative of higher average values for the population.

Other measures of central tendency are the *median* and the *mode.* The median is that value for which 50% of the observations, when arranged in order of magnitude, lie on each side. It is useful when the distributions of the samples are unknown or when they are skewed. The mode is the value of most frequent occurrence.

Measures of Dispersion

In most situations, the population values will not have the same value, and some (or most) observations will deviate from the mean. The average square deviations, or *population variance,* σ^2, can be quantified by computing the average square deviation of each population value from the population mean. Mathematically, this is represented by

$$\sigma^2_y = (1/N) \sum_{i=1}^{N} (y_i - \mu)^2 \qquad\qquad [3]$$

A data set with little variability will have most of the observations located near the mean, and σ^2_y will be low. Deviations from the sample mean for a more variable set of measurements would be relatively large. There are several measures of dispersion that compute the variance from the sample mean.

Estimates of the population variance can be calculated by the *sample variance* s^2 based on a sample size n:

$$s^2 = s_y{}^2 = \sum_{i=1}^{n} (y_i - \bar{y})^2 / (n-1) \qquad\qquad [4]$$

The quantity in the numerator on the right-hand side is known as the *sum of squares* of the deviations. As before, the summation is taken over all values of n. The sample variance is also known as the mean square.

The *standard deviation,* s_y, which is often designated as s, can be calculated from the sample variance by

$$s_y = (s_y{}^2)^{1/2} \qquad\qquad [5]$$

Large values of s correspond to samples that are heterogeneous (i.e., dissimilar), whereas small values of s indicate that the samples were mostly close to the estimated mean. Therefore, the lower the value of s, the higher the precision. The units of s are the same as those of the observations.

Central-Limit Theorem

Although normal distributions of soil physical parameters occur frequently, many random variables are not normally distributed, and it would be inappropriate to use a normal distribution as the model. In spite of this, if the samples are sufficiently large, a normal distribution can often be used to approximate the sampling distribution of $\bar{y}$ because of the *central-limit theorem: If random samples of n measurements are repeatedly drawn from a population with finite mean μ and a standard deviation σ, then, when n is large, the relative frequency histogram for the sample means will be approximately normal with mean μ and standard deviation $\sigma/\sqrt{n}$. The approximation becomes more precise as n increases.*

The central-limit theorem concerns the sampling distribution of averages and has the following properties:

1. The mean of the sampling distribution of averages is the same as the mean of the underlying population.
2. The variance of the sampling distribution of averages is equal to the variance of the underlying population divided by n.
3. If n is sufficiently large, then the sampling distribution of averages is almost symmetrical and unimodal.

One interesting and practical aspect of the central-limit theorem is that no matter what the shape of the original probability distribution, the sampling distribution of the mean approaches a normal distribution. Furthermore, for most probability distributions, a normal distribution is approached quickly as n increases.

The estimated *standard error of the mean,* $s_{\bar{y}}$, is the ratio of the standard deviation of the sampling distribution of $\bar{y}$ and the square root of the number of observations used in calculating the mean. The working equation is

$$s_{\bar{y}} = s_y / n^{1/2} = s / \sqrt{n} \qquad [6]$$

The standard error is a measure of the variability associated with the sample mean. It also has the same units as the observations.

Another statistical parameter often used to characterize the variability of a soil parameter is the *coefficient of variation,* CV. Mathematically, CV is defined as the sample standard deviation divided by the sample mean and is usually expressed as a percentage of the sample mean, or as

$$CV = 100(s/\bar{y}) \qquad [7]$$

The CV is a relative measure of variation, in contrast to the standard deviation. It is independent of the unit of measurement used.

Some soil physical parameters are inherently more variable than others. Warrick and Nielsen (1980) presented a table of various soil parameters and grouped their variability into categories of low, medium, and high. Parameters having CV values less than 10% were considered to be in the low-variability category and included bulk density and saturated water content. Soil parameters having CV values greater than 50% were considered to be in the high-variability group and included saturated hydraulic conductivity, unsaturated hydraulic conductivity, apparent diffusion coefficient, pore water velocity, and electrical conductivity.

Other Useful Classical Statistical Parameters

Confidence limits can be placed around the mean by adding to and subtracting from its estimate a quantity associated with its precision, the s:

$$CL = \bar{y} \pm t_{\alpha/2}s_{\bar{y}} = \bar{y} \pm t_{\alpha/2}(s_y^2 / n)^{1/2} \qquad [8]$$

where CL is the confidence limit and $t_{\alpha/2}$ is the two-tailed Student's t variate with $(n - 1)$ degrees of freedom at the α probability level. The upper confidence limit is obtained when the positive sign is used in equation [8]; the lower confidence limit is obtained when the negative sign is used.

The number of samples, n, required to estimate a parameter within a specified difference from the mean within $(1 - \alpha)$ 100% of the time is sometimes used to make sampling decisions. Values of n can be calculated from

$$n = t_{\alpha/2}^2 s^2 / d^2 \qquad [9]$$

where d is the difference from the mean or how close we wish to estimate the mean, and t is determined with $(n - 1)$ degrees of freedom. Equation [9] shows that the variability and the number of samples are directly related. The assumptions are that the samples are independent and normally distributed. Independence means that the observed values in one sample are independent of the values that appear in another sample. "Normally distributed" means that the variables have Gaussian symmetry; that is, they have bell-shaped frequency distributions that can be approximated by using a normal curve. These statistics do not in themselves provide information about the type of frequency distribution associated with each parameter.

Linear Correlation

Correlation analysis is used to measure the strength of the relationship between two variables, for example, X and Y, by means of a single number called the *correlation coefficient*. We denote the linear correlation coefficient by ρ and its magnitude can be calculated from

$$\rho = \sigma_{xy}/(\sigma_x\sigma_y)^{0.5} \qquad [10]$$

where σ_{xy} is the covariance of X and Y, and σ_x and σ_y are the population standard deviations of X and Y, respectively. To estimate ρ we use the sample correlation r, which is defined as

$$r = s_{xy} / s_x s_y = [n\sum XY - (\sum X)(\sum Y)]/\{[n\sum X^2 - (\sum X)^2] $$
$$[n\sum Y^2 - (\sum Y)^2]\}^{0.5} \qquad [11]$$

This equation for r is defined so that r will always assume a value between -1 and $+1$. A value of $r = +1$ will occur when all the points lie in a straight line sloping up to the right. A value of $r = -1$ will occur when all the data points lie exactly in a straight line sloping down to the right. Therefore, a perfect relationship exists between X and Y when $r = +/-1$. If r is near zero, the linear relationship is weak or nonexistent.

Various relationships of X and Y are shown in the scatter diagrams in Figure A.1. If the data points closely follow a straight line that slopes up to the right, then we have a high positive correlation between the two variables. On the other hand, if the points follow closely a straight line sloping down to the right, we have a high negative correlation. The correlation between the two variables decreases numerically as the scattering of points from a straight line increases. If the scatter of points follows a strictly random pattern, we have a zero correlation and conclude that no *linear* relationship exists between X and Y.

In evaluating values of r, there are several important points to consider. First, it is important to remember that the correlation coefficient between two variables is a measure of their association,

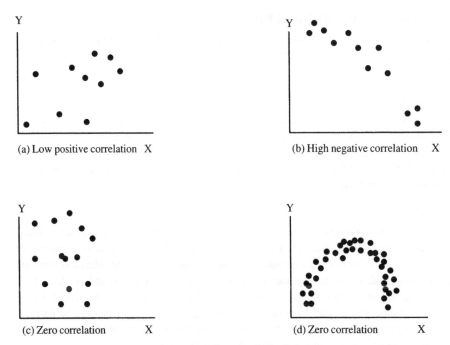

Figure A.1. Scatter diagrams showing various degrees of correlation.

that is, their linear relationship. A value of r = 0 implies a lack of linearity but not necessarily a lack of association. Therefore, if a strong quadratic relationship exists between X and Y as indicated in Figure A.1d, the zero correlation that may result indicates a nonlinear relationship. Second, one must be careful in interpreting r between different data sets. For example, values of r equal to 0.3 and 0.6 only mean that we have two positive correlations, one somewhat stronger than the other. It is incorrect to conclude that the r value of 0.6 of one data set indicates a linear relationship twice as good as that indicated by the r value of 0.3 in the other data set. And third, it would be wrong to conclude that there is necessarily a cause-and-effect relationship between the variables. One must determine from the nature of the study whether a significant correlation coefficient implies that one variable has a direct or indirect effect on the other or whether both variables are being influenced by other variables.

Continuous Frequency Distributions

Every soil property has a frequency (or probability) distribution (Figure A.2). Given the frequency distribution, we can determine statistical parameters such as measures of central tendency and measures of dispersion and even the probability that a randomly drawn sample will be within specified limits. Two continuous probability distributions frequently found in the examination of soil physical properties will be examined.

Normal Probability Distribution

Probably, the most well known continuous frequency distribution is the bell-shaped curve, called the normal or Gaussian distribution. The equation for this function is

$$f(y_i, \mu, \sigma) = [1/\sigma(2\pi)^{1/2}]\exp[-(y_i - \mu)^2/2\sigma^2] \qquad [12]$$

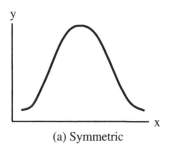

(a) Symmetric

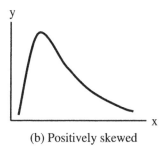

(b) Positively skewed

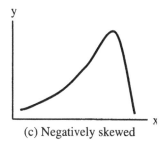

(c) Negatively skewed

Figure A.2. Important shapes of frequency distributions.

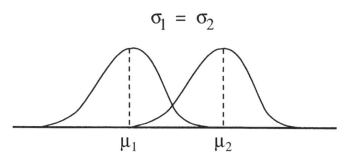

$$\sigma_1 = \sigma_2$$

μ_1 μ_2

Figure A.3. Normal curves with unequal means but equal standard deviations.

where f(y) represents a frequency of occurrence of the random variable Y. The mean μ falls at the center of the bell curve (Figure A.3). The median, for which half of the population values are smaller and half are larger, also falls at the central point. The mode, for which f(y) is a maximum, coincides with both the mean and the median. The μ value determines where the center of the bell is located. The value of σ controls the spread; for example, a large σ gives an f(x) curve that is low and wide, whereas small values of σ give narrow, tall peaks (Figure A.4). Two distributions with different means and different variances are shown in Figure A.5. The area under the probability function is unity, which is equivalent to stating that we are 100% certain any value in the population is included somewhere in the distribution. A typical frequency distribution of approximately normally distributed data is shown in Figure A.6.

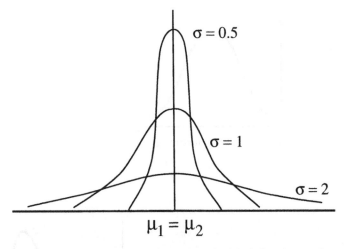

Figure A.4. Normal curves with unequal standard deviations but equal means.

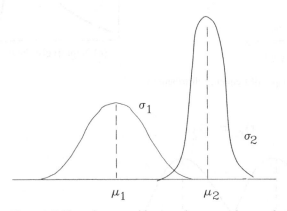

Figure A.5. Normal curves with unequal means and unequal standard deviations.

The cumulative normal probability function, F(y), is calculated from

$$P(Y \leq y) = F(y) = \int_{-\infty}^{y} f(y) \, dy \qquad [13]$$

As y becomes large, F(y) approaches 1. A plot of the cumulative frequency versus the magnitude of Y should be a straight line if the data are normally distributed (Figure A.7).

Lognormal Probability Distribution

Frequently, cumulative probability plots of water and solute transport parameters in soil such as Ksat, pore water velocity, solute dispersion coefficients, and aggregate diameters are not normally distributed but skewed to the right. These soil properties typically have a wide range of values, with the mean near the origin (Figure A.8). The few large values of the mass transport coefficients are due to the wide range in pore size and distribution of pores. One way to statistically characterize these data is to take the $\log_{10}$ or ln of each Y. The logarithmically transformed data are then

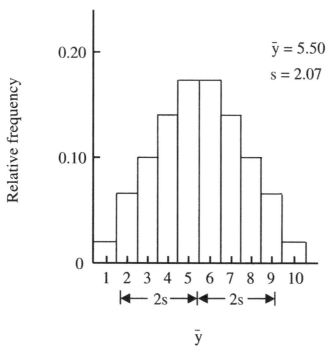

Figure A.6. A typical distribution of approximately normally distributed data.

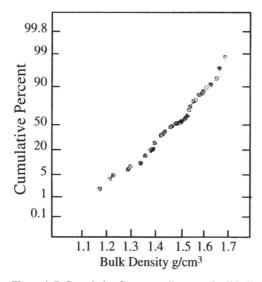

Figure A.7. Cumulative frequency diagram of soil bulk density.

treated as if they approximate the normal distribution. If the cumulative frequency is plotted against the value of the variable on probability graph paper and the result is curvilinear, then perhaps the experimental data are not normally distributed. Frequently, a transformation of the data can be made so that the transformed data are normally distributed. The most well known of these

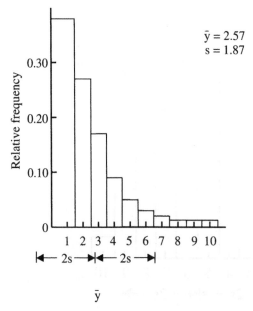

Figure A.8. A typical distribution of lognormally distributed data.

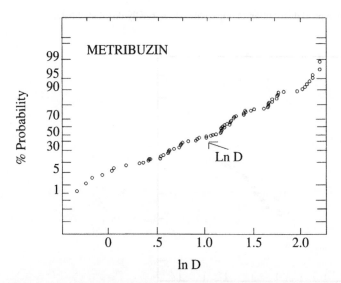

Figure A.9. Cumulative probability of the natural logarithm of the dispersion coefficient of the herbicide metribuzin in soil.

transformed distributions in soil physics is the lognormal distribution. The lognormal distribution can be used whenever the logarithm of the random variable Y is normally distributed (Figure A.9).

The lognormal frequency distribution of a random variable X, given by equation [14], has a form similar to equation [12] if $Y = \ln X \sim N(\mu, \sigma^2)$:

$$f(x, \mu, \sigma^2) = \exp[-(\ln x - \mu)^2/2\sigma^2]/[x\sigma(2\pi)]^{1/2} \qquad [14]$$

where ln Y is the natural logarithm of Y. Since the lognormal distribution is skewed, the mean, median, and mode have different values. Numerically, the mean of x is equal to $\exp(\mu + \sigma^2/2)$, the median is equal to $\exp(\mu)$, and the mode is equal to $\exp(\mu - \sigma^2)$.

Spatial Statistics as a Measure of Variability

Independence is a convenient assumption that makes much of the statistical theory tractable. However, models involving spatial dependence are more realistic in nature. Data that are spatially dependent are present in all directions and become weaker as sampling locations become more dispersed.

When soil properties are not purely random in the landscape but have some spatial arrangement, the assumption of independence between samples is violated. Therefore, classical statistical methods of analysis seriously bias the estimates of variances, with the direction and magnitude of the bias depending on the nature of the correlations. According to the *first law of geostatistics,* observations (or samples) that are close to each other have similar magnitudes, while observations farther apart have magnitudes differing by a greater order of magnitude. Thus, the closely spaced variables may be spatially correlated. This law has been successfully used to model physical and spatial phenomena.

Geostatistical data are measurements taken at fixed locations. The locations are generally spatially continuous, the underlying process is continuous in space, and the data are sometimes called random field data. Geostatistical data often exhibit small-scale variations that may be modeled as spatial correlation. The spatial variability is modeled as a function of distance between sampling sites.

Examples of continuous geostatistical data include rainfall recorded at geo-referenced weather stations, soil permeabilities at known sampling locations within a watershed, crop yields within a field, soil nutrient concentrations within a field, and concentrations of pollutants at geo-referenced monitoring stations.

Exploring the spatial correlation structure is a primary goal in analyzing geostatistical data. If the data are spatially correlated, then how does this affect our knowledge of adjacent unsampled values of a soil parameter?

For spatially uncorrelated data, the best estimate of a parameter at an unsampled position is the population mean. Samples from any position are equally valuable in estimating the population mean, and an estimate of uncertainty in the mean is given by the standard error.

For spatially correlated data, however, a measurement taken near the location of an unsampled point should better represent the magnitude of the property at that point than would the mean. Values at points farther away have less effect and give less information.

Spatial prediction is any prediction that incorporates spatial dependence. One of the simpler methods is kriging (discussed in a later section). Spatial prediction involves two steps. First, the covariance or semivariogram of the spatial process is modeled. Second, this model of spatial dependence is used in solving the kriging system at the specified set of spatial points, resulting in predicted values and associated standard errors.

Regionalized Variables

When a variable is distributed in space, it is known to be *regionalized.* A regionalized variable is a function f(s) that takes a value at every point s with coordinates (x, y, z) in three-dimensional space. The regionalized function with the spatial distribution varies from one location to another with apparent continuity, but the changes cannot be represented by any easily used mathematical function. Examples of regionalized variables include depth of the water table, distribution of soil physical properties in a landscape, rainfall, and crop yield. Almost all variables encountered in the earth sciences can be regarded as regionalized variables (Journel and Huijbregts 1991; Vieira et al. 1983; Nielsen et al. 1983). Some common characteristics of regionalized variables are localization, anisotropy, and continuity.

Localization: A regionalized variable is numerically defined by a value that is associated with a sample of specific size, shape, and orientation. These geometric characteristics of the sample are called the geometric support. A geometric field is that larger area or volume from which the samples are taken. The geometric field and the geometric support do not necessarily comprise volumes but may instead be areas, lines, or time intervals. When the size of the geometric support tends to zero, we have a point, or punctual, sample, and the geometric support is immaterial.

Anisotropy: Some regionalized variables are anisotropic; that is, changes in magnitude are gradual in one direction and rapid or irregular in another.

Continuity: The spatial variation of a regionalized variable may be extremely large or very small, depending on the phenomenon. Despite the complexity of the fluctuations, an average continuity is generally present. A regionalized variable possesses two characteristics: a local, random, erratic aspect and a general or average structured aspect that is based on a certain mathematical function, for example, a line or a polynomial. Representation of the spatial variability must take these two aspects of randomness and structure into account.

Mathematically, a regionalized variable is a function $Z(x)$ that gives the magnitude at point x (in a one-, two-, or three-dimensional space) of a soil property Z. Frequently, the variables have a spatial behavior too complex to be described by the usual analytical functions.

Suppose that soil water contents were measured along a transect. The water contents are expressed on a percentage by weight basis and were collected at 10 m intervals along the x axis. Assume that the following two different series of soil water contents were found:

Position (m):	10	20	30	40	50	60	70	80	90
Series A:	16	14	12	10	10	12	14	16	18
Series B:	12	16	18	10	12	10	16	14	14

An analysis of the data in both series shows that the arithmetic mean is 13.56%, the standard deviation is 2.79%, and the percentage coefficient of variation is 20.6%. These statistical parameters as well as the frequency diagram are the same for both transects. Therefore, the classical statistical parameters do not differentiate between the two series. However, a careful analysis of the individual data indicates that series A has a spatial continuity that series B does not have. Therefore, classical statistics is not completely adequate for the study of regionalized variables, because the information concerning the geographic locations of the samples is not accounted for.

Regionalized data may be decomposed into a trend component and a stochastic (or random) component according to

regionalized data = global trend + stochastic component

The global trend may reflect the changes in the mean of the variable with location and the effects of covariate variables. The stochastic component may include spatial correlation between neighboring observations and random sampling variations.

Assumptions in Spatial Modeling

The terms *stationarity* and *isotropy* are often used to describe assumptions under which inference is performed. Exploratory and diagnostic techniques are often used to determine whether the assumptions are reasonable for a particular data set since from a statistical point of view these assumptions are hard to check.

Most of the techniques used for estimating spatial correlation require the data to be stationary. This requires that the mean of the variable not change over the region of interest. We also refer to stationarity of variance; that is, the variance of the function is constant over the region of interest. Techniques such as kriging and spatial regression models allow us to remove trends while modeling the spatial correlation in the data.

Isotropy refers to a spatial process that evolves the same way in all directions. That is, the correlation between two observations depends only upon their distance from one another and not upon the direction in which they are separated. Directional variograms are used to examine whether isotropy is present.

A geostatistical evaluation of regionalized data is performed using two main tools: the correlogram and the semivariogram. For the correlogram, strong stationarity is required of the following:

1. The mean exists and is constant for all locations x.
2. The covariance between the observations exists and is a unique function of the separation distance h. Frequently, h is called the lag. This is known as second-order stationarity.

Weaker assumptions for the variogram include the following:

1. The mean exists for all space.
2. For all vectors, the variance of $Z(x + h) - Z(x)$ is defined and is a unique function of h and not on location.

Autocorrelation and the Correlogram

One of the oldest methods of estimating space or time dependence between neighboring observations is through autocorrelation. The autocorrelation coefficient, $\rho(h)$, of a regionalized variable, $Z(x)$, is defined by

$$\rho(h) = \text{Cov}[Z(x), Z(x + h)]/\sigma^2 = C(h)/C(0) \qquad [15]$$

where the covariance, Cov, is for any two values of Z at a distance h apart, and $C(0) = \text{Var}[Z(x)] = \sigma^2$ is finite. The sample covariance function, $C(h)$, is defined as

$$C(h) = E[Z(x + h) - Z(x)] - m^2$$
$$C(h) = [1/n(h)]\sum[Z(x) - m][Z(x + h) - m] \qquad [16]$$

where n(h) is the number of pairs of sample points a distance h apart and m is the arithmetic mean. The distance between neighbors is called the lag. Sometimes, it is easier to calculate the sample covariance from the following equation:

$$C(h) = \{1/[n(h) -1]\}\sum[Z(x)Z(x + h) - m^2] \qquad [17]$$

According to equation [16], values of $C(h)$ are computed by subtracting the mean from all of the measured values, multiplying by the difference of the adjacent values, and summing over all pairs of observations. For example, in series A for h = 10 m (lag 1), we compute

$$C(10) = [(16.0 - 13.56)(14.0 - 13.56) + \ldots$$
$$+ (16.0 - 13.56)(12.0 - 13.56)]/7 = 5.06$$

where 7 is the number of lag 1 pairs of observations minus 1. The autocorrelation coefficient for lag 1 (i.e., this position) is

$$\rho(10) = 5.06/(2.79)^2 = 0.649$$

This process can be repeated for lags of 2, 3, . . . , n – 1.

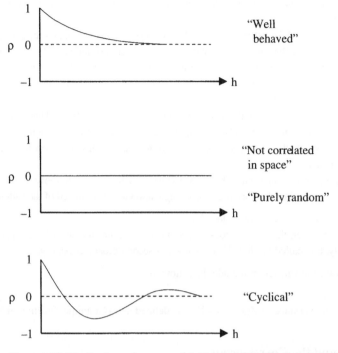

Figure A.10. Idealized correlograms, $\rho(h)$. (Adapted from Warrick et al. 1986.)

Ideally, the number of data pairs used in the calculations should be large for each value of h (typically greater than 50), but in practice this is often difficult, especially for small data sets and increasing values of h.

The correlogram is an ordered set of correlation coefficients $\rho(h_1)$, $\rho(h_2)$, . . . , for a common variable where each couple or pair of measurements is separated by distance h—that is, a plot of the autocorrelation coefficient $\rho(h)$ as a function of distance or lag number h (Figure A.10). The correlogram is a measure of the strength of the linear association between pairs of observations. It is useful in defining the separation distance between samples beyond which there is no correlation between pairs of values.

The autocorrelation coefficient, $\rho(h)$, is dimensionless and varies between +1 and –1, indicating positive and negative correlations, respectively. The maximum $\rho(h)$ is 1, which occurs at zero lag (h = 0), and the values tend to decrease for larger lags. This means that as h increases, the covariance decreases; that is, the correlation between the observations decreases. Therefore, samples taken close together are more alike and samples somewhat separated are less alike and samples remote from each other are not correlated at all. Eventually, for large lags or distances, $\rho(h)$ is approximately zero, and no correlation or spatial dependence exists. In practice, as $h \to \infty$, $C(h) \to 0$. In addition, we put $C(h) = 0$ when $|h| \geq a$, where a is a distance beyond which $C(h)$ can be considered to be equal to zero. This distance is called the *range* and represents the transition from the state in which a spatial correlation exists ($|h| < a$) to the state in which there is no correlation ($|h| > a$).

The autocorrelation coefficient is related to the correlation coefficient r used in linear regression techniques. The square of r (R^2) gives the fraction of the total variation in $Z(x)$ that can be explained by $Z(x + h)$. The autocorrelation coefficient gives the same information. The set of all possible h's produces the correlogram.

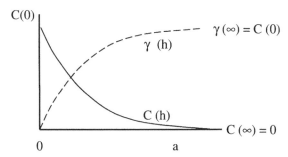

Figure A.11. The relationship between the covariance and the semivariogram.

Properties of the Covariance

There are three main properties of the covariance:

$$C(0) = Var[Z(x)] \geq 0 \tag{18}$$

$$C(h) = C(-h) \tag{19}$$

$$C(h) \leq C(0) \tag{20}$$

The first property states that the variance cannot be negative; the second states that the covariance is an even function; and the third property is Schwarz's inequality.

The autocorrelation coefficient can also be defined mathematically as

$$\rho(h) = C(h)/C(0) = 1 - \gamma(h)/C(0) \tag{21}$$

where $\gamma(h)$ is one-half of the variogram, called the semivariogram. Therefore, $\gamma(h) = C(0) - C(h)$. Graphically, the relationship between $C(h)$ and $\gamma(h)$ is shown in Figure A.11.

This graph indicates that the covariance and the variogram are two equivalent tools for characterizing the autocovariance between two variables $Z(x)$ and $Z(x + h)$ separated by distance h.

Correlation Length

The distance over which a significant correlation exists is called the correlation length, scale, or range. Commonly, the correlation length l is defined for a one-dimensional transect by the exponential relation

$$\rho(h) = r_0 \exp(-x/l) \tag{22}$$

where $r_0 = 1$ and ρ is decreased by e^{-1} (i.e., 0.3679) at distance l. Sampling at intervals less than l may be unnecessary because the observations are spatially related to each other. Sampling at intervals greater than l does not allow meaningful interpolation between neighboring observations.

Semivariance and the Semivariogram

Probably the most important method used to quantify spatial dependence is a type of correlogram known as the semivariogram. A semivariogram expresses the degree of spatial variation as a function of distance of separation. The main goal of a variogram analysis is to construct a variogram that best estimates the autocorrelation structure of the underlying stochastic process.

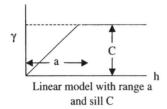

Linear model with range a
and sill C

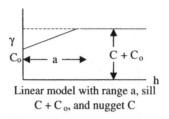

Linear model with range a, sill
$C + C_o$, and nugget C

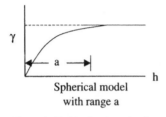

Spherical model
with range a

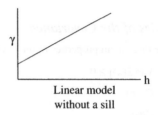

Linear model
without a sill

Figure A.12. Idealized semivariograms. (After Warrick et al. 1986.)

Mathematically, the experimental semivariogram can be calculated from the equation

$$\gamma(h) = [(1/2) \mid n(h) \mid] \sum_{i=1}^{n(h)} [(z(x) - z(x+h)]^2 \qquad [23]$$

where the 2 is there for mathematical convenience. The parameter n(h) is the set of all pairs of points separated by a distance h. The value of n(h) depends on the number of pairs of observations used in the analysis separated by a distance, or lag vector, h. Even though used interchangeably, the term $\gamma(h)$ is the semivariogram and the term $2\gamma(h)$ is called the variogram.

The semivariogram is a graph describing the expected squared difference in value between pairs of samples with a given relative orientation (Figure A.12). Usually, the distance between the pairs of samples along the horizontal axis, or the lag number, is plotted against the value of the semivariogram along the vertical axis. By definition, h starts at zero, since it is impossible to take two samples on top of each other. The γ axis also starts at zero, since it is an average of squared values. The graph of the experimental semivariogram computed using equation [23] will display a series of discrete points corresponding to each value of h. A continuous function or model must be fitted to these points.

Consider the case when h is equal to zero. The difference between the two values of Z must be zero, so that γ must always pass through the origin of the graph. Now, suppose we take the two samples a little distance apart. We would now expect a little difference between the magnitude of the two values, so that the semivariogram will show some difference between the two values. As the samples become farther apart, the differences between magnitude should increase.

In the ideal case, the sample values become independent of one another when the distance of separation becomes large. The semivariogram value will then become more or less constant, since it will be calculating the difference between sets of independent samples. This so-called ideal

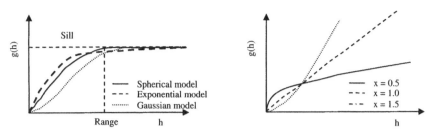

Figure A.13. Shapes of three semivariogram models.

shape for the semivariogram is to geostatistics what the normal distribution is to classical statistics. It is a model semivariogram and is usually called the spherical model. The distance at which samples become independent of one another is denoted by the symbol a and is called the *range* of influence of a sample. Measurements separated by distances closer than a are spatially correlated to each other. The value of γ at which the graph levels off is denoted by the symbol C and is called the *sill* of the semivariogram. The sill, if one exists, is nearly equal to the ordinary sample variance, that is, $C \approx s^2$, if we are concerned with point samples. The value of γ at the intercept (h = 0) must be zero since two samples measured at exactly the same position must have the same value. This is called the *nugget effect,* C_0. In practice, there often appears to be a discontinuity, that is, a small range of influence. This may be due to a microregionalization on a scale much smaller than the spacing of the data points, measurement error, and/or differences in variance due to unmeasured values at small distances. When the range is smaller than the shortest sampling distance, we have a *pure nugget effect,* the samples are completely random, and classical statistical methods can be applied.

Theoretical Semivariogram Models

There are several possible mathematical functions to model shapes of semivariograms (Figure A.13), but only a few are commonly used (Marx and Thompson 1987; Deutsch and Journel 1998). The most common single models are discussed below and are classified according to their behavior at or near the origin and the presence or absence of a sill.

Models with a Sill and Linear Behavior at the Origin

As h increases from zero, $\gamma(h)$ increases linearly up to a certain corresponding value of h, that is, the range a, after which it remains relatively constant. The value of $\gamma(h)$ at this point is approximately equal to the variance of the observations and is called the sill C. We will consider two models.

Spherical Model

The equations for the spherical model are

$$\gamma(h) = C_0 + C_1[(3h/2a) - (h^3/2a^3)] \qquad (0 \le h \le a) \qquad [24a]$$
$$\gamma(h) = C_0 + C_1 \qquad (h \ge a) \qquad [24b]$$

where $C_0 + C_1$ is the sill of the semivariogram, C_0 is the nugget effect, and a is the range. The spherical model is obtained by first selecting the nugget effect C_0 and the sill value $C_0 + C_1$. Then, a line intercepting the y axis at C_0 and the tangent to the points near the origin will reach the sill at a distance $a' = 2/3a$, where a' is the practical range. Thus, the range is $3/2a'$. The spherical model behaves linearly up to approximately a/3.

Exponential Model

The equation for the exponential model is

$$\gamma(h) = C_o + C_1[1 - \exp(-3h/a)] \tag{25}$$

This model rises more slowly from the origin than the spherical model and asymptotically approaches the sill. For the exponential model, $h = a = a'/3$, that is, one-third of the practical range. The parameter a is obtained by taking a tangent to the points near the origin, intercepting the y axis at C_0.

Model with a Sill and Parabolic Behavior at the Origin
Gaussian Model

The equation for the Gaussian model is

$$\gamma(h) = C_o + C_1[1 - \exp(-3h^2/a^2)] \tag{26}$$

The sill is asymptotically approached and a practical range of a.

The effective range of the Gaussian and exponential models is taken to be the distance where the semivariance is 95% of the sill. The major distinguishing feature of the Gaussian and exponential models is the shape in the neighborhood of the origin, that is, $h \approx 0$. This is why small lags are important in determining an appropriate model for the semivariogram.

Models without a Sill

These models correspond to phenomena with an infinite capacity for dispersion, for which neither the variances of the data nor the covariances can be defined (Journel and Huijbregts 1991). We consider three models.

Power Model

The equation for the power model is

$$\gamma(h) = C_o + ph^x \tag{27}$$

where x is a dimensionless quantity with typical values greater than 0 but less than 2 and positive slope p. There is no sill for this model.

Linear Model

The equation for the linear model without a sill is

$$\gamma(h) = C_o + ph \qquad (0 \le h \le a) \tag{28}$$

where p is the slope of the semivariogram line. This model is a special case of the power model where $x = 0$.

Logarithmic Model

The equation for the logarithmic model without a sill is

$$\gamma(h) = C_o + p \log(h) \tag{29}$$

This model gives a linear semivariogram if plotted against the logarithm of the distance.

For a given set of spatial data, a semivariogram may not seem to fit any one of the theoretical models presented above. In such a case, the covariance structure of the spatial process may be the

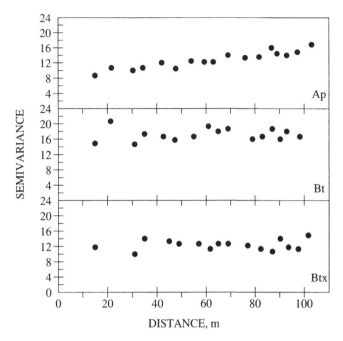

Figure A.14. Semivariograms for clay content in the Ap, Bt, and Btx horizons (from Wood 1988). Only the clay content in the Ap horizon was spatially dependent.

sum of two or more covariances. This is common where there are correlations of different scales, which leads to more than one spatial structure and to nested models. The sum of two models does not resemble any single theoretical model but has the characteristic shape of both models at different portions of the semivariogram. Thus, a combination of models may be needed instead of a single model.

Burrough (1991) listed the following rules of thumb for estimating variograms:

1. Make sure that there are sufficient pairs of points at all distance classes needed for the study.

2. Do not estimate the variogram for lags that exceed one-half the longest dimension of the area, because longer lags will be based on a few points located at the extremes of the area and may suffer from edge effects.

3. For irregularly spaced data, choose distance classes as small as possible so that local variation is not smoothed away.

4. Don't be afraid to transform the data if possible.

An example of the linear model fitted to the clay distribution in a Captina profile is shown in Figure A.14 (Wood 1988). The silt content in the same profile was not spatially correlated and, therefore, exhibited no spatial structure (Figure A.15).

All the data points are not equally valuable in determining the reliability of the model for the variogram. We note that the greater the distance between the pairs, the fewer the number of pairs of observations. Thus, the rule that we go by is that the fewer the pairs, the less reliable our model. This means that the most reliable points on the graph are those for small distances, and the reliability drops off slowly and regularly as the distance of separation increases. Also, a good model fit to the semivariogram near the origin (for small lags) is most important for kriging.

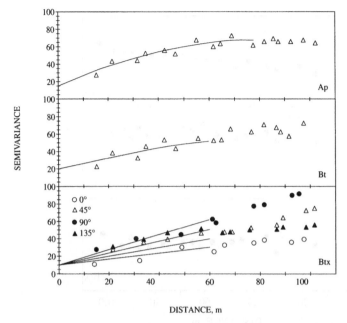

Figure A.15. Semivariograms for silt content in the Ap, Bt, and Btx horizons (from Wood 1988). The silt content in the Btx horizon was anisotropic.

Kriging

Kriging is a weighted interpolation scheme named after D. G. Krige, who developed it for estimating the gold content of ore in South Africa. The technique of kriging provides a means of interpolating values of points not physically sampled using knowledge about the spatial relationships of the data set. The spatial dependency of the data is obtained from the semivariogram model, and the error of the prediction depends only on the variogram, the number and configuration of the data points, and the size of the block for which the estimate is made. Therefore, kriging is based on regionalized variable theory and is superior to other means of interpolation because it provides an optimal interpolation estimate for a given coordinate location as well as a variance estimate for the interpolation value.

According to Clark (1984), the advantages of kriging are as follows:

1. Given the assumptions, no trend, and a model for the semivariogram, kriging always produces the best linear unbiased estimator.

2. If the proper models are used for the semivariogram, and the system is set up correctly, there is always a unique solution to the kriging system.

3. If the value at a location that has been sampled is estimated, the kriging system will return the sample value as the estimator and a kriging variance of zero. In other words, you already know that value. This is usually referred to as an exact interpolator.

4. If regular sampling is used, and hence the same sampling/block setup at many different positions within the field, it is not necessary to recalculate the kriging system each time.

Kriging requires that several choices be made. First, spatial analysis in the form of the semivariogram has to be computed. Second, the choice of the model for the semivariogram has to be made. The choice of the model requires knowledge of any anisotropy in the data set, with special care given to the fit of the model at small lag distances. Third, computation of the weights in each local average must be made.

There are three types of kriging: punctual (or point or ordinary), block, and universal. The simpler types are punctual and block kriging.

Punctual Kriging

Punctual kriging is concerned with the estimation of the exact values for points (Burgess and Webster 1980a). It is a linear interpolation procedure that allows prediction of unknown values of a random function from observations at known' locations. Also, it assumes a constant mean and variance across the prediction region; that is, no trend is present. A random function model of spatial correlation is used to calculate a weighted linear combination of available samples for prediction of a nearby unsampled location. The weights in the linear combination are chosen to return an unbiased estimate of the unknown value that has minimum prediction variance. A variogram is needed to specify the expectation under which the prediction variance is calculated.

If the values of $Z(x)$ are known at n experimental points $x_1, x_2, \ldots, x_n$ in a field, the problem is to estimate a quantity Z^* at all locations x_0 where values of Z have not been measured; the estimation may be any linear combination of the measured variables. To estimate Z^* at x_0, consider a linear-weighted sum of the n measured $Z(x_i)$ values to be

$$Z^*(x_0) = \sum \lambda_i Z(x_i) \tag{30}$$

where $Z^*(x_0)$ is the estimated value of Z at x_0, and the sets of weights λ_i give the best possible estimation. The extent of the neighborhood around x_0 determines the n points used in the estimation. The weights must be determined before the interpolation can be solved. Points near x_0 have more weight than distant points, points that are clustered carry more weight than lone points, and those points between the point to be interpolated and more distant points screen the distant points so that the latter have less weight. There are numerous ways to distribute the weights, including the (1) inverse of the distance of separation, (2) inverse of the square of the distances, and (3) inverse of the number of values. In kriging, the weights must be such that the best estimator is (1) unbiased, that is, has no systematic over- and underestimation, and (2) optimal, that is, has minimum squared error. Three conditions are required: (1) linearity, (2) lack of bias, and (3) minimal mean square prediction error, $E[Z^*(x_0) - Z(x_0)]^2$.

The first condition requires that $Z(x_0)$ is linear in $Z(x_1), \ldots, Z(x_n)$.

The second condition requires that the estimator $E^*(x_0)$ be written as

$$E[Z^*(x_0) - Z(x_0)] = 0 \tag{31}$$

where E is the expected value of the function in the brackets, and $Z(x_0)$ is the value of the random function Z at x_0. Thus, substituting [30] into [31] gives

$$E\{[\sum \lambda_i Z(x_i)] - Z(x_0)\} = 0 \tag{32}$$

Taking the expectation of each value and equating it to the mean, m, which is assumed constant over space, yields

$$\sum \{\lambda_i E[Z(x_i)] - E[Z(x_0)]\} = m(\sum \lambda_i - 1) = 0 \tag{33}$$

Therefore, the estimation will be unbiased if

$$\sum \lambda_i = 1 \quad \text{and} \quad I = 1, n \tag{34}$$

The weights are chosen so that the estimate $Z^*(x_0)$ is unbiased, and the prediction variance σ^2 is less than for any other linear combination of the observed values.

The third condition requires the estimator $Z^*(x_0)$ have minimum variance of estimation. This can be written as

$$\sum \lambda^* \gamma(x_i, x_j) + \Phi = \gamma(x_i, x_0) \qquad (j = 1, n, \text{ and for all } i) \tag{35}$$

and the prediction variance is

$$\sigma^2_E = \sum \lambda_j^* \gamma(x_j, x_0) + \Phi \qquad (i = 1, 2, \ldots, n) \tag{36}$$

The quantity $\gamma(x_i, x_j)$ is the semivariance of Z between the sampling points x_i and x_j; $\gamma(x_i, x_0)$ is the semivariance between the sampling point x_i and the unsampled point x_0. Both of these quantities are obtained from the fitted variogram. The quantity Φ is a Lagrange multiplier required for the minimization. These systems of equations are best solved with matrix algebra as presented in Vieira et al. (1983).

Block Kriging

Equations [30] and [35] give estimates of the variable and its prediction variance at unsampled locations for areas or volumes of soil that are the same size as that of the original sampling support. Block kriging may be more appropriate than punctual interpolation whenever average values of properties are more meaningful than single-point values, especially where spatial dependence is weak (Burgess and Webster 1980b). It also gives an error term for each value estimated, providing a measure of reliability for the interpolations.

The average value of Z over a block B is given by

$$Z(x_B) = \int_B Z(x) \, dx / \text{area } B$$

and is estimated by

$$Z(x_B) = \sum \lambda_i^* Z(x_i) \tag{37}$$

with equation [34] satisfied. In block kriging, the weights λ_i are calculated using average semivariances between the data points and all points in the block.

The minimum prediction variance for the block B is

$$\sigma_B^2 = \sum \lambda_j \gamma(x_j, x_B) + \Phi_B - \gamma(x_j, x_B) \tag{38}$$

Given the semivariogram, predictions can be made of the value of an attribute Z at any location within the map unit for blocks of land having a minimum area of the sample support or larger.

Universal Kriging

Universal kriging is an adaptation of ordinary kriging that accommodates trends. Universal kriging can be used to produce local estimates in the presence of a trend and to estimate the underlying trend itself. In universal kriging, a polynomial form for the trend in the mean is specified, and coefficients for the polynomial are estimated.

In kriging of large areas, the assumption that there is no drift is difficult to obtain. One approach is to subdivide the area into regions. Another approach is to account for the drift by fitting a polynomial to the data. This introduces some uncertainty in that the polynomial function is generally unknown and the semivariogram should be constructed without drift. Thus, fitting of the semivariogram requires that the drift function be known, but the coefficients in the model can be estimated optimally only if the variogram is known. The problem is circular. Some investigators have fitted linear and quadratic functions to the drift with least squares and substituted this into the kriging equations.

In summary, kriging takes the following into account:

1. the distances between the estimated point and the surrounding data points,
2. the distances between the data points themselves with the ij terms, and the structure of the variable through the variogram.

With a regular variable, kriging gives higher weights to the closest data points. When the phenomenon is irregular, the weights given to the closest experimental points are dampened. When x_0 coincides with the data point x_i, $\lambda_i = 1$, $\lambda_j = 0$, $Z^*(x_0) = Z(x_i)$, and the variance is zero. Thus, kriging estimation is based on the identical spatial structure rather than some predefined procedure of weighting (Isaaks and Srivastava 1989; Upchurch and Edmonds 1991).

Most soil properties tend to be correlated over space. This means that estimates for a property at an unsampled location will be similar to those nearby, rather than the average of the area. The primary application of geostatistics is for estimating values at locations where measurements have not been made. Geostatistical techniques can be applied to distances of a few molecules or many kilometers.

Cokriging and Multivariate Geostatistics

Up to this point, we have considered the spatial correlation of only one variable, Z, with itself. However, the techniques can also be applied to two or more properties to determine, for example, to what extent variable Z at location x depends upon variable W. The spatial distributions of two variables are quantified by coregionalization with a technique known as cross correlation.

Cokriging uses both variables simultaneously to predict values of both variables at locations where data are not collected. It is particularly useful when one variable is easier to measure than others. For example, the saturated hydraulic conductivity of soil is more difficult to measure than the particle size distribution, but these variables are usually related. The estimation of one variable using information about both itself and another cross-correlated and perhaps easier-to-measure variable is more useful than the kriging of that variable by itself. This estimation is accomplished by cokriging.

Cited References

Burgess, T. M., and R. Webster. 1980a. Optimal interpolation and isarithmic mapping of soil properties. I. The semi-variogram and punctual kriging. *Journal of Soil Science* 31:315–331.

_____. 1980b. Optimal interpolation and isarithmic mapping of soil properties. II. Block kriging. *Journal of Soil Science* 31:333–341.

Burrough, P. A. 1991. Sampling designs for quantifying map unit composition. In M. J. Mausbach and L. P. Wilding (eds.), *Spatial Variabilities of Soils and Landforms*. Soil Science Society of America Special Publication 28:89–125.

Clark, I. 1984. *Practical Geostatistics*. Elsevier Applied Science Publishers. New York.

Deutsch, C. V., and A. G. Journel. 1998. GSLIB: *Geostatistical Software Library and User's Guide*. Oxford University Press. New York.

Dixon, W. J. 1986. Extraneous values. In A. Klute et al. (eds.), *Methods of Soil Analysis,* part 1, *Physical and Mineralogical Methods,* pp. 83–90. 2d ed. Agronomy Monograph 9. Soil Science Society of America. Madison.

Goovaerts, P. 1997. *Geostatistics for Natural Resources Evaluation*. Oxford University Press, New York.

Isaaks, E. H., and R. M. Srivastava. 1989. *Applied Geostatistics*. Oxford University Press. New York.

Journel, A. G., and C. J. Huijbregts. 1991. *Mining Geostatistics.* Academic Press, New York.

Kempthorne, O., and R. R. Allmaras. 1986. Errors and variability of observations. In A. Klute et al. (eds.), *Methods of Soil Analysis,* part 1, *Physical and Mineralogical Methods,* pp. 1–31. 2d ed. Agronomy Monograph 9. Soil Science Society of America. Madison.

Marx, D., and K. Thompson. 1987. Practical aspects of agricultural kriging. *Arkansas Agricultural Experiment Station Bulletin* 903.

Nielsen, D. R., P. J. Wierenga, and J. W. Biggar. 1983. Spatial soil variability and mass transfers from agricultural soils. In *Chemical Mobility and Reactivity in Soil Systems.* American Society of Agronomy. Madison.

Petersen, R. G., and L. D. Calvin. 1986. Sampling. In A. Klute et al. (eds.), *Methods of Soil Analysis,* part 1, *Physical and Mineralogical Methods,* pp. 33–51. 2d ed. Agronomy Monograph 9. Soil Science Society of America. Madison.

Upchurch, D. R., and W. J. Edmonds. 1991. Statistical procedures for specific objectives. In M. J. Mausbach and L. P. Wilding (eds.), *Spatial Variabilities of Soils and Landforms,* pp. 49–71. SSSA Special Publication No. 28. Soil Science Society of America. Madison.

Vieira, S. R., J. L. Hatfield, D. R. Nielsen, and J. W. Biggar. 1983. Geostatistical theory and application to variability of some agronomical properties. *HILGARDIA* 51:1–75.

Warrick, A. W., D. E. Myers, and D. R. Nielsen. 1986. Geostatistical methods applied to soil science. In A. Klute et al. (eds.), *Methods of Soil Analysis,* part 1, *Physical and Mineralogical Methods,* pp. 53–82. 2d ed. Agronomy Monograph 9. Soil Science Society of America. Madison.

Warrick, A. W., and D. R. Nielsen. 1980. Spatial variability of soil physical properties in the field. In D. Hillel (ed.), *Applications of Soil Physics,* pp. 319–344. Academic Press. New York.

Wood, L. S. 1988. Spatial variability of the physical properties of Captina soils. M.S. thesis. Department of Agronomy, University of Arkansas, Fayetteville.

Appendix B
Review of Mathematics

Scientists and engineers use mathematics as a tool to solve problems encountered in the design and analysis of soil systems and to quantify physical, chemical, and biological processes. Mathematics is especially helpful in quantifying soil physical properties and processes. This appendix reviews several important rules and laws from selected topics in mathematics that are often used in soil physics, other sciences, and engineering. Topics include algebra, functions, trigonometry, analytical geometry, linear algebra, calculus, differential equations, initial and boundary conditions, and numerical analysis. By necessity, the review is brief. It is recommended that for additional, more complete reviews, the student consult one or more of the modern mathematical textbooks on these topics.

Algebra

Definition and Characteristics

A relation is a law or rule that describes how one variable quantity is related to another. Symbolically, this can be expressed as

$$Y = f(x), \quad y = F(x), \quad y = g(x), \quad \theta_v = f(h), \quad C = f(x, t),$$

or in other equations similar to these that state that y is a function of x; that is, each value of x has a value of y that corresponds to it according to some law or rule. The symbol x is called the independent variable, and y is called the dependent variable. Independent variables often include spatial and temporal parameters but may also include soil temperature, soil water content, soil water pressure, solute concentration, time, depth in the profile, clay content, etc. As an example, if the variable θ_v represents the volume of water retained by a soil, then, in some cases,

$$\theta_v = f(h) = \theta_s e^{-ah}$$

where h is the soil water pressure, and θ_s and a are parameters that may have some physical significance. This relationship, which shows that θ is a function of a, h, and θ_s, has been described by several other mathematical equations.

If there exists more than one value of the dependent variable for each assigned value of the independent variable, the relation is said to be *multivalued*. An example of a multivalued relation is that of water retention in a soil exhibiting hysteresis. A function is, by definition, a *single-valued*

379

relation when for each value of the independent variable there is only one assigned value of the dependent variable. Examples include the rate laws of Darcy, Fick, and Fourier, where for a given value of the potential gradient (i.e., driving force) and transport coefficient, only one value is calculated for the flux density.

Often, the independent variable is constrained within a certain interval, called the *domain*. This may be expressed as follows: $a < x < b$, $a \leq x < b$, $a < x \leq b$, or $a \leq x \leq b$. These expressions are equivalent to the statement that x can be any number within the interval of numbers from a to b. The absence of an equal bar on the inequality sign, however, denotes that the adjacent number is excluded from the interval. The dependent variable is usually constrained to a similar interval, called its *range*. The domain and range are important in the mathematical formulation of the initial and boundary conditions of a soil system. Initial conditions are those equations that describe the state of the system at $t = 0$. Boundary conditions are those equations that define the behavior of the system at its boundaries, e.g., the soil surface and the bottom of the root zone.

Types of Elementary Functions

There are two basic types of elementary functions, polynomials and rational functions, which are formed by applying the arithmetic operations of addition, subtraction, multiplication, and division to the independent variable. *Polynomial* functions of *degree n* have the form

$$f(x) = a_0 + a_1x + a_2x^2 + a_3x^3 + \ldots + a_nx^n$$

and are continuous for all values of x in the range $-\infty < x < \infty$.

A special case of the polynomial functions is the straight line of degree one.

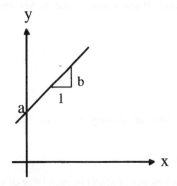

The general equation for a line is

$$y = f(x) = a \pm bx$$

where a is the y intercept and b is the slope. The sign of the slope, b, indicates whether the line slants up or down to the right or is horizontal or vertical. The slope is usually measured as

    b = rise/run

      = vertical change/horizontal change

      = y value change/x value change

      = $\Delta y/\Delta x$

      = $(y_2 - y_1)/(x_2 - x_1)$

The units of the y intercept are the same as those for y, whereas the units of the slope are the ratio of the units of y to the units of x.

The linear equation can be expressed in three forms:

Point-slope: $y - y_1 = b(x - x_1)$
Slope intercept: $y = a + bx$
Two intercept: $1 = (x/c) + (y/a)$

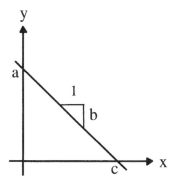

In these equations, b is the slope, a is the y intercept, and c is the x intercept.

The general second-degree polynomial equation is given by

$$f(x) = a + bx + cx^2$$

which represents a parabola. The solution of this quadratic equation is

$$x = [-b \pm (b^2 - 4ac)^{0.5}]/2a$$

If $b^2 - 4ac$ is negative, there are no x intercepts and the two roots are called complex. If $b^2 - 4ac > 0$, then the parabola intercepts the x axis twice. If $b^2 - 4ac = 0$, then there is only one x intercept.

Cubic and higher-order equations are most often solved by trial and error or with numerical analysis.

Rational functions (ratios of polynomials) have the form

$$f(x) = (a_0 + a_1x + a_2x^2 + \ldots)/(b_0 + b_1x + b_2x^2 + \ldots + b_nx^n)$$

where both the numerator and denominator are polynomials and are continuous. The rational function is also continuous except where the denominator vanishes. Polynomials are frequently used in statistical regression models because they can approximate numerous functions.

Algebraic functions tend to be more complicated and may have forms such as

$$y = x^{0.5} \qquad \text{or} \qquad y = (1 + x)^{0.5}$$

Another group of elementary functions are the *transcendental* functions. This group includes the trigonometric, exponential, and logarithmic functions. We will examine them in the sections that follow.

Trigonometry

Trigonometry is the study of triangles. The primary functions in trigonometry involve the ratios between the sides of a right triangle. The right triangle is shown below, where θ is the angle measured in degrees or radians; x and y are the lengths of the sides, perpendicular to each other; and r is the hypotenuse of the triangle.

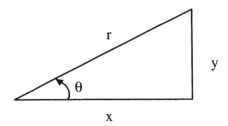

The six trigonomic functions are defined by using a circle with the equation $x^2 + y^2 = r^2$ and an angle in standard position with its vertex at the center of the circle and its initial side along the positive portion of the x axis. The trig functions are defined as

sine: $\sin \theta = y/r$
cosine: $\cos \theta = x/r$
tangent: $\tan \theta = y/x$
cotangent: $\cot \theta = x/y$
secant: $\sec \theta = r/x$
cosecant: $\csc \theta = r/y$

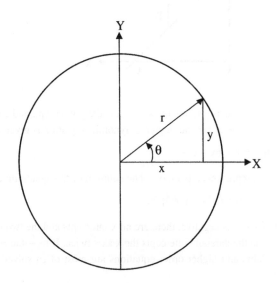

The trig functions $\sin \theta$ and $\cos \theta$ are periodic functions with a period of 2π. This means that they perfectly reproduce themselves every interval of $\theta = 2\pi$; for example, $\sin(\theta \pm 2\pi) = \sin \theta$.

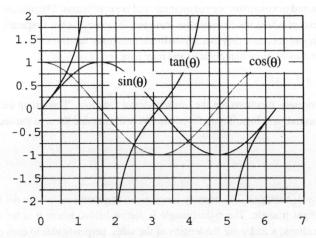

Graphs of the trigonometric functions for $\sin \theta$, $\cos \theta$, and $\tan \theta$.

There are two important units in trigonometry: degrees and radians.

Degrees: A circle is divided into 360 equal angles. Each of these angles is 1 degree (1°). Each degree is further subdivided into 60 minutes (60′), and each minute is subdivided into 60 seconds (60″).

Radians: The value of an angle in radians is the length of arc of a circle divided by the radius of the circle. The units of many practical applications in soil physics have angles measured in radians.

The relation between these two types of angles is as follows:

$$1 \text{ radian} = 360°/2\pi = 180°/\pi = 57.29577°$$
$$1 \text{ degree} = 2\pi \text{ radians}/360 = \pi \text{ radians}/180 = 0.01745 \text{ radians}$$

It is often convenient to use the inverse trigonometric function, which designates the angle in a specified interval for which the trig function has the specified value. Thus, the inverse trig function to $y = \sin \theta$ is $\theta = \arcsin y$, which is read as the "arcsine of y" and stands for the angle whose sin is y (or inverse sin is y). The arccos y, arctan y, etc., are similarly defined. The arc functions are multivalued and cyclic and generally defined over a limited range of y that is repeated indefinitely.

The following trigonometric identities are useful in solving problems.

$$\sin^2 \theta + \cos^2 \theta = 1$$
$$\sin 2\theta = 2 \sin \theta \cos \theta$$
$$\cos 2\theta = \cos^2 \theta - \sin^2 \theta$$
$$\sin(a \pm b) = \sin(a) \cos(b) \pm \sin(b) \cos(a)$$
$$\cos(a \pm b) = \cos(a) \cos(b) \mp \sin(a) \sin(b)$$
$$\tan(a \pm b) = [\tan(a) \pm \tan(b)]/[1 \mp \tan(a) \tan(b)]$$

Note the sign reversals $\pm$ and $\mp$ mean either $(+, -)$ or $(-, +)$, respectively.

Exponential and Logarithmic Functions

The *exponential* function has the form

$$y = a^x$$

where x is a real number and the parameter a is a positive number called the *base*. The value of the base cannot be equal to 1. Then, we write

$$x = \log_a y$$

which is called the logarithm of y to the base a. The number $y = a^x$ is called the *antilogarithm* of x to the base a. Graphs of $y = a^x$ and $x = \log_a y$ are identical.

The equation $y = \log_a x$ is called the inverse function of the exponential function $y = a^x$. The graph of $y = a^x$ is a mirror image of $y = \log_a x$ about $y = x$. In general, the inverse of $y = f(x)$, when it exists, is written as $x = f^{-1}(y)$.

There are two special values of the base a that because of its importance in mathematics are worthy of special consideration. The first is $\log_{10} y$, which is known as common log to a base 10 and is often written as log y, and the second is ln y, or natural log to the base e ($e = 2.71828 \ldots$). If any other base is used, it should be designated as $\log_b y$.

Logarithms of positive numbers less than 1 are negative, the logarithm of 1 is 0 for all bases, and logarithms of numbers greater than 1 are positive. Several characteristics of logarithms are given below:

$$\ln x^a = a \ln x$$
$$\ln(xy) = \ln x + \ln y$$
$$\ln(x/y) = \ln x - \ln y$$
$$\ln x = 2.303 \log x$$
$$\log_b = 1$$
$$\ln 1 = 0$$
$$\ln e^a = a$$

The symbol e has the value 2.71828 . . . and is a fundamental constant like the number π. The exponential function $y = e^x$ changes at a rate proportional to itself, is not changed by differentiation, and can be differentiated repeatedly, always reproducing itself. No other function remains unchanged by differentiation except constant multiples of this one. Many physical, chemical, and biological processes appear to behave exponentially.

Exponents

Several characteristics of exponents are given below:

$$x^0 = 1$$
$$x^{-a} = 1/x^a$$
$$x^a x^b = x^{a+b}$$
$$(xy)^a = x^a y^a$$
$$(x^a)^b = x^{ab}$$
$$x^a/x^b = x^{a-b}$$

Partial Fractions

A rational fraction $P(x)/Q(x)$ where both $P(x)$ and $Q(x)$ are polynomials and the degree of $P(x)$ is less than the degree of $Q(x)$ can be resolved into partial fractions for the following five cases.

Case 1: $Q(x)$ factors into n different linear terms; that is, $Q(x) = (x - a_1)(x - a_2) \ldots (x - a_n)$. Then

$$\frac{P(x)}{Q(x)} = \sum_{i=1}^{n} \frac{A_i}{x - a_i}$$

Case 2: $Q(x)$ factors into n identical terms; that is, $Q(x) = (x - a)^n$. Then

$$\frac{P(x)}{Q(x)} = \sum_{i=1}^{n} \frac{A_i}{(x - a)^i}$$

Case 3: $Q(x)$ factors into n different quadratic terms; that is, $Q(x) = (x^2 + a_1 x + b_1)(x^2 + a_2 x + b_2) \ldots (x^2 + a_n x + b_n)$. Then

$$\frac{P(x)}{Q(x)} = \sum_{i=1}^{n} \frac{A_i x + B}{x^2 + a_i x + b_i}$$

Case 4: $Q(x)$ factors into n identical quadratic terms; that is, $Q(x) = (x^2 + ax + b)^n$. Then

$$\frac{P(x)}{Q(x)} = \sum_{i=1}^{n} \frac{A_i x + B}{(x^2 + ax + b)^i}$$

Case 5: $Q(x)$ factors into a combination of the above. The obvious partial fractions are obtained from the expansions of the above.

Geometry

Shapes of some common objects found in soil science include the following:

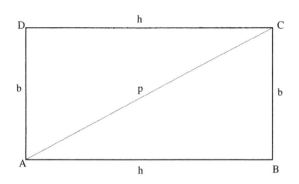

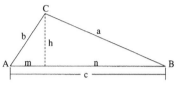

Triangle
Area = ch/2
Perimeter = a + b + c

Rectangle
Area = bh
Perimeter = 2b + 2h

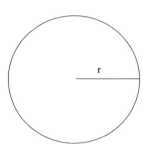

Circle
Area = πr^2
Perimeter = $2\pi r$

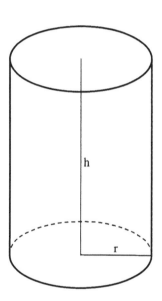

Right Circular Cylinder
Volume = $\pi r^2 h$
Lateral surface area = $2\pi rh$
Total surface area = $2\pi r(r + h)$

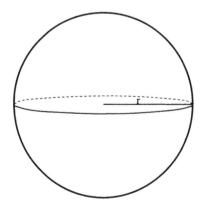

Sphere
Volume = $(4/3)\pi r^3$
Surface area = $4\pi r^2$

Coordinate Systems

In some applied problems, it is more convenient to use other coordinate systems instead of the rectangular (or Cartesian) coordinate system. Three other coordinate systems often used in science and engineering applications are

polar (r, θ)
cylindrical (r, θ, z)
spherical (r, θ, φ)

The *polar* coordinate system is restricted to a two-dimensional plane. The relationships between the polar coordinates r, θ and the rectangular coordinates x, y are given below.

$x = r \cos \theta$
$y = r \sin \theta$
$r = (x^2 + y^2)^{0.5}$

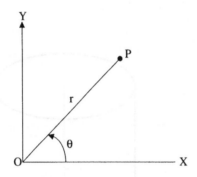

The distance between two points $P_1(x_1, y_1, z_1)$ and $P_2(x_2, y_2, z_2)$ in three dimensions is

$d = [(x_2 - x_1)^2 + (y_2 - y_1)^2 + (z_2 - z_1)^2]^{0.5}$

The *cylindrical* coordinate system is three-dimensional and is represented by the following relationships between the cylindrical coordinates r, θ, z and the rectangular coordinates x, y, z.

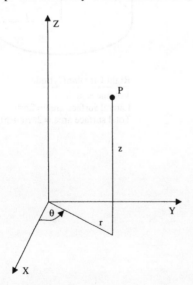

$$x = r \cos \theta \qquad r = (x^2 + y^2)^{0.5}$$
$$y = r \sin \theta \qquad \theta = \tan^{-1}(y/x)$$
$$z = z \qquad z = z$$

Cylindrical coordinates are just polar coordinates in the (x, y) plane with z for the third variable. Plant roots and fungal mycelia can often be approximated with cylindrical geometry.

The *spherical* coordinate system is also three-dimensional and can be represented by

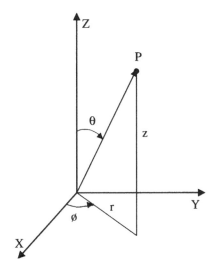

$$x = r \sin \theta \cos \phi \qquad r = (x^2 + y^2 + z^2)^{0.5}$$
$$y = r \sin \theta \sin \phi \qquad \phi = \tan^{-1}(y/x)$$
$$z = r \cos \theta \qquad \theta = \cos^{-1}[z/(x^2 + y^2 + z^2)^{0.5}]$$

Note that the r and θ in the spherical coordinate system are defined differently from the cylindrical and polar coordinates r and q. Seed and some bacteria can often be approximated with spherical geometry.

Taylor Series for Functions of One Variable

Many problems cannot be solved in terms of the so-called elementary functions. In such cases, an infinite series with only as many terms as necessary is used to obtain the accuracy desired. A power series with a nonzero radius of convergence can be used to represent an analytic function. One of the most important of the power series is the Taylor series. A Taylor series in general means a series of powers of $(x - a)$ where a is some constant. The general form is

$$f(x) = f(a) + f'(a)(x - a) + f''(a)(x - a)^2/2! + \ldots + f^{n-1}(a)(x - a)^{n-1}/(n - 1)! + R_n$$

where the parameter $(x - a)$ is a distance from the center, a, and R_n is the remainder after n terms, which is given by either of the following forms:

Lagrange form: $R_n = f^n(\xi)(x - a)^n/n!$

where ξ is in the range (a, x); or

Cauchy's form: $R_n = f^n(\xi)(x - \xi)^{n-1}(x - a)/(n - 1)!$

A Taylor series with center a = 0 is referred to as a Maclaurin series. Maclaurin power series representations of four frequently used functions are given below.

$$e^x = 1 + x/1! + x^2/2! + \ldots = \sum_{k=0}^{\infty} x^k/k!$$

$$\ln(1+x) = x - x^2/2 + x^3/3 - x^4/4 + \ldots -1 < x \le 1$$

$$\sin x = 1 - x^3/3! + x^5/5! - \ldots = \sum_{k=0}^{\infty} (-1)^k x^{2k+1}/(2k+1)!$$

$$\cos x = 1 - x^2/2! + x^4/4! - \ldots = \sum_{k=0}^{\infty} (-1)^k x^{2k}/2K!$$

The Taylor power series can be integrated and differentiated term by term within its circle of convergence.

Binomial Theory

The binomial equation is used to expand an algebraic expression of the form $(a + x)^n$:

$$(a + x)^n = a^n + na^{n-1}x + n(n-1)/2! \ a^{n-2}x^2 + \ldots \qquad (x^2 < a^2)$$

If n is zero, then the result is 1.0. If n is positive, the expansion contains $(n + 1)$ terms. If n is negative or a fraction, an infinite series expansion results.

Matrices and Systems of Equations: Linear Algebra

One of the primary objectives of linear algebra is to solve a set of n linear, algebraic equations for n unknowns $(x_1, \ldots, x_n)$. One of the best ways to solve these types of equations is with matrix algebra.

A matrix is defined as a rectangular array of numbers. The numbers in the array are called entries, or *elements,* of the matrix. If a matrix has m rows and n columns, we say that the size is m by n, written as m $\times$ n. An n $\times$ n matrix is called a *square* matrix or a matrix of *order* n. A 1 $\times$ 1 matrix is a constant or scalar.

The element in the ith row and j^{th} column of an m $\times$ n matrix **A** is written as a_{ij} and is represented as $\mathbf{A} = \{a_{ij}\}_{m,n}$. For an n $\times$ n square matrix, the elements $a_{11}, a_{22}, \ldots, a_{nn}$ are called the main diagonal elements.

Properties of a Matrix

Suppose that **A, B,** and **C** are m $\times$ n matrices and k_1 and k_2 are scalars. Then we write for matrix addition and scalar multiplication

$$\mathbf{A} + \mathbf{B} \qquad = \mathbf{B} + \mathbf{A}$$
$$\mathbf{A} + (\mathbf{B} + \mathbf{C}) = (\mathbf{A} + \mathbf{B}) + \mathbf{C}$$
$$(k_1 k_2)\mathbf{A} \qquad = k_1(k_2 \mathbf{A})$$
$$1\mathbf{A} \qquad = \mathbf{A}$$
$$k_1(\mathbf{A} + \mathbf{B}) \qquad = k_1 \mathbf{A} + k_1 \mathbf{B}$$
$$(k_1 + k_2)\mathbf{A} \qquad = k_1 \mathbf{A} + k_2 \mathbf{A}$$

Matrix multiplication is an important procedure in practical problems. Let **A** be a matrix having m rows and p columns, and let **B** be a matrix having p rows and n columns. The product **AB** is the m $\times$ n matrix. This is illustrated by

$$\mathbf{A}_{m \times p} \cdot \mathbf{B}_{p \times n} = \mathbf{C}_{m \times n}$$

Note that the product **AB** is defined only when the number of columns in matrix **A** is the same as the number of rows in matrix **B**. The dimensions of the **C** matrix are m × n. Unlike matrix addition, matrix multiplication is, in general, not commutative; that is, **BA** ≠ **AB**. Two additional matrix multiplication laws are

$$\mathbf{A}(\mathbf{BC}) \quad = (\mathbf{AB})\mathbf{C}$$
$$\mathbf{A}(\mathbf{B} + \mathbf{C}) = \mathbf{AB} + \mathbf{AC} \qquad \text{if both } \mathbf{B} \text{ and } \mathbf{C} \text{ are p} \times \text{n}$$

In general, when a matrix **A** of m rows and n columns, called the m × n matrix **A**, is multiplied by an n × p matrix **B**, the result is an m × p matrix **C**, written **AB** = **C**. The elements of **C** = $\{c_{ik}\}_{m,p}$ are defined as

$$c_{ik} = \sum_{j=1}^{n} a_{ij} b_{jk}$$

where i = 1 to m, k = 1 to p. In matrix multiplication, the order of multiplication is important. The multiplication **BA** cannot be performed unless p = m, and even then the result will not be **C**, except in special cases.

In the soil profile, a set of linear equations may be representative of soil properties arranged by depth interval or horizon. The set, or system, of equations that we have to solve may have the form

$$
\begin{array}{l}
a_{11}x_1 + a_{12}x_2 + \ldots + a_{1n}x_n = c_1 \\
a_{21}x_1 + a_{22}x_2 + \ldots + a_{2n}x_n = c_2 \\
\ldots \qquad \ldots \qquad \ldots \qquad \ldots \; = \ldots \\
a_{n1}x_1 + a_{n2}x_2 + \ldots + a_{nn}x_n = c_n
\end{array}
\quad \text{or} \quad
\begin{vmatrix}
a_{11} & a_{12} & \ldots & a_{1n} \\
a_{21} & a_{22} & \ldots & a_{2n} \\
\ldots & \ldots & \ldots & \ldots \\
a_{n1} & an_2 & \ldots & a_{nn}
\end{vmatrix}
\begin{vmatrix}
x_1 \\
x_2 \\
\ldots \\
x_n
\end{vmatrix}
=
\begin{vmatrix}
c_1 \\
c_2 \\
\ldots \\
c_n
\end{vmatrix}
$$

where the x's are various soil properties. In matrix algebra, these equations are written as **Ax** = **c**.

This is also an example of matrix multiplication. Note how each element of **c** is a sum of the products of the elements of a row of **A** times the elements of the single column of **x**. For this to work, the number of elements in a row of **A** must equal the number of elements in the column of **x**.

In setting up a problem like this, we may also wish to perform matrix addition. Suppose we have the m × n matrices **A** and **B** and the n × 1 vectors **x** and **y** such that

$$(\mathbf{A} + \mathbf{B})\mathbf{x} = \mathbf{y}$$

We can simplify the equation to **Cx** = **y** where **C** is the m × n matrix with elements c_{ij} such that

$$c_{ij} = a_{ij} + b_{ij}$$

where i = 1 to m, j = 1 to n. In this case, **A**, **B**, and **C** must all have the same size, or dimensions, of m × n, but the order of addition does not matter.

In order to solve the first problem of n equations in n unknowns, **Ax** = **c**, for n unknowns **A** must always be of dimensions n × n for vectors **x** and **c** of dimensions n × 1. Not only that, but none of the rows of **A** can be linear combinations of any of the other rows, or the problem cannot be solved for all of the $(x_1, \ldots, x_n)$ unknowns. If, for example, the first row were the simple sum of the second and third rows, then we have at best n – 1 equations with n unknowns. In that case, we cannot find a numerical answer for at least one of the unknowns; it has to be expressed as a linear combination of the other unknowns.

A set of equations may have (1) no solution, (2) a unique solution, or (3) an infinite number of solutions. In order to have at least one solution, the equations must be consistent. This means that any linear dependence among the rows of **A** must also exist among the corresponding elements of **c**. Consistent equations have either a unique solution or an infinite number of solutions. If the rank r(A) equals the number of unknowns, then the solution is unique and is given by

$$\mathbf{x} = \mathbf{A}^{-1}\mathbf{c}$$

when **A** is square (n × n) or

$$\mathbf{x} = \mathbf{A}_1^{-1}\mathbf{c}$$

where $\mathbf{A}_1$ is a full-rank submatrix of **A** when **A** is rectangular (m × n).

To solve $\mathbf{Ax} = \mathbf{c}$ for x, we must construct the inverse of **A**, called $\mathbf{A}^{-1}$. The inverse can only be constructed for an n × n matrix, also called a square matrix. It has the property that when multiplied by **A** on either side, it generates the *identity* matrix, **I**:

$$\mathbf{A}^{-1}\mathbf{A} = \mathbf{AA}^{-1} = \mathbf{I}$$

where $i_{ij} = 1$, i = j; and $i_{ij} = 0$, i ≠ j.

The identity matrix **I** has elements of 1 on the diagonal and zero elsewhere. It performs nearly the same function as 1 in scalar arithmetic. Any matrix left- or right-multiplied by **I** remains unchanged.

Therefore, to get **x**, we left-multiply both sides of the equation by $\mathbf{A}^{-1}$:

$$\mathbf{Ax} = \mathbf{c} \rightarrow \mathbf{A}^{-1}\mathbf{Ax} = \mathbf{A}^{-1}\mathbf{c} \rightarrow \mathbf{Ix} = \mathbf{A}^{-1}\mathbf{c} \rightarrow \mathbf{x} = \mathbf{A}^{-1}\mathbf{c}$$

The inverse of a matrix is obtained in general by (1) replacing every element of the matrix and its cofactor, (2) transposing the resulting matrix, and (3) dividing by the determinant of the original matrix.

The *determinant* of a matrix is a scalar computed from the elements according to well-defined rules. Determinants are defined only for square matrices and are denoted by |A|, where A is a square matrix. For example, if A is a 2 × 2 matrix,

$$\mathbf{A} = \begin{vmatrix} a_{11} & a_{12} \\ a_{21} & a_{22} \end{vmatrix}$$

then $|\mathbf{A}| = a_{11}a_{22} - a_{12}a_{21}$. The determinants of higher-order matrices are determined by expanding the determinants as linear functions of determinants of 2 × 2 submatrices. First, it is convenient to define the *minor* and the *cofactor* of an element in a matrix. Let **A** be a square matrix of order n. For any element a_{rs} in **A**, a square matrix of order (n – 1) is formed by eliminating the row and column containing the element a_{rs}. We label this matrix $\mathbf{A}_{rs}$. Then $|\mathbf{A}_{rs}|$ is called the minor of the element a_{rs}. The product $M_{rs} = (-1)^{i+j}|\mathbf{A}_{rs}|$ is called the cofactor of a_{rs}. Each element in a square matrix has its own minor and cofactor. The determinant of a square matrix of order n is expressed in terms of the elements and any row or column and their cofactors:

$$|\mathbf{A}| = \sum_{i=1}^{n} a_{ij}M_{ij}$$

using row i for illustration.

Except for the simplest cases of 2 × 2 or 3 × 3 matrices, inverse matrices are usually generated in practice by numerical methods or computers. Matrix addition, multiplication, and inversion are common tools in many popular spreadsheet programs. To become adept in working with these techniques, the reader is encouraged to take a course in linear algebra, advanced engineering mathematics, and/or numerical methods.

Differential Calculus

Calculus, discovered in the 17th century by Newton and independently by Leibniz, has its origins in efforts to accurately describe the motion of bodies. Calculus is where advanced mathematics begins. Despite this, the mathematical manipulations used in elementary calculus are often simple to apply.

Definition of Limit

The limit of a function f(x) describes the behavior of the function close to the particular x value. Suppose a function f(x) is defined in an open interval containing the point a, except possibly at point a itself. Then, we write that

$$\lim_{x \to a} f(x) = L$$

which means that as x approaches a, the function f(x) approaches the number L. If the function does not approach L as x approaches a, then the limit does not exist.

The concept of the limit allows us to give a precise formulation of instantaneous rates of change, which are expressed as derivatives of the function f(x).

Definition of a Derivative

Suppose a function f is defined on an open interval containing a point x. Then the derivative of f at x, whenever it exists, is given by

$$\lim_{\Delta x \to 0} \frac{f(x + \Delta x) - f(x)}{\Delta x}$$

Note that this expression depends on the chosen point x, so that the derivative of a function is also a function. In mathematical notation, this "derived function" is denoted in a variety of ways: $f'(x)$, $df(x)/dx$, y', dy/dx, where $y = f(x)$. Although it is important to understand the fundamental definition of the derivative in order to employ it in modeling rates of processes, it is much too cumbersome for use in the computation of derivatives. For this reason, rules of differentiation have been formulated using the definition. The following are the derivatives of some of the more commonly used elementary functions in soil physics:

$$d(a)/dx = 0 \text{ (a is a constant)}$$
$$d(x^p)/dx = px^{p-1} \text{ (p is a real number)}$$
$$d(e^x)/dx = e^x$$
$$d(\ln x)/dx = 1/x \text{ (where } x > 0)$$
$$d[\sin(x)]/dx = \cos(x)$$
$$d[\cos(x)]/dx = -\sin(x)$$

The following rules of differentiation show how the derivative of various combinations of functions depends on the derivatives of the individual functions f(x) and g(x):

$$d(f + g)/dx = f' + g'$$
$$d[af(x)]/dx = af' \text{ (where a is a real number)}$$
$$d[f(x)g(x)]/dx = fg' + f'g$$
$$d[f(x)/g(x)]/dx = (gf' - fg')/g^2$$
$$d\{f[g(x)]\}/dx = f'[g(x)]g'(x)$$

The last of these rules of differentiation, known as the *chain rule,* shows how to differentiate the composition of two functions.

If the process of differentiation is repeated, higher-order derivatives are obtained. Some notations for higher-order derivatives are $f''(x)$, d^2y/dx^2, y''. For example, $d^2(\sin x)/dx^2 = -\sin x$. If $x(t)$ is the location of an object moving on a line at time t, then $x'(t)$ is the velocity and $x''(t)$ is the acceleration of the object at time t.

Two of the most common applications of differentiation are optimization and approximation. The applicability of differentiation to optimization problems is a result that if a continuous, differentiable function $f(x)$ on a closed interval [a, b] attains its maximum (or minimum) value on [a, b] at a point c in the open interval (a, b), then the slope or $f'(c)$ equals zero.

In practice, this property leads us to consider the equation $f'(x) = 0$. Solving $f'(x) = 0$ is like a "primary election," considerably narrowing the field of candidates at which a desired maximum value can occur.

In incremental form, the slope of a curve $y = f(x)$ is the ratio of the change in y to the change in x as the change in x becomes infinitesimally small. Using the definition, the first derivative can be written as

$$\frac{dy}{dx} = \lim_{\Delta \to 0} \frac{\Delta y}{\Delta x} = \lim_{\Delta x \to 0} \frac{f(x + \Delta x) - f(x)}{\Delta x}$$

This equation is another form of the definition of the derivative. Alternative notations of the first derivative are $f'(x)$, $d[f(x)]/dx$, dy/dx, and y'.

The derivative of a function $f(x)$ at point P is shown graphically in the figure below.

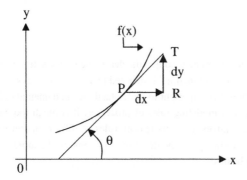

The second derivative of the function with respect to x can be written as

$$\frac{d^2y}{dx^2} = \lim_{\Delta x \to 0} \frac{\Delta y'}{\Delta x} = y'' = \frac{[df(x + \Delta x)/dx] - [df(x)/dx]}{\Delta x}$$

Integral Calculus

The inverse of differentiation is called integration or the antiderivative. Suppose a curve, given by $y = f(x)$, is continuous in the closed interval [a, b], and this interval is divided into n equal parts such that $\Delta x = (b - a)/n$. Then the definite integral of $f(x)$ with respect to x, $F(x)$, between the limits $x = a$ and $x = b$ is given by

$$\int_a^b f(x)\, dx = \lim_{\Delta x \to 0} \sum_{i=0}^{n-1} f(x_i)\Delta x = \int f(x)dx\, |_a^b$$

$$= F(x)\, |_a^b - F(a)$$

Geometrically, the definite integral of a positive function $f(x)$ with respect to x between limits $x = a$ and $x = b$ is the area under the curve bounded by $f(x)$, the x axis, with the vertical lines through the endpoints of a and b.

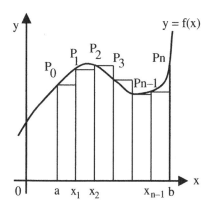

If the integral has no limits, it is called an *indefinite* integral. Similarly, if the integral has limits, it is a *definite* integral. Limits of the integral are almost always needed in practical situations.

Some indefinite integrals of elementary functions are listed below, with the symbol C serving as the constant of integration.

$$\int 1 \, dx \quad = x + C$$
$$\int af(x) \, dx \quad = a\int f(x) \, dx$$
$$\int x^n \, dx \quad = [x^{n+1}/n + 1] + C \quad (\text{if } n \neq -1)$$
$$\int x^{-1} \, dx \quad = \ln x + C$$
$$\int e^{ax} \, dx \quad = (1/a)e^{ax} + C$$
$$\int \sin x \, dx \quad = -\cos x + C$$
$$\int \cos x \, dx \quad = \sin x + C$$
$$\int (u + v) \, dx = \int u \, dx + \int v \, dx$$
$$\int u \, dv \quad = uv - \int v \, du$$

where C is an arbitrary constant of integration, u and v are variables that depend on x, and n is a constant. Because of the infinite number of values of C, there are an infinite number of integrals, differing only by the relative position of the antiderivative to the x axis.

Numerical Integration

Frequently, in soil physics experiments, tabulated values are obtained that represent the results of a time or spatial distribution of a parameter. In particular, we may be interested in determining the total mass present or the area under the curve.

In calculus, the definite integral is limit of what are called Riemann sums. Mathematically, this is represented by the fundamental theorem of integral calculus as

$$I(x) = \int_a^b f(x) \, dx$$

The integration of this expression results in

$$I(x) = F(b) - F(a)$$

where $F(x)$ is any antiderivative of $f(x)$. While many integrals can be determined by this method, many integrals in experimental soil physics cannot be evaluated in this way because the integrals do not have antiderivatives expressible in terms of elementary functions. Thus, additional methods are needed for evaluating such integrands. The central strategy behind most ideas for approximating equations is to replace $f(x)$ by an approximating function whose integral can be evaluated. These formulas are known as Newton-Cotes integration formulas and are based on replacing a complicated function or tabulated data with some approximating function that is easy to integrate. We will present only two of the simplest numerical integration methods of approximation.

Midpoint Rule

One way of approximating a definite integral is by constructing rectangular elements under the graph of $f(x)$ and adding their areas. Suppose that $y = f(x)$ is continuous on $[a, b]$ and that this interval is partitioned into n subintervals of equal width $\Delta x = (b - a)/n$. A simple and fairly accurate approximation rule consists of summing the areas of the n rectangular elements whose lengths are calculated at the midpoints of each subinterval. Graphically, this is shown below.

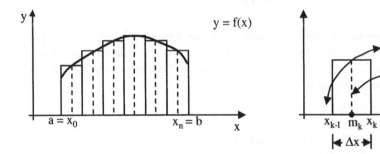

Illustration of the midpoint rule.

We let m_k be the midpoint of the rectangular element $[x_{k-1}, x_k]$; that is, $m_k = (x_{k-1} + x_k)/2$. The area of rectangle k is

$$A_k = \text{width} \times \text{height} = f(m_k)\Delta x = [f(x_{k-1} + x_k)/2]\Delta x_k$$

By identifying the endpoints $a = x_0$ and $b = x_n$ and summing the areas of the n rectangles, we obtain

$$A_n = \int_a^b f(x) \ dx \approx [f(x_0 + x_1)/2]\Delta x + [f(x_1 + x_2)/2]\Delta x + \ldots + [f(x_{n-1} + x_n)/2]\Delta x$$

If we replace Δx by $(b - a)/n$, the *midpoint rule* can be summarized as

$$A_n = \int_a^b f(x) \ dx \approx [(b-a)/n]\{[f(x_0 + x_1)/2]\Delta x + [f(x_1 + x_2)/2]\Delta x + \ldots + [f(x_{n-1} + x_n)/2]\Delta x\}$$

The accuracy of the approximation can be improved by increasing the number of subintervals.

Trapezoidal Rule

The more popular method for estimating the integral is the *trapezoidal rule*. The trapezoidal rule is derived by first assuming that the function is linear between the boundaries; that is,

$$f(x) = f(a) + \{[f(b) - f(a)]/(b - a)\}(x - a) \qquad (a \le X < b)$$

where a and b are the lower and upper boundaries, respectively. The integral of f(x) over [a, b] is the area of the shaded trapezoid shown graphically below. Geometrically, this area is given by the product of the width and the averaged height, that is, the average of the function values at the endpoints. The area of the trapezoid is

$$A_k = \Delta x \{[f(x_{k-1}) + f(x_k)]/2\}$$

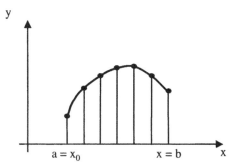 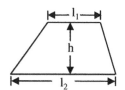

Illustration of the trapezoidal rule.

Improvement of the accuracy of the trapezoidal rule can be obtained when f(x) is not linear on [a, b] by dividing the interval into smaller subintervals and applying the equation for the one trapezoid on each subinterval. Again, we let the number of subintervals be denoted by n and the width of each subinterval by Δx. The value of Δx is calculated by

$$\Delta x = (b - a)/n$$

The endpoints of the subintervals are given by

$$x_j = a + j\Delta x$$

where j = 0, 1, 2, 3, . . ., n.

The trapezoidal rule is summarized as

$$A_n = \int_a^b f(x) \ dx \approx [(b-a)/2n] \ [f(x_0) + 2f(x_1) + 2f(x_2) + \ldots + 2f(x_{n-1}) + f(x_n)]$$

The points $x_0, x_1, x_2, \ldots, x_n$ are called the numerical integration node points.

The choice of n is important for the ease of calculations. In general, it is best to double n since each A value will include all of the earlier function values used in the preceding A, which ensures that all previously computed information is used in the new calculations, thereby making the method more accurate.

Ordinary Differential Equations

In soil physics, we often wish to describe or model the physical behavior of soils in mathematical terms. The description starts with identifying the variables that are responsible for changing the system and making reasonable assumptions about the boundaries. These assumptions may include empirical rate laws that are related to soil systems such as those of Darcy, Fick, Fourier, and Newton. The mathematical construct of these assumptions is often a differential equation whose solution should be representative of known physical behavior of the soil.

An equation containing the derivatives or differentials of one or more dependent variables, with respect to one or more independent variables, is said to be a differential equation. Differential equations are classified according to three properties: type, order, and linearity. We follow the presentation of Zill and Cullen (1992).

Classification by Type

If an equation contains only ordinary derivatives of one or more dependent variables, with respect to a single independent variable, then it is said to be an *ordinary differential equation* (ODE). For example,

$$dy/dx - 5y \qquad = 1$$
$$(x + y)\, dx - 4y\, dy \quad = 0$$
$$d^2y/dx^2 - 2dy/dx + 6y = 0$$

are ODEs because the only independent variable is x. By ODE, we mean a relation that involves one or several derivatives of an unspecified function y of x with respect to x; the relation may also involve y itself, given functions of x, and constants.

A differential equation involving the partial derivatives of one or more dependent variables and two or more independent variables is called a *partial differential equation* (PDE). For example,

$$\partial C/\partial t \quad = -\partial J/\partial z$$
$$\partial C/\partial t \quad = D\partial^2 C/\partial x^2$$

are PDEs because z, x, and t are independent variables. The rate laws in soil physics are included in this group of differential equations.

Classification by Order

The *order* of the highest derivative in a differential equation is called the order of the equation. For example,

$$d^2y/dx^2 + 5(dy/dx)^3 - 4y = x$$

is a second-order differential equation. Since the differential equation $x^2\, dy + y\, dx = 0$ can be put in the form

$$x^2\, dy/dx + y = 0$$

by dividing by the differential dx, it is an example of a first-order differential equation. The equation

$$a^2 \partial^4 u/\partial x^4 + \partial^2 u/\partial t = 0$$

is a fourth-order ordinary PDE.

Classification by Linearity

A differential equation is said to be linear if it has the following form:

$$a_n(x)\, d^n y/dx^n + a_{n-1}(x)\, d^{n-1}y/dx^{n-1} + \ldots + a_1(x)\, dy/dy + a_0(x)y = g(x)$$

Linear differential equations are characterized by two properties: (1) the dependent variable y and all its derivatives are of the first degree (i.e., the power of each term involving y is 1); and (2) each coefficient depends on only the independent variable x. An equation that is not linear is said to be nonlinear. Nonlinear equations, with the exception of some first-order equations, are usually difficult or impossible to solve analytically in terms of standard elementary functions. Examples of nonlinear differential equations are those of the transient-state flow equations above the water table where hydraulic conductivity depends upon the hydraulic head.

Solution of a Differential Equation

Any function f defined on some interval I that, when substituted into a differential equation, reduces the equation to an identity is said to be a solution of the equation on the interval. In practice, this means that a solution of a differential equation is a function $y = f(x)$ that possesses at least n derivatives and satisfies an equation of the form

$$F[x, f(x), f'(x), y', \ldots, f^n(x)] = 0$$

for every x in the interval I. An nth-order ODE will yield an n-parameter family of solutions. A differential equation is called *homogeneous* if any linear combination of solutions is also a solution.

Solutions of differential equations can be distinguished as explicit or implicit solutions. A relation $G(x, y) = 0$ is said to define a solution of a differential equation implicitly on an interval I provided it defines one or more explicit solutions on I. Solutions of an nth-order ODE result in an n-parameter family of solutions. A given differential equation usually possesses an infinite number of possible solutions.

A *general* solution of a differential equation involves a number of arbitrary constants equal to the order of the equation. If conditions are specified, the arbitrary constants may be calculated. When an n-parameter family of solutions gives every solution of a differential equation on some interval I, it is called a general, or complete, solution.

A *particular* solution is any solution free of arbitrary parameters that satisfies the differential equation. One way to obtain a particular solution is to use specific values of the parameters in a family of solutions, for example, the initial and/or boundary conditions.

A *singular* solution is any solution that cannot be obtained from an n-parameter family of solutions by assigning values to the parameters.

Solution of First-Order Ordinary Differential Equations

We consider two types of first-order differential equations. A first-order differential equation is *separable* if it can be expressed as

$$M(x) \, dx + N(y) \, dy = 0$$

The solution is obtained by integrating each of the terms.

A linear, first-order differential equation has the form

$$y' + h(x)y = g(x)$$

and the solution of

$$y(x) = (1/u)\int ug(x) \, dx + C/u$$

where $u(x)$ is called the integrating function and has the form

$$u(x) = e^{\int h(x)dx}$$

Solution of Second-Order Ordinary, Linear, Homogeneous Differential Equations with Constant Coefficients

The general form of a second-order, linear, homogeneous differential equation with constant coefficients is

$$y'' + Ay' + By = 0$$

To find a solution, we first solve the characteristic equation

$$m^2 + Am + B = 0$$

If the solutions to the characteristic equation $m_1 \neq m_2$ and both solutions are real, the general solution is

$$y(x) = c_1 e^{m_1 x} + c_2 e^{m_2 x}$$

If $m_1 = m_2$, the general solution is

$$y(x) = c_1 e^{m_1 x} + c_2 x e^{m_2 x}$$

If the components of m are complex, i.e., $m_1 = a + ib$ and $m_2 = a - ib$, then the general solution is

$$y(x) = [c_1 \sin(bx) + c_2 \cos(bx)] e^{ax}$$

Solution of Second-Order Ordinary, Linear, Nonhomogeneous Differential Equations with Constant Coefficients

The nonhomogeneous version of the second-order linear differential equation is

$$y'' + Ay' + By = g(x)$$

The general solution is found by finding the solution $y_h(x)$ to the homogeneous equation and adding to it a particular solution $y_p(x)$ using the table of particular solutions below. The solution is expressed as

$$y(x) = y_h(x) + y_p(x)$$

Particular solutions are given in the table below.

$g(x)$	$y_p(x)$	Provisions
a	C	
ax + b	Cx + D	
$ax^2 + bx + c$	$Cx^2 + Dx + E$	
e^{ax}	Ce^{ax}	if m_1 or $m_2 \neq a$
	Cxe^{ax}	if m_1 or $m_2 = a$
b sin(ax)	C sin(ax) + D cos(ax)	if $m_{1,2} \neq ai$
	Cx sin(ax) + Dx cos(ax)	if $m_{1,2} = \pm ai$
b cos(ax)	(same as above)	

Note: a, b, C, D, and E are constants whose values must be determined from the initial and boundary conditions.

Partial Differential Equations

A PDE differs from an ODE in that the state variable depends on more than one independent variable. For example, a PDE may model the transport of soil water in both space and time. PDE models may also be independent of time but depend on several spatial variables.

When differentiation is applied to multivariable functions it is called *partial* differentiation. Suppose soil temperature can be represented by T(z, t), then $\partial T / \partial t$ is the derivative of T with respect to t, holding z constant. It represents the slope of a line in a plane of constant z tangent to the surface of T(z, t) intersecting with the plane. The rate of change of T with t at a fixed depth z is

$$\frac{\partial T}{\partial t} = \lim_{h \to 0} \frac{T(t + h, z) - T(t, z)}{h}$$

Partial differentiation is like a controlled experiment: the rate of change of the function with respect to one of the variables is measured while holding all other variables fixed.

PDEs are classified according to their order and to other properties. For example, like ODEs, they are classified as linear or nonlinear. If the state variable and all of its derivatives are linear, then the PDE is linear. Otherwise, it is nonlinear. For example, Laplace's and Fourier's equations and Fick's second law are classified as linear. Linear PDEs have an algebraic structure to their solution sets, whereas nonlinear equations do not. As a result, nonlinear PDEs are more difficult to solve, and their solution sets are more difficult to analyze.

Initial and Boundary Conditions

Initial and boundary conditions are perhaps the most important set of conditions needed to describe the transport of heat, gas, water, and solutes in soil. Initial conditions describe the values of the solute concentration at some starting time equal to zero. Boundary conditions specify the interaction between the length, area, or volume under study and its external environment. The initial and boundary conditions represent the limits of integration and are necessary to obtain a solution to the transport equations. Mathematical solutions of elementary processes are of two types: analytical or numerical. In general, there are three types of boundary conditions used in transport modeling, and these are expressed in terms of location and time.

First Type of Boundary Condition

The boundary condition often called the Dirichlet boundary condition is when a fixed or constant value of the dependent variable is prescribed at one or both ends of a length (or time); for example,

$$C(0, t) = C_o$$
$$C(L, t) = C_L$$

where in the case of contaminant transport C_o and C_L are different solute concentrations that remain constant over time. If $C_o = 0$, then we have

$$C(0, t) = C_o = 0$$

which is called a homogeneous boundary condition of the first type. Dirichlet's boundary conditions are imposed on the boundary if the variable is independent of the flow conditions in the soil profile.

Second Type of Boundary Condition

The second type is often called the flux or Neumann boundary condition and concerns the fixed gradient of the concentration; for example,

$$-D\partial C/\partial z = J_o \quad (z = 0)$$
$$\partial C/\partial z = \alpha_L \quad (z = L)$$

There are two special cases. First, if the right-hand side is zero (i.e., J_o or α_L is zero) and no flow occurs, we have

$$\partial C/\partial z = 0$$

which is called the homogeneous boundary condition of the second type. The second special case occurs if a prescribed nonzero flux occurs at the boundary.

Third Type of Boundary Condition

The third type combines the first two types of boundary conditions and accounts for transport across the interface. It is known as the variable-flux boundary. If the flux of water is constant with a constant input concentration, then

$$-D\partial C/\partial z + vC = vC_0$$

Again, if the right-hand side is zero, we have a homogeneous boundary condition of the third type.

Finite Differences

Often, for problems of heat flow, unsaturated water flow, or contaminant transport, there may be no analytic solutions or neat equations describing the result. In such cases, we use numerical methods on a computer. Perhaps the simplest of these numerical methods to understand and to program are finite differences, derived from Taylor series expansions (DuChateau and Zachmann 1989). Some numerical solutions are so simple, they can even be done in a spreadsheet. But in the interests of accuracy, we will only discuss the methods that require some ability to program in a computer language such as C, BASIC, or FORTRAN. The example given here will be in FORTRAN but can be converted to other computer languages. The section was written by Dr. Don Baker.

Finite differences can be explained and used in a cookbook manner, if one is careful. If the reader has no other experience in these methods, he or she should keep in mind that this is a limited discussion. Such important issues as stability, convergence, iteration methods, implicitness, discretization errors, and non-Darcian flow will not be covered here. Thus, the reader must understand that a great deal more study is necessary before these methods can be used successfully in many cases.

Due to the need for brevity, we will discuss how to numerically solve only one type of problem: conduction of heat flow in soil. This type of problem is one of the easiest because it only involves a conduction/diffusion-type equation and is usually linear with many analytic solutions available (Carslaw and Jaeger 1959).

Development of the Numerical Equations

Consider the previous definition for the first ordinary or partial derivative (equations [1a] and [1b] below). Suppose that we have a temperature, $T(z, t)$, distributed along a grid of points in the soil profile, $z_0, z_1, \ldots, z_{i-1}, z_i, z_{i+1}, \ldots, z_{np-1}, z_{np}$, with np segments of fixed size, $\Delta z = z_{i+1} - z_i$, and that we want to evaluate the first partial derivative with respect to z, $\partial T(z, t)/\partial z$, from z_0 to z_{np}. We can simply look at the definition, equation [1b], and estimate the derivatives with equation [2]. This method is only an estimate and has error components linearly related to Δz, Δz^2, Δz^3, and on up, but the largest error component is directly related to the size of Δz. Thus, as Δz gets larger, so does the error of the approximation.

$$\partial T(z)/dz = \lim_{\Delta z \Rightarrow 0}[T(z + \Delta z) - T(z)]/\Delta z \qquad [1a]$$

$$\partial T(z, t)/\partial z = \lim_{\Delta z \Rightarrow 0}[T(z + \Delta z, t) - T(z, t)]/\Delta z \qquad [1b]$$

$$\begin{aligned}
\partial T(z, t)/\partial z &\approx \{T[z_0 + (i + 1)\Delta z, t] - T(z_0 + i\Delta z, t)\}/\Delta z \\
&= [T(z_{i+1}, t_j) - T(z_i, t_j)]/(z_{i+1} - z_i) \\
&= (T_{i+1}{}^j - T_i{}^j)/(z_{i+1} - z_i) \qquad [2]
\end{aligned}$$

Notice how we have changed the notation to make it more compact. If Δz is fixed over the range from z_0 to z_{np}, we can write each $z_i = z_0 + i*\Delta z$. If we have a Δt fixed over the range of interest, we can write $t_j = t_0 + j*\Delta t$. Each value of $T(z_i, t_j)$ can be given in notation as T_i^j, even if the Δz and Δt are not fixed. These values of T are called finite or discrete, because they occur, or are known, only at discrete positions, (z_i, t_j), in time and space.

If we want to estimate the second partial derivative [3], we apply the difference rules as shown below. The $1/\Delta z$ in front of the difference term should be expressed as $\Delta z = z_{i+1/2} - z_{i-1/2}$, where the $i + 1/2$ and $i - 1/2$ indicate the midpoints between z_{i+1} and z_i and between z_i and z_{i-1}. If all the z_i are fixed and Δz apart, then the differences between adjacent points are all Δz. If not, then $1/\Delta z$ can be expressed by $2/(z_{i+1} - z_{i-1})$. The error of this approximation goes as the size of Δz^2. Since the soil physics problems we discuss involve only first and second partial derivatives, [2] and [3] are the two basic formulas needed. But if these are blindly applied, without a full knowledge of stability and convergence criteria, they will likely fail.

$$\partial^2 T(z, t)/\partial z^2 = \lim_{\Delta z \Rightarrow 0} \{[\partial T(z + \Delta z, t)/\partial z] - [\partial T(z, t)/\partial z]\}/\Delta z$$

$$\approx (1/\Delta z)|[(T_{i+1}^j - T_i^j)/(z_{i+1} - z_i)] - [(T_i^j - T_{i-1}^j)/(z_i - z_{i-1})] \tag{3}$$

Example of One-Dimensional Heat Flow in Soil

Suppose that we want to estimate the solution of the transient-state Fourier's heat flow equation [4] in the vertical direction, where the space step, Δz, and time step, Δt, are fixed. Using equations [2] and [3], and also assuming that the thermal diffusivity, D_T, is constant in time and space, the explicit finite difference formula [5] can approximate it. It is called explicit because when all the discrete variables for T at the $j + 1$ time level (i.e., $t_{j+1} = t_0 + (j + 1)*\Delta t$) are placed on the left and all the variables at the j time level are placed on the right, the numerical solution for each T_i^{j+1} is separate and distinct and can be solved in a spreadsheet.

$$\partial T/\partial z = D_T(\partial^2 T/\partial z^2) \tag{4}$$

where T is soil temperature (°C or K), t is time (s), D_T is thermal diffusivity (m²/s), and z is the vertical distance, positive in the upward direction (m).

$$(T_i^{j+1} - T_i^j)/\Delta t = (D_T/\Delta z^2)(T_{i+1}^j - 2T_i^j + T_{i-1}^j) \tag{5}$$

If one knows the initial temperature distribution, $T(z, t = t_0)$, then one has the initial values of $T_i^{j=0}$. One can use equation [5] to solve for all of the values at $T_i^{j=1}$, and from those the values at $T_i^{j=2}$, and so on. So in this notation, the values at the j time level are always known and the values at the $j + 1$ time level are to be solved. This method of numerical approximation can be done in a spreadsheet.

Unfortunately, there are limitations on the sizes of Δt and Δz. If they are chosen incorrectly, method [5] becomes unstable and simply blows up. For this reason, we will use a more difficult form (equation [6]), called an implicit method. It is unconditionally stable with all Δz and Δt. If Δz and Δt are chosen poorly or too large, it may give an incorrect answer, but it will not explode. Therefore, we have

$$T_i^{j+1} - T_i^j = r(T_{i+1}^{j+1} - 2T_i^{j+1} + T_{i-1}^{j+1}) \tag{6}$$

where $r = (D_T \cdot \Delta t)/\Delta z^2$.

If we gather all the terms at the $j + 1$ time level on the left and all those at the j time level on the right, we see what looks like three unknowns, $(T_{i-1}, T_i, T_{i+1})^{j+1}$, in one equation [7]. Because of our use of compact notation, we actually have a large number of equations. We let $i = 0$ to np, where the boundary conditions are $T_0 = T(z_0)$ and $T_{np} = T(z_{np})$ for all time $t > t_0$. That gives us equations for $i = 1$ to $np - 1$, with unknown values of T_i^{j+1} for $i = 1$ to $np - 1$, making the solution possible.

$$[-rT_{i-1} + (1 + 2r)T_i - rT_{i+1}]^{j+1} = T_i^j \qquad [7]$$

Suppose that we have constant boundary conditions, T_0 and T_{np}. Those values are always known and go on the right-hand side, producing equations [8] and [9]. Even if T_0 and T_{np} vary with time as known functions, they go on the right-hand side because they are known, but to be mathematically consistent with the implicit method, they should be expressed at the $j + 1$ time level. An example of this situation might be a soil temperature that is considered to be constant at $z = -2$ m but to vary with time and weather at the soil surface, $z = 0$ m, known from datalogger measurements.

$$[(1 + 2r)T_1 - rT_2]^{j+1} \qquad = T_1^j + rT_0 \qquad [8]$$

$$[-rT_{np-2} + (1 + 2r)T_{np-1}]^{j+1} = T_{np-1}^j + rT_{np} \qquad [9]$$

Equations [7] to [9] can be expressed as a matrix equation in linear algebra, equation [10], multiplying an $(np - 1) \times (np - 1)$ matrix by an $(np - 1) \times 1$ matrix to get an $(np - 1) \times 1$ matrix. Notice how the boundary conditions fit on the first and last rows. Remember how in matrix multiplication a row of the matrix on the left multiplies a column of the matrix on the right of the multiplication. See how this gives back equations [7] to [9].

$$\begin{vmatrix} (1+2r) & -r & 0 & & & & \\ -r & (1+2r) & -r & \cdots & & & \\ 0 & -r & (1+2r) & & & & \\ & & & \cdots & & & \\ & & & (1+2r) & -r & 0 \\ & & & -r & (1+2r) & -r \\ & & & 0 & -r & (1+2r) \end{vmatrix} \begin{vmatrix} T_1^{j+1} \\ T_2^{j+1} \\ T_3^{j+1} \\ \cdots \\ T_{np-3}^{j+1} \\ T_{np-2}^{j+1} \\ T_{np-1}^{j+1} \end{vmatrix} = \begin{vmatrix} T_1 + rT_0 \\ T_2^j \\ T_3^j \\ \cdots \\ T_{np-3}^j \\ T_{np-2}^j \\ T_{np-1}^j + rT_{np} \end{vmatrix} \quad [10]$$

In the even more compact matrix notation, equation [10] can be expressed as equation [11], where the bold letters indicate the entire matrix and the exponent, -1, indicates the inverse matrix. It shows that if we have the inverse of the $(np - 1) \times (np - 1)$ matrix, $\mathbf{A}$, we can solve for all the T_i^{j+1} with a simple matrix multiplication.

$$\mathbf{A}\mathbf{T}_{j+1} = \mathbf{T}_j \Rightarrow \mathbf{A}^{-1}\mathbf{A}\mathbf{T}_{j+1} = \mathbf{A}^{-1}\mathbf{T}_j \Rightarrow \mathbf{T}_{j+1} = \mathbf{A}^{-1}\mathbf{T}_j \qquad [11]$$

The matrix $\mathbf{A}$ is called tridiagonal because it has only three diagonals of $-r$, $(1 + 2r)$, and $-r$. It is called diagonally dominant because $|1 + 2r| > |-r| + |-r|$; the absolute value of each element of the diagonal is greater than the sum of the absolute values of the elements on the same row off the diagonal. This second property ensures that the solution to equation [11] will exist. The first allows an easy solution with a well-known algorithm for solving tridiagonal systems of equations, subroutine tridia in Listing 1. If the second property is not true, then the algorithm will likely fail.

Suppose we wish to solve equation [12], at time $t_1 = 24$ h (or 86,400 s). Let the soil temperature at $z = -2$ m be $T_0 = 15°C$, the temperature at the soil surface, $z = 0$ m, be $T_{np} = 25°C$, the initial soil temperature between the boundaries be 16°C, and the thermal diffusion $D_T = 6 \times 10^{-7}$ m²/s. Listing 1, FORTRAN program fdtemp.for, shows how to set up and use subroutine tridia in each time step to solve for the next set of T_i. The problem becomes to solve

$$\partial T/\partial t = D_T(\partial^2 T/\partial z^2) \qquad [12]$$

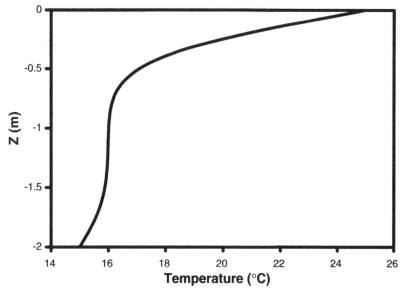

Figure B.1. Numerical simulation of the temperature distribution in the soil profile for Fourier's equation [12] after 24 hours.

with $D_T = 6 \times 10^{-7}$ m²/s; boundary conditions $T(z_0, t) = 15°C$, $T(z_{np}, t) = 25°C$, $z_0 = -2$m, $z_{np} = 0$, $t \geq t_0 = 0$; and initial conditions $T(z, t_0) = 18°C$, $z_0 < z < z_{np}$.

This program is set up to calculate 80 segments of $\Delta z = 0.025$ m, for time steps of $\Delta t = 1$ s, and then write out the results to the file outtem.dat. It begins by initializing all the variables, using the array tmj(i) for the T_i^j discrete values, and tm(i) for the T_i^{j+1} discrete values. Notice how the tridiagonal **A** matrix is set up in the aa(i), bb(i), and cc(i) arrays, and the T^j matrix is set up in the dd(i) array, with additions to the proper elements to account for the boundary conditions. When the subroutine tridia is called, it changes the aa, bb, and cc arrays and puts the answer for T^{j+1} in the dd array. Figure B.1 shows the temperature distribution in the soil 24 hours after the initial conditions. Notice how the temperature at the boundary temperatures has diffused into the soil from the ends.

Listing 1. FORTRAN Program fdtemp.for, an implicit finite-difference method to solve the heat flow equation (equation [12])

```
c       fdtemp

        implicit double precision (a-h, k, o-z)
        parameter (iz=100)
        dimension aa(0:iz), bb(0:iz), cc(0:iz), dd(0:iz)
        dimension tmj(0:iz), tm(0:iz)

        open (1, file='outtem.dat', status='unknown')
        zl = 2.d0
        tl = 3600.d0*24.d0
        np = 80
        dt = 1.0d0
```

```
dift = 6.d-7
t0 = 15.d0
tnp = 25.d0
tini = 16.d0
t = 0.d0
dz = zl/dble(np)
rz = dift*dt/(dz*dz)
rz2 = 1.d0 + 2.d0*rz
do i = 1, np-1
    tmj(i) = tini
    tm(i) = tini
end do
tm(0) = t0
tm(np) = tnp
do while (t .le. tl)
    t = t + dt
    do i = 1, np-1
        aa(i) = -rz
        bb(i) = rz2
        cc(i) = -rz
        dd(i) = tmj(i)
    end do
    dd(1) = dd(1) + rz*t0
    dd(np-1) = dd(np-1) + rz*tnp

    call tridia (np-1, aa, bb, cc, dd)

    do i = 1, np-1
        tm(i) = dd(i)
        tmj(i) = tm(i)
    end do

end do

do i = 0, np
    z = dble(i)*dz - zl
    write (1, '(f6.3,f7.3)') z, tm(i)
end do
close (1)

end

subroutine tridia (nn, a, b, c, d)

implicit double precision (a-h, o-z)
parameter (iz=100)
dimension a(0:iz), b(0:iz), c(0:iz), d(0:iz)
```

```
do n = 2, nn
    r = a(n)/b(n-1)
    b(n) = b(n) - r*c(n-1)
    d(n) = d(n) - r*d(n-1)
end do
d(nn) = d(nn)/b(nn)
do n = nn-1, 1, -1
    d(n) = (d(n) - c(n)*d(n+1))/b(n)
end do

return
end
```

Cited References

Carslaw, H. S., and J. C. Jaeger. 1959. *Conduction of Heat in Solids.* Oxford University Press. London.

DuChateau, P., and D. Zachmann. 1989. *Applied Partial Differential Equations.* Harper and Row. New York.

Zill, D. G., and M. R. Cullen. 1992. *Advanced Engineering Mathematics.* PWS-KENT Publishing Co., Wadsworth. Boston.

Additional References

CRC. 1995. *Standard Mathematical Tables.* Student ed. Chemical Rubber Company. Boca Raton, FL.

Logan, J. D. 1998. *Applied Partial Differential Equations.* Springer-Verlag. New York.

Potter, M. C. 1990. *Fundamentals of Engineering.* Great Lakes Press. Okemos, MI.

Waite, T. D., and N. J. Freeman. 1977. *Mathematics of Environmental Processes.* D. C. Heath and Co. Lexington, MA.

Appendix C
Review of Physics

Several parameters have been developed to describe the physical state of a system. The review below groups these parameters into translational motion, force, work and energy, and hydrostatics and dynamics. Since only a brief discussion is given for each selected parameter, the student is encouraged to consult textbooks on classical physics to obtain a broader background.

Translational Motion

Speed and Velocity

Speed can be determined by dividing the distance traveled by the length of time taken to travel that distance. It is a scalar quantity, since only magnitude is given. If the distance traveled is designated as Δs and the time needed to travel this distance is Δt, then the average speed can be calculated as

$$\text{average speed} = v = (s_2 - s_1)/(t_2 - t_1) = \Delta s/\Delta t \qquad [1]$$

We define Δs in three dimensions as $(x_2 - x_1, y_2 - y_1, z_2 - z_1)$. Speed can be positive or negative and has units in the SI system of meters per second, or m/s.

Velocity is a vector quantity having both speed and direction as its magnitudes. The instantaneous velocity is defined as

$$v = \lim_{\Delta t \to 0} \Delta s / \Delta t = ds / dt \qquad [2]$$

As the time interval is made smaller and smaller, the limit of the average velocity approaches ds/dt. The ratio is evaluated as the change in time approaches zero. The velocity is the tangent of the curve of distance traveled, s, versus time, t, at a given point.

Velocity is used whenever the rates of transport of soil particles, water, solutes, and energy are calculated. This is because we are interested not only in the rate of the process but also in the direction. Rates of processes are often keys to understanding soil behavior, and therefore, much emphasis is given in this book to calculations of rates of the various processes in soil.

Uniform Velocity

If a body traverses equal distances in equal time intervals, then it moves with uniform velocity. A plot of the distance traveled versus the time elapsed that results in a constant slope indicates that the velocity is constant. In this case, the value of $\Delta s/\Delta t$ will be constant.

Acceleration

The change in velocity divided by the corresponding time interval is known as acceleration. Mathematically, the average acceleration can be calculated from

$$\text{average acceleration} = a = (v_2 - v_1)/(t_2 - t_1) = \Delta v/\Delta t \qquad [3]$$

Equation [3] indicates that in the case of uniform velocity, the acceleration is zero. The units of acceleration in the SI system are meters per square second, or m/s^2.

The instantaneous time rate of change of velocity also is a vector quantity. Mathematically, acceleration is defined as

$$a = \lim_{\Delta t \to 0} \Delta v / \Delta t = dv / dt \qquad [4]$$

Acceleration can be positive, zero, or negative. A negative acceleration can indicate that an object has either (1) a decreasing velocity in the positive direction or (2) an increasing velocity in the negative direction.

Uniform Acceleration

Uniform acceleration occurs when the velocity increases at a specified and constant rate. When the acceleration is constant in magnitude and direction and when the motion is along the line of a, we have the relation

$$v_{ave} = 0.5(v_0 + v) \qquad [5]$$

where v_{ave} is the average velocity over the time interval 0 to t and v_0 is the velocity at t equal to zero. Under these conditions, the velocity is

$$v = v_0 + at \qquad [6]$$

Gravitational Acceleration

Every body falling freely near the surface of the earth has an almost constant downward acceleration of $g = 9.8 \ m/s^2$.

Force

Mass and Weight

The property a body has of resisting any change in its state of rest or of uniform motion in a straight line is called inertia. Inertia is related to what can be loosely called the amount of matter. The quantitative measure of inertia is mass. Mass is an invariant quantity with units of kilograms.

The weight of a body is the gravitational force exerted on the body. The weight of a body, which varies depending on g, has units of newtons (N). Thus, mass and weight are distinctly different quantities.

Newton's First Law of Motion

A body at rest remains at rest and a body in motion continues to move at a constant speed along a straight line unless there is a resultant force acting on the body. From this law, we have force defined as anything that changes or tends to change the state of motion of an object.

Newton's Second Law of Motion

The acceleration of a body takes place in the direction of the resultant force acting upon it. The acceleration is directly proportional to the resultant force and inversely proportional to the mass of the body. Therefore,

$$\sum F = \text{mass} \times \text{acceleration} = ma \tag{7}$$

where $\sum F$ is the vector sum of all forces acting on the body, m is the mass of the body, and a is the acceleration. The units of force in the SI system are $kg/m \cdot s^2$ or N.

For the special case where the resultant force is zero, then acceleration equals zero. This implies that the body's velocity is constant in magnitude and direction.

Newton's Third Law of Motion

For every action there is an equal and opposite reaction, and the two actions are directed along the same straight line. Therefore, no force occurs by itself.

Newton's first and second laws of motion state that the tendency of any body is to stay in the state that it finds itself. The natural state is either for the body to remain at rest or to move at constant speed in a straight line. Therefore, the acceleration would be zero and the force would also be zero. A force in the form of a push or pull is necessary either to start the body moving from rest or to change the velocity while it is moving. We also know that if a body is denser or has more matter, and therefore has more mass, it requires more force than a lighter body to bring it up to the same acceleration as the lighter body.

Newton's Law of Universal Gravitational Acceleration

The force of attraction varies directly with the product of the masses and inversely as the square of the distance between them. Mathematically, the equation is

$$F = GM_1M_2/R^2 \tag{8}$$

where G is the universal gravitational constant (6.67×10^{-11} N/kg), M is the masses of the bodies, and R is the distance between their centers. The subscripts refer to the two bodies.

Friction

The magnitude of friction depends mainly on two conditions:

1. The force of friction is directly proportional to the normal force. The normal force is the force action perpendicular to the two surfaces, pressing one body against the other. The greater the normal force, the deeper the individual "hills" will penetrate the "valleys," the more opposition to motion, and the greater the force of friction.

2. Friction depends upon the material making up the sliding surface. A wooden surface offers more friction with the same normal force than an ice surface, for in the former, the hills and valleys are more pronounced. The factor determining the amount of friction due to the nature of the material is called the coefficient of friction. The coefficient of friction has a range of values from 0 to 1 and therefore is a decimal value.

Work and Energy

Work

Work is force times distance. In equation form, work is defined as

$$\text{work} = W = Fs \tag{9}$$

The units of work in the SI system are newton-meters (N·m) or joules (J).

Energy

Energy is the ability to do work. The two types of energy that we will consider are due to motion (kinetic) and to its position (potential). Mathematically, they are defined as

$$\text{kinetic energy} = KE = (1/2)mv^2 \tag{10}$$

$$\text{potential energy} = PE = mgh \tag{11}$$

where m is the mass, v is the velocity, g is the gravitational acceleration, and h is the distance above a reference level. In all cases, the force and the distance through which the force acts must be parallel. In this case, the potential energy mgh is the work that a mass, m, would do falling a distance h, through a gravitational field g.

Power

Power is the time rate of doing work. Therefore, we write

$$\text{power} = \text{work/time} = Fs/t = Fv \tag{12}$$

The units of power are joules per second or watts.

Hydrostatics and Hydrodynamics

Fluid

The term *fluid* applies to a substance that does not have a fixed shape but is able to flow and take the shape of its container.

Streamline

A streamline, or flow line, is an imaginary line in a fluid, taken at an instant of time, such that the velocity vector at each point of the line is tangential to it.

Steady-State Flow

In steady-state flow, the fluid velocity at a given location is independent of time. Generally, the velocity can vary from location to location.

Incompressible Fluid

A fluid is considered to be incompressible if its density is constant, regardless of pressure.

Pressure

Pressure is force per unit area. In equation form, we write

$$P = F/A \tag{13}$$

where P has units of newtons per square meter (N/m^2) or pascals (Pa). Thus, an object weighing very little can exert a large pressure if the force acts only on a small surface area.

Archimedes's Principle

One of the earliest discoveries about fluids is Archimedes's principle, which describes the buoyancy, or "lift," when a body is wholly or partially immersed in a fluid. If a body is submerged in a fluid, there is a buoyant force action on the body that is equal to the weight of the displaced fluid. A body floats if its weight exactly equals the weight of the displaced fluid, for the net force on the body is zero. A body sinks if the buoyant force on it is less than its weight. However, there is still a buoyant force on a body even if it is not floating. The buoyant force is the equivalent loss in weight supplied by the fluid and is exerted upward.

Bernoulli's Equation

For steady-state flow of a nonviscous incompressible fluid, Bernoulli's equation relates the pressure P, the fluid speed, v, and the height, z, at any two points on the same streamline. In basic fluid mechanics, Bernoulli's equation, which represents an energy balance between two points along a streamline, is

$$v_1^2/2g + z_1 + p_1/\gamma = v_2^2/2g + z_2 + p_2/\gamma + h_{L,1-2} \qquad [14]$$

where v is the average flow velocity, g is the gravitational constant, z is the elevation from a reference level, p is a pressure, γ is the unit weight of water, and $h_{L,1-2}$ represents the energy loss between locations 1 and 2. Each complete term has the units of length (i.e., energy per unit of flowing fluid) usually designated as a "head," and the locations are designated with the subscripts 1 and 2. Thus, $v^2/2g$ is termed the velocity head and represents the kinetic energy, z is the elevation head and represents the potential energy, and p/γ is the pressure head and represents the stored energy. The sum of the kinetic, potential, and stored energy is the total energy. Bernoulli's equation states that if an incompressible fluid is in laminar flow, then the direction of the total energy line is the direction of the flow of the fluid.

Usually in soils, v is small so that Bernoulli's equation [14] becomes

$$z_1 + p_1/\gamma = z_2 + p_2/\gamma + h_{L,1-2} \qquad [15]$$

Density

The density of a body is defined as its mass per unit volume. The units of density are kilograms per cubic meter (kg/m^3).

Specific Gravity

Specific gravity is the ratio of the weight of a body to the weight of an equal volume of water at some reference temperature. It is often convenient to compare substances with one another. Pure water is often used as a standard substance. Then the specific gravity is defined as

$$\text{Specific gravity} = \frac{\text{weight density of substance}}{\text{weight density of water}} \qquad [16]$$

where weight density is defined as the weight per unit volume of a substance (kg/m^3).

Index

Printed and bound by CPI Group (UK) Ltd, Croydon, CR0 4YY

16/04/2025

14658420-0004